THE FRONTIERS COLLECTION

THE FRONTIERS COLLECTION

Series Editors
A.C. Elitzur   L. Mersini-Houghton   T. Padmanabhan   M. Schlosshauer
M.P. Silverman   J.A. Tuszynski   R. Vaas

The books in this collection are devoted to challenging and open problems at the forefront of modern science, including related philosophical debates. In contrast to typical research monographs, however, they strive to present their topics in a manner accessible also to scientifically literate non-specialists wishing to gain insight into the deeper implications and fascinating questions involved. Taken as a whole, the series reflects the need for a fundamental and interdisciplinary approach to modern science. Furthermore, it is intended to encourage active scientists in all areas to ponder over important and perhaps controversial issues beyond their own speciality. Extending from quantum physics and relativity to entropy, consciousness and complex systems—the Frontiers Collection will inspire readers to push back the frontiers of their own knowledge.

More information about this series at http://www.springer.com/series/5342

For a full list of published titles, please see back of book or springer.com/series/5342

Anthony Aguirre · Brendan Foster
Zeeya Merali
Editors

QUESTIONING THE FOUNDATIONS OF PHYSICS

Which of Our Fundamental Assumptions Are Wrong?

Editors
Anthony Aguirre
Department of Physics
University of California
Santa Cruz, CA
USA

Zeeya Merali
Foundational Questions Institute
New York, NY
USA

Brendan Foster
Foundational Questions Institute
New York, NY
USA

ISSN 1612-3018   ISSN 2197-6619   (electronic)
THE FRONTIERS COLLECTION
ISBN 978-3-319-38353-8   ISBN 978-3-319-13045-3   (eBook)
DOI 10.1007/978-3-319-13045-3

Springer Cham Heidelberg New York Dordrecht London
© Springer International Publishing Switzerland 2015
Softcover reprint of the hardcover 1st edition 2015

Printed on acid-free paper

Springer International Publishing AG Switzerland is part of Springer Science+Business Media (www.springer.com)

Preface

This book is a collaborative project between Springer and The Foundational Questions Institute (FQXi). In keeping with both the tradition of Springer's Frontiers Collection and the mission of FQXi, it provides stimulating insights into a frontier area of science, while remaining accessible enough to benefit a non-specialist audience.

FQXi is an independent, nonprofit organization that was founded in 2006. It aims to catalyze, support, and disseminate research on questions at the foundations of physics and cosmology.

The central aim of FQXi is to fund and inspire research and innovation that is integral to a deep understanding of reality, but which may not be readily supported by conventional funding sources. Historically, physics and cosmology have offered a scientific framework for comprehending the core of reality. Many giants of modern science—such as Einstein, Bohr, Schrödinger, and Heisenberg—were also passionately concerned with, and inspired by, deep philosophical nuances of the novel notions of reality they were exploring. Yet, such questions are often overlooked by traditional funding agencies.

Often, grant-making and research organizations institutionalize a pragmatic approach, primarily funding incremental investigations that use known methods and familiar conceptual frameworks, rather than the uncertain and often interdisciplinary methods required to develop and comprehend prospective revolutions in physics and cosmology. As a result, even eminent scientists can struggle to secure funding for some of the questions they find most engaging, while younger thinkers find little support, freedom, or career possibilities unless they hew to such strictures.

FQXi views foundational questions not as pointless speculation or misguided effort, but as critical and essential inquiry of relevance to us all. The Institute is dedicated to redressing these shortcomings by creating a vibrant, worldwide community of scientists, top thinkers, and outreach specialists who tackle deep questions in physics, cosmology, and related fields. FQXi is also committed to engaging with the public and communicating the implications of this foundational research for the growth of human understanding.

As part of this endeavor, FQXi organizes an annual essay contest, which is open to everyone, from professional researchers to members of the public. These contests are designed to focus minds and efforts on deep questions that could have a profound impact across multiple disciplines. The contest is judged by an expert panel and up to 20 prizes are awarded. Each year, the contest features well over a hundred entries, stimulating ongoing online discussion for many months after the close of the contest.

We are delighted to share this collection, inspired by the 2012 contest, "Questioning the Foundations: Which of Our Basic Physical Assumptions Are Wrong?" In line with our desire to bring foundational questions to the widest possible audience, the entries, in their original form, were written in a style that was suitable for the general public. In this book, which is aimed at an interdisciplinary scientific audience, the authors have been invited to expand upon their original essays and include technical details and discussion that may enhance their essays for a more professional readership, while remaining accessible to non-specialists in their field.

FQXi would like to thank our contest partners: The Gruber Foundation, Sub-Meta, and Scientific American. The editors are indebted to FQXi's scientific director, Max Tegmark, and managing director, Kavita Rajanna, who were instrumental in the development of the contest. We are also grateful to Angela Lahee at Springer for her guidance and support in driving this project forward.

2014   Anthony Aguirre
                                                                                        Brendan Foster
                                                                                         Zeeya Merali

Contents

Chapter 1
Introduction

Anthony Aguirre, Brendan Foster and Zeeya Merali

> *Our conceptions of Physical Reality can never be definitive; we*
> *must always be ready to alter them, to alter, that is, the*
> *axiomatic basis of physics, in order to take account of the facts*
> *of perception with the greatest possible logical completeness.*
> (Einstein, A: Maxwell's influence on the evolution of the idea of
> physical reality. In: Thomson, J. J., ed.: James Clerk Maxwell: a
> commemoration volume, pp. 66–73. Cambridge University
> Press (1931).)
>
> Albert Einstein (1931)
>
> *Scientific development depends in part on a process of*
> *non-incremental or revolutionary change. Some revolutions are*
> *large, like those associated with the names of Copernicus,*
> *Newton, or Darwin, but most are much smaller, like the*
> *discovery of oxygen or the planet Uranus. The usual prelude to*
> *changes of this sort is, I believe, the awareness of anomaly, of an*
> *occurrence or set of occurrences that does not fit existing ways of*
> *ordering phenomena. The changes that result therefore require*
> *'putting on a different kind of thinking-cap', one that renders the*
> *anomalous lawlike but that, in the process, also transforms the*
> *order exhibited by some other phenomena, previously*
> *unproblematic.* (Kuhn, T.S.: The Essential Tension (1977).)
>
> Thomas S. Kuhn (1977)

Over the course of history, we can identify a number of instances where thinkers have sacrificed some of their most cherished assumptions, ultimately leading to scientific revolutions. We once believed that the Earth was the centre of the universe; now, we know that we live in a cosmos littered with solar systems and extra-solar

A. Aguirre (✉)
Department of Physics, University of California, Santa Cruz, CA, USA
e-mail: aguirre@scipp.ucsc.edu

B. Foster · Z. Merali
Foundational Questions Institute, New York, NY, USA
e-mail: foster@fqxi.org

Z. Merali
e-mail: merali@fqxi.org

© Springer International Publishing Switzerland 2015
A. Aguirre et al. (eds.), *Questioning the Foundations of Physics*,
The Frontiers Collection, DOI 10.1007/978-3-319-13045-3_1

planets. Cosmologists today are even questioning whether our universe is itself unique or one of many parallel cosmoses.

Such paradigm shifts can be forced by experiment, an internal inconsistency in accepted physics, or simply a particular philosophical intuition. Based, in part, on the theoretical insight that the speed of light in a vacuum should be a constant, in the early twentieth century, Einstein developed his special theory of relativity, which threw out the common-sense belief that time and space are absolute. With his general theory of relativity, Einstein went on to claim that space and time are stitched together creating a four-dimensional fabric pervading the universe and that gravity manifests as this fabric warps and bends around massive cosmic objects. Around the same time, at the other extremity of scale, physicists realised that in order to explain perplexing experimental results they must formulate a new set of rules for the behaviour of subatomic entities—quantum physics—that muddies the boundaries between what we define to be particles and what we traditionally think of as waves. Inherently probabilistic, quantum theory also forces us to relinquish some of our deepest-held intuitions and to accept that, at its core, reality may be indeterministic.

But those revolutions in our understanding raised as many questions as they answered. Almost a century on, the time appears ripe for reassessing our current assumptions. Relativity and quantum theory together form the cornerstones of modern physics but they have brought us to an impasse. Both theories have been corroborated by experiments; yet physicists have failed to bring the two descriptions together into one overarching framework of "quantum gravity", suggesting that one or other, or even both, must be modified.

Astronomical observations also mock our understanding of the contents of the universe. By monitoring galaxies, astronomers have surmised that most of the mass of the universe resides in some unknown form, dubbed "dark matter", that is detectable only through its gravitational pull on visible matter. Furthermore, at the end of the twentieth century, cosmologists were blind-sided by the discovery that the universe is expanding at an accelerated rate, without any clear cause. This propulsive push is now attributed to "dark energy", but the origin and nature of this entity remains a mystery.

The world's biggest experiment at the Large Hadron Collider, at the CERN laboratory, has recently helped to verify the standard model of particle physics with unprecedented precision. Yet, this success has left physics with many unanswered questions. The standard model cannot explain the nature of dark matter, or why certain known particles have their observed masses and properties. In fact, if the standard model is correct, it is difficult to understand how we even came to exist, since it predicts that equal amounts of matter and antimatter should have been produced during the big bang, and that this matter and antimatter should subsequently have annihilated leaving nothing behind to form stars, galaxies, or people.

It seems clear that we are lacking some fundamental insight. In order to understand the origin of the universe, its contents and its workings—and our own existence—it is likely that we will once again need to give up one or more of the notions that lie at the base of our physical theories and which we currently hold sacred.

So which of our current underlying preconceptions–tacit or explicit—need rethinking? That is the question that we posed in the 2012 FQXi contest: "Questioning the Foundations: Which of Our Basic Physical Assumptions Are Wrong?" This was one of our broadest and most ambitious essay topics and it drew over 270 entries from Africa, Asia, Australasia, Europe, and North and South America. It also generated record levels of discussion on our online forums. This volume brings together the top 18 prize-winning entries.

Our first prize winner, Robert Spekkens, questions the distinction between a theory's kinematics—that is, the specification of the space of physical states it allows—and its dynamics—which encompasses the description of how these states may evolve. Though this conceptual separation has traditionally been central to the way that physicists build theories, whether classical or quantum, in Chap. 2, Spekkens argues that it is a convention that should be abandoned. In its stead, he champions underpinning new theories with a "causal structure" that explicitly relates variables in terms of how they have been influenced by, or could in turn affect, other variables.

Chapters 3 and 4 also deal with causation. George Ellis scrutinizes the implicit assumption that causation flows from the bottom up—that is, from micro to macro scales—instead positing that complexity in biology, and even the arrow of time, emerge from a top-down causal flow from macroscopic scales downwards. Benjamin Dribus meanwhile rejects the traditional spacetime manifold invoked by relativity in favour of a new central principle based on considering causal order.

The tenets upon which relativity are built are examined in more detail in Chaps. 5 and 6. In particular, Israel Perez questions Einstein's assumption that there are no preferred reference frames in the universe. In their essay, Sean Gryb and Flavio Mercati propose unstitching time from space in Einstein's fabric and argue that the fundamental description of reality must be based on shape.

Daryl Janzen also tackles physicists' accepted conceptions of time. In Chap. 7, he argues that by rethinking time in cosmological contexts, we may get a better handle on cosmic expansion and the origin of dark energy. Chapter 8 also deals with current mysteries in cosmology. Olaf Dreyer derives observable consequences that relate to both dark energy and dark matter by reformulating these problems in a framework in which particles are described as emergent excitations of the background, rather than as existing on a background.

Connecting cosmology and quantum mechanics in Chap. 9, Steven Weinstein challenges the orthodox view that physical facts at one point in space must be held independent from those at another point. In so doing, he argues, we may better understand the surprising homogeneity of the universe on cosmic scales and also the origin of quantum entanglement—the spooky property that appears to link distant quantum particles so that measurements of one influence the properties of its partners.

Chapters 10–12 deal specifically with aspects at the foundations of quantum theory. Angelo Bassi, Tejinder Singh and Hendrik Ulbricht question the principle of quantum linear superposition (that is, the consensus notion that the actual state of a quantum particle is the sum of its possible states). Although this has been experimentally confirmed for relatively small particles and molecules, they note that superposition breaks down for macroscopic objects; tables are never seen in two places at once, for instance. The team proposes experiments to test whether quantum theory is an approximation to a stochastic non-linear theory. In his essay, Giacomo D'Ariano searches for new quantum-information principles at the foundations of physics based on epistemological and operational rules. In Chap. 12, Ken Wharton argues that aspects of quantum physics would feel less paradoxical and may be open to explanation if we let go of the intuitive implicit belief that the universe is effectively a computer, processing itself in the same time-evolved manner that we use when performing calculations.

The challenge of devising a theory of quantum gravity that will unite quantum theory with Einstein's general theory of relativity occupies the authors of Chaps. 13–16. Debates over the best approach for developing such a unified theory often focus on whether quantum theory or our general-relativistic view of spacetime is more fundamental. Giovanni Amelino–Camelia argues that when quantum mechanical effects dominate, the assumption that spacetime exists becomes a hindrance and should be thrown out. By contrast, Torsten Asslemeyer–Maluga reviews both options in Chap. 14—that either spacetime must be quantized or that spacetime emerges from something deeper—and then presents an alternative view in which spacetime defines the quantum state. Sabine Hossenfelder also makes the case for a third way, arguing that the final theory need not be either classical or quantized. In Chap. 16, Michele Arzano opens a new avenue for approaching a potential theory of quantum gravity by scrutinizing the founding principles of quantum field theory that determine the structure of the quantum fields.

To close the volume, we include award-winning entries that looked at the philosophical stance of reductionism. In Chap. 17, Julian Barbour argues that while reductionism has been a successful approach in science, in order to understand quantum mechanics and other mysteries such as the arrow of time, we may require a more holistic approach. Ian Durham defends reductionism in Chap. 18, but questions the paradigm that modern science simply consists of posing questions and then testing them. Finally, in Chap. 19, Sara Walker examines the merits of reductionism for tackling perhaps the biggest unanswered question of all—the origin of life—by challenging the edict that "all life is just chemistry".

In summary, the volume brings together an eclectic mix of approaches for addressing current mysteries that range from the peculiarities of the subatomic quantum scale to those that span cosmic distances, examining our beliefs about time, causation, and even the source of the spark of life, along the way. The winners include experts in physics, mathematics, astronomy, astrobiology, condensed-matter physics, aerospace engineering, and cosmology and each provides ample food for thought for the basis of our next scientific revolution.

Chapter 2
The Paradigm of Kinematics and Dynamics Must Yield to Causal Structure

Robert W. Spekkens

Abstract The distinction between a theory's kinematics and its dynamics, that is, between the space of physical states it posits and its law of evolution, is central to the conceptual framework of many physicists. A change to the kinematics of a theory, however, can be compensated by a change to its dynamics without empirical consequence, which strongly suggests that these features of the theory, considered separately, cannot have physical significance. It must therefore be concluded (with apologies to Minkowski) that henceforth kinematics by itself, and dynamics by itself, are doomed to fade away into mere shadows, and only a kind of union of the two will preserve an independent reality. The notion of causal structure seems to provide a good characterization of this union.

Proposals for physical theories generally have two components: the first is a specification of the space of physical states that are possible according to the theory, generally called the *kinematics* of the theory, while the second describes the possibilities for the evolution of the physical state, called the *dynamics*. This distinction is ubiquitous. Not only do we recognize it as a feature of the empirically successful theories of the past, such as Newtonian mechanics and Maxwell's theory of electromagnetism, it persists in relativistic and quantum theories as well and is even conspicuous in proposals for novel physical theories. Consider, for instance, some recent proposals for how to unify quantum theory and gravity. Fay Dowker describes the idea of causal histories as follows [1]:

> The hypothesis that the deep structure of spacetime is a discrete poset characterises causal set theory at the kinematical level; that is, it is a proposition about what substance is the subject of the theory. However, kinematics needs to be completed by dynamics, or rules about how the substance behaves, if one is to have a complete theory.

She then proceeds to describe the dynamics. As another example, Carlo Rovelli describes the basics of loop quantum gravity in the following terms [2]:

> The kinematics of the theory is well understood both physically (quanta of area and volume, discrete geometry) and from the mathematical point of view. The part of the theory that is not yet fully under control is the dynamics, which is determined by the Hamiltonian constraint.

R.W. Spekkens (✉)
Perimeter Institute for Theoretical Physics, Waterloo, Ontario N2L 2Y5, Canada
e-mail: rspekkens@perimeterinstitute.ca

© Springer International Publishing Switzerland 2015
A. Aguirre et al. (eds.), *Questioning the Foundations of Physics*,
The Frontiers Collection, DOI 10.1007/978-3-319-13045-3_2

In the field of quantum foundations, there is a particularly strong insistence that any well-formed proposal for a physical theory must specify both kinematics and dynamics. For instance, Sheldon Goldstein describes the de Broglie-Bohm interpretation [3] by specifying its kinematics and its dynamics [4]:

> In Bohmian mechanics a system of particles is described in part by its wave function, evolving, as usual, according to Schrödinger's equation. However, the wave function provides only a partial description of the system. This description is completed by the specification of the actual positions of the particles. The latter evolve according to the "guiding equation," which expresses the velocities of the particles in terms of the wave function.

John Bell provides a similar description of his proposal for a pilot-wave theory for fermions in his characteristically whimsical style [5]:

> In the beginning God chose 3-space and 1-time, a Hamiltonian H, and a state vector $|0\rangle$. Then She chose a fermion configuration n (0). This She chose at random from an ensemble of possibilities with distribution D (0) related to the already chosen state vector $|0\rangle$. Then She left the world alone to evolve according to [the Schrödinger equation] and [a stochastic jump equation for the fermion configuration].

The distinction persists in the Everett interpretation [6], where the set of possible physical states is just the set of pure quantum states, and the dynamics is simply given by Schrödinger's equation (the appearance of collapses is taken to be a subjective illusion). It is also present in dynamical collapse theories [7, 8], where the kinematics is often taken to be the same as in Everett's approach—nothing but wavefunction— while the dynamics is given by a stochastic equation that is designed to yield a good approximation to Schrödinger dynamics for microscopic systems and to the von Neumann projection postulate for macroscopic systems.

While proponents of different interpretations of quantum theory and proponents of different approaches to quantizing gravity may disagree about the correct kinematics and dynamics, they typically agree that any proposal must be described in these terms.

In this essay, I will argue that the distinction is, in fact, conventional: kinematics and dynamics only have physical significance when considered jointly, not separately.

In essence, I adopt the following methodological principle: any difference between two physical models that does not yield a difference at the level of empirical phenomena does not correspond to a physical difference and should be eliminated. Such a principle was arguably endorsed by Einstein when, from the empirical indistinguishability of inertial motion in free space on the one hand and free-fall in a gravitational field on the other, he inferred that one must reject any model which posits a physical difference between these two scenarios (the strong equivalence principle).

Such a principle does not force us to operationalism, the view that one should only seek to make claims about the outcomes of experiments. For instance, if one didn't already know that the choice of gauge in classical electrodynamics made no difference to its empirical predictions, then discovery of this fact would, by the lights of the principle, lead one to renounce real status for the vector potential in favour of only the electric and magnetic field strengths. It would not, however, justify a blanket rejection of *any* form of microscopic reality.

As another example, consider the prisoners in Plato's cave who live out their lives learning about objects only through the shadows that they cast. Suppose one of the prisoners strikes upon the idea that there is a third dimension, that objects have a three-dimensional shape, and that the patterns they see are just two-dimensional projections of this shape. She has constructed a hidden variable model for the phenomena. Suppose a second prisoner suggests a different hidden variable model, where in addition to the shape, each object has a property called colour which is completely irrelevant to the shadow that it casts. The methodological principle dictates that because the colour property can be varied without empirical consequence, it must be rejected as unphysical. The shape, on the other hand, has explanatory power and the principle finds no fault with it. Operationalism, of course, would not even entertain the possibility of such hidden variables.

The principle tells us to constrain our model-building in such a way that every aspect of the posited reality has some explanatory function. If one takes the view that part of achieving an adequate explanation of a phenomenon is being able to make predictions about the outcomes of interventions and the truths of counterfactuals, then what one is seeking is a *causal* account of the phenomenon. This suggests that the framework that should replace kinematics and dynamics is one that focuses on causal structure. I will, in fact, conclude with some arguments in favour of this approach.

Different Formulations of Classical Mechanics

Already in classical physics there is ambiguity about how to make the separation between kinematics and dynamics. In what one might call the *Newtonian* formulation of classical mechanics, the kinematics is given by configuration space, while in the *Hamiltonian* formulation, it is given by phase space, which considers the canonical momentum for every independent coordinate to be on an equal footing with the coordinate. For instance, for a single particle, the kinematics of the Newtonian formulation is the space of possible positions while that of the Hamiltonian formulation is the space of possible pairs of values of position and momentum. The two formulations are still able to make the same empirical predictions because they posit different dynamics. In the Newtonian approach, motion is governed by the Euler-Lagrange equations which are second-order in time, while in the Hamiltonian approach, it is governed by Hamilton's equations which are first order in time.

So we can change the kinematics from configuration space to phase space and maintain the same empirical predictions by adjusting the dynamics accordingly. It's not possible to determine which kinematics, Newtonian or Hamiltonian, is the *correct* kinematics. Nor can we determine the correct dynamics in isolation. The kinematics and dynamics of a theory can only ever be subjected to experimental trial as a pair.

On the Possibility of Violating Unitarity
in Quantum Dynamics

Many researchers have suggested that the correct theory of nature might be one that shares the kinematics of standard quantum theory, but which posits a different dynamics, one that is not represented by a unitary operator. There have been many different motivations for considering this possibility. Dynamical collapse theorists, for instance, seek to relieve the tension between a system's free evolution and its evolution due to a measurement. Others have been motivated to resolve the black hole information loss paradox. Still others have proposed such theories simply as foils against which the predictions of quantum theory can be tested [9].

Most of these proposals posit a dynamics which is *linear* in the quantum state (more precisely, in the density operator representing the state). For instance, this is true of the prominent examples of dynamical collapse models, such as the proposal of Ghirardi et al. [7] and the continuous spontaneous localization model [8]. This linearity is not an incidental feature of these models. Most theories which posit dynamics that are nonlinear also allow superluminal signalling, in contradiction with relativity theory [10]. Such nonlinearity can also lead to trouble with the second law of thermodynamics [11].

There is an important theorem about linear dynamics that is critical for our analysis: such dynamics can always be understood to arise by adjoining to the system of interest an auxiliary system prepared in some fixed quantum state, implementing a unitary evolution on the composite, and finally throwing away or ignoring the auxiliary system. This is called the *Stinespring dilation theorem* [12] and is well-known to quantum information theorists.[1]

All proposals for nonunitary but linear modifications of quantum theory presume that it is in fact possible to distinguish the predictions of these theories from those of standard quantum mechanics. For instance, the experimental evidence that is championed as the "smoking gun" which would rule in favour of such a modification is *anomalous decoherence*—an increase in the entropy of the system that cannot be accounted for by an interaction with the system's environment. Everyone admits that such a signature is extremely difficult to detect if it exists. But the point I'd like to make here is that *even if* such anomalous decoherence were detected, it would not vindicate the conclusion that the dynamics is nonunitary. Because of the Stinespring dilation theorem, such decoherence is also consistent with the assumption that there are some hitherto-unrecognized degrees of freedom and that the quantum system under investigation is coupled unitarily to these.[2]

---

[1] It is analogous to the fact that one can simulate indeterministic dynamics on a system by deterministic dynamics which couples the system to an additional degree of freedom that is probabilistically distributed.

[2] A collapse theorist will no doubt reject this explanation on the grounds that one cannot solve the quantum measurement problem while maintaining unitarity. Nonetheless, our argument shows that someone who does not share their views on the quantum measurement problem need not be persuaded of a failure of unitarity.

So, while it is typically assumed that such an anomaly would reveal that quantum theory was mistaken in its *dynamics*, we could just as well take it to reveal that quantum theory was correct in its dynamics but mistaken in its *kinematics*. The experimental evidence alone cannot decide the issue. By the lights of our methodological principle, it follows that the distinction must be purely conventional.

Freedom in the Choice of Kinematics
for Pilot-Wave Theories

The pilot-wave theory of de Broglie and Bohm supplements the wavefunction with additional variables, but it turns out that there is a great deal of freedom in how to choose these variables. A simple example of this arises for the case of spin. Bohm, Schiller, and Tiomno have proposed that particles with spin should be modeled as extended rigid objects and that the spinor wavefunction should be supplemented not only with the positions of the particles (as is standardly done for particles without spin), but with their orientation in space as well [13]. In addition to the equation which governs the evolution of the spinor wavefunction (the Pauli equation), they propose a guidance equation that specifies how the positions and orientations evolve over time.

But there is another, more minimalist, proposal for how to deal with spin, due to Bell [14]. The only variables that supplement the wavefunction in his approach are the particle positions. The particles follow trajectories that are different from the ones they would follow if they did not have spin because the equations of motion for the particle positions depend on the spinor wavefunction.

The Bohm, Schiller and Tiomno approach and the Bell approach make exactly the same experimental predictions. This is possible because our experience of quantum phenomena consists of observations of macroscopic variables such as pointer positions rather than direct observation of the properties of the particle.

The non-uniqueness of the choice of kinematics for pilot-wave theories is not isolated to spin. It is generic. The case of quantum electrodynamics (QED) illustrates this well. Not only is there a pilot-wave theory for QED, there are multiple viable proposals, all of which produce the same empirical predictions. You could follow Bohm's treatment of the electromagnetic field, where the quantum state is supplemented by the configuration of the electric field [15]. Alternatively, you could make the supplementary variable the magnetic field, or any other linear combination of the two. For the charges, you could use Bell's discrete model of fermions on a lattice (mentioned in the introduction), where the supplementary variables are the fermion numbers at every lattice point [5]. Or, if you preferred, you could use Colin's continuum version of this model [16]. If you fancy something a bit more exotic, you might prefer to adopt Struyve and Westman's minimalist pilot-wave theory for QED, which treats charges in a manner akin to how Bell treats spin [17]. Here, the variables that are taken to supplement the quantum states are *just* the electric field strengths. No variables for the charges are introduced. By virtue of Gauss's law, the field nonetheless carries an image of all the charges and hence it carries an image of the pointer positions. This image is what we infer when our eyes detect the fields. But the charges are an illusion. And, of course, according to this model the stuff of which we are made is not charges either: we are fields observing fields.

The existence of many empirically adequate versions of Bohmian mechanics has led many commentators to appeal to principles of simplicity or elegance to try to decide among them. An alternative response is suggested by our methodological principle: any feature of the theory that varies among the different versions is not physical.

Kinematical Locality and Dynamical Locality

I consider one final example, the one that first set me down the path of doubting the significance of the distinction between kinematics and dynamics. It concerns different notions of locality within realist models of quantum theory. Unlike a purely operational interpretation of quantum theory, a realist model seeks to provide a causal explanation of the experimental statistics, specifically, of the correlations that are observed between control variables in the preparation procedure and outcomes of the measurement procedure. It is presumed that it is the properties of the system which passes between the devices that mediates the causal influence of the preparation variable on the measurement outcome [18]. We refer to a full specification of these properties as the system's *ontic state*.

It is natural to say that a realist model has *kinematical locality* if, for any two systems A and B, every ontic state λ_{AB} of the composite is simply a specification of the ontic state of each component,

$$\lambda_{AB} = (\lambda_A, \lambda_B).$$

In such a theory, once you have specified all the properties of A and of B, you have specified all of the properties of the composite AB. In other words, kinematical locality says that there are no holistic properties.[3]

It is also natural to define a dynamical notion of locality for relativistic theories: a change to the ontic state λ_A of a localized system A cannot be a result of a change to the ontic state λ_B of a localized system B if B is outside the backward light-cone of A. In other words, against the backdrop of a relativistic space-time, this notion of locality asserts that all causal influences propagate at speeds that are no faster than the speed of light.

Note that this definition of dynamical locality has made reference to the ontic state λ_A of a *localized* system A. If A is part of a composite system AB with holistic properties, then the ontic state of this composite, λ_{AB}, need not factorize into λ_A and λ_B therefore we cannot necessarily even define λ_A. In this sense, the dynamical notion of locality already presumes the kinematical one.

---

[3] The assumption has also been called *separability* [19].

It is possible to derive Bell inequalities starting from the assumption of kinematical and dynamical locality together with a few other assumptions, such as the fact that the measurement settings can be chosen freely and the absence of retrocausal influences. Famously, quantum theory predicts a violation of the Bell inequalities. In the face of this violation, one must give up one or more of the assumptions. Locality is a prime candidate to consider and if we do so, then the following question naturally arises: is it possible to accommodate violations of Bell inequalities by admitting a failure of the dynamical notion of locality while salvaging the kinematical notion?

It turns out that for any realist interpretations of quantum theory wherein the ontic state encodes the quantum state, termed "ψ-ontic" models[4] in Ref. [19], there is a failure of *both* sorts of locality. In such models, kinematical locality fails simply by virtue of the existence of entangled states. This is the case for all of the interpretations enumerated in the introduction: Everett, collapse theories, de Broglie-Bohm. Might there nonetheless be some alternative to these interpretations that *does* manage to salvage kinematical locality?

I've told the story in such a way that this seems to be a perfectly meaningful question. But I would like to argue that, in fact, it is not.

To see this, it suffices to realize that it is *trivial* to build a model of quantum theory that salvages kinematical locality. For example, we can do so by a slight modification of the de Broglie-Bohm model. Because the particle positions can be specified locally, the only obstacle to satisfying kinematical locality is that the other part of the ontology, the universal wavefunction, does not factorize across systems and thus must describe a holistic property of the universe. This conclusion, however, relied on a particular way of associating the wavefunction with space-time. Can we imagine a different association that would make the model kinematically local? Sure. Just put a copy of the universal wavefunction at every point in space. It can then pilot the motion of every particle by a local causal influence. Alternatively, you could put it at the location of the center of mass of the universe and have it achieve its piloting by a superluminal causal influence—remember, we are allowing arbitrary violations of dynamical locality, so this is allowed. Or, put it under the corner of my doormat and let it choreograph the universe from there.

The point is that the failure of dynamical locality yields so much leeway in the dynamics that one can easily accommodate any sort of kinematics, including a local kinematics. Of course, these models are not credible and no one would seriously propose them,[5] but what this suggests to me is *not* that we should look for nicer models, but rather that the question of whether one can salvage kinematical locality was not an interesting one after all. The mistake, I believe, was to take seriously the distinction between kinematics and dynamics.

---

[4] Upon learning this terminology, a former student, Chris Granade, proposed that the defining feature of these types of model—that the ontic state encodes the quantum state—should be called "ψ-ontology". I and other critics of ψ-ontic approaches have since taken every opportunity to score cheap rhetorical points against the ψ-ontologists.

[5] Norsen has proposed a slightly more credible model but only as a proof of principle that kinematical locality can indeed be achieved [20].

Summary of the Argument

A clear pattern has emerged. In all of the examples considered, we seem to be able to accommodate wildly different choices of kinematics in our models without changing their empirical predictions simply by modifying the dynamics, and vice-versa. This strikes me as strong evidence for the view that the distinction between kinematics and dynamics—a distinction that is often central to the way that physicists characterize their best theories and to the way they constrain their theory-building—is purely conventional and should be abandoned.

From Kinematics and Dynamics to Causal Structure

Although it is not entirely clear at this stage what survives the elimination of the distinction between kinematics and dynamics, I would like to suggest a promising candidate: the concept of *causal structure*.

In recent years, there has been significant progress in providing a rigorous mathematical formalism for expressing causal relations and for making inferences from these about the consequences of interventions and the truths of counterfactuals. The work has been done primarily by researchers in the fields of statistics, machine learning, and philosophy and is well summarized in the texts of Spirtes et al. [21] and Pearl [22]. According to this approach, the causal influences that hold among a set of classical variables can be modeled by the arrows in a directed acyclic graph, of the sort depicted in Figs. 2.1 and 2.2, together with some causal-statistical parameters describing the strengths of the influences.

The causal-statistical parameters are conditional probabilities $P(X|\text{Pa}(X))$ for every X, where $\text{Pa}(X)$ denotes the causal parents of X, that is, the set of variables that have arrows into X. If a variable X has no parents within the model, then one simply specifies $P(X)$. The graph and the parameters together constitute the causal model.

It remains only to see why this framework has some hope of dispensing with the kinematics-dynamics distinction in the various examples I have presented.

The strongest argument in favour of this framework is that it provides a way to move beyond kinematical and dynamical notions of locality. John Bell was someone who clearly endorsed the kinematical-dynamical paradigm of model-building, as the quote in the introduction illustrates, and who recognized the distinction among notions of locality, referring to models satisfying kinematical locality as theories of "local beables" [23]. In his most precise formulation of the notion of locality, however—which, significantly, he called *local causality*—he appears to have transcended the paradigm of kinematics and dynamics and made an early foray into the new paradigm of causal structure.

Consider a Bell-type experiment. A pair of systems, labeled A and B, are prepared together and then taken to distant locations. The variable that specifies the choice of

Fig. 2.1 The causal graph associated with Bell's notion of local causality

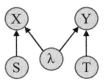

measurement on A (respectively B) is denoted S (respectively T) and the variable specifying the measurement's outcome is denoted X (respectively Y). Bell interprets the question of whether a set of correlations $P(XY|ST)$ admits of a locally causal explanation as the question of whether the correlations between X and Y can be entirely explained by a common cause λ, that is, whether they can be explained by a causal graph of the form illustrated in Fig. 2.1. From the causal dependences in this graph, we derive that the sorts of correlations that can be achieved in such a causal model are those of the form

$$P(XY|ST) = \sum_{\lambda} P(X|S\lambda) \, P(Y|T\lambda) \, P(\lambda).$$

Such correlations can be shown to satisfy certain inequalities, called the Bell inequalities, which can be tested by experiments and are found to be violated in a quantum world.

If we think of the variable λ as the ontic state of the composite AB, then we see that we have not needed to specify whether or not λ factorizes as (λ_A, λ_B). Bell recognized this fact and emphasized it in his later writing: "It is notable that in this argument nothing is said about the locality, or even localizability, of the variable λ [24]." Indeed, whether λ is localized in the common past of the two measurement events and effects them by means of intermediary influences that propagate subluminally, or whether it is localized under my doormat and influences them superluminally, or whether it is not even localized at all, is *completely irrelevant*. All that is needed to prove that $P(XY|ST)$ must satisfy the Bell inequalities is that whatever the separate kinematics and dynamics might be, together they define the effective causal structure that is depicted in Fig. 2.1. By our methodological principle, therefore, only the effective causal structure should be considered physically relevant.[6]

We see that Bell's argument manages to derive empirical claims about a class of realist models without needing to make any assumptions about the separate nature of their kinematics and dynamics. This is a remarkable achievement. I propose that it be considered as a template for future physics.

---

[6] This analysis also suggests that the concepts of space and time, which are primitive within the paradigm of kinematics and dynamics, ought to be considered as secondary concepts that are ultimately defined in terms of cause-effect relations. Whereas in the old paradigm, one would consider it to be part of the definition of a cause-effect relation that the cause should be temporally prior to the effect, in the new paradigm, what it means for one event to be temporally prior to another is that the first could be a cause of the second.

Fig. 2.2 Causal graphs for
Hamiltonian (*left*) and
Newtonian (*right*)
formulations of classical
mechanics

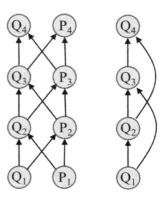

It is not as clear how the paradigm of causal structure overcomes the conventionality of the kinematics-dynamics distinction in the other examples I've presented, but there are nonetheless some good reasons to think that it is up to the task.

Consider the example of Hamiltonian and Newtonian formulations of classical mechanics. If we let Q_i denote a coordinate at time t_i and P_i its canonically conjugate momentum, then the causal models associated respectively with the two formulations are depicted in Fig. 2.2. The fact that Hamiltonian dynamics is first-order in time implies that the Q and P variables at a given time are causally influenced directly only by the Q and P variables at the previous time. Meanwhile, the second-order nature of Newtonian dynamics is captured by the fact that Q at a given time is causally influenced directly by the Qs at *two* previous times. In both models, we have a causal influence from Q_1 to Q_3, but in the Newtonian case it is direct, while in the Hamiltonian case it is mediated by P_2. Nonetheless, the kinds of correlations that can be made to hold between Q_1 and Q_3 are the same regardless of whether the causal influence is direct or mediated by P_2.[7] The consequences for Q_3 of interventions upon the value of Q_1 also are insensitive to this difference. So from the perspective of the paradigm of causal structure, the Hamiltonian and Newtonian formulations appear less distinct than they do if one focusses on kinematics and dynamics.

Empirical predictions of statistical theories are typically expressed in terms of statistical dependences among variables that are observed or controlled. My guiding methodological principle suggests that we should confine our attention to those causal features that are relevant for such dependences. In other words, although we can convert a particular claim about kinematics and dynamics into a causal graph, not all features of this graph will have relevance for statistical dependences. Recent work that seeks to infer causal structure from observed correlations has naturally gravitated towards the notion of equivalence classes of causal graphs, where the equivalence relation is the ability to produce the same set of correlations. One could also try to

---

[7] There is a subtlety here: it follows from the form of the causal graph in the Newtonian model that Q_1 and Q_4 are conditionally independent given Q_2 and Q_3, but in the Hamiltonian case, this fact must be inferred from the causal-statistical parameters.

characterize equivalence classes of causal models while allowing for restrictions on the forms of the conditional probabilities and while allowing not only observations of variables but interventions upon them as well. Such equivalence classes, or something like them, seem to be the best candidates for the mathematical objects in terms of which our classical models of physics should be described.

Finally, by replacing conditional probabilities with quantum operations, one can define a quantum generalization of causal models—*quantum causal models* [25, 26]—which appear promising for providing a realist interpretation of quantum theory. It is equivalence classes of causal structures here that are likely to provide the best framework for future physics.

The paradigm of kinematics and dynamics has served us well. So well, in fact, that it is woven deeply into the fabric of our thinking about physical theories and will not be easily supplanted. I have nonetheless argued that we must abandon it. Meanwhile, the paradigm of causal structure is nascent, unfamiliar and incomplete, but it seems ideally suited to capturing the nonconventional distillate of the union of kinematics and dynamics and it can already claim an impressive achievement in the form of Bell's notion of local causality.

Rest in peace kinematics and dynamics. Long live causal structure!

Acknowledgments My thanks to Howard Wiseman and Travis Norsen for valuable discussions, especially those on the subject of kinematical locality. Research at Perimeter Institute is supported by the Government of Canada through Industry Canada and by the Province of Ontario through the Ministry of Research and Innovation.

References

1. F. Dowker. Causal sets and the deep structure of spacetime, ed. by A. Ashtekar. 100 Years of Relativity, Space-Time Structure: Einstein and Beyond, pp. 445–464 (2005).
2. C. Rovelli. Loop quantum gravity. Living Rev. Relativ. 11 (2008).
3. D. Bohm, A suggested interpretation of the quantum theory in terms of "hidden" variables I. Phys. Rev. **85**(2), 166 (1952).
4. S. Goldstein. Bohmian mechanics. The Stanford Encyclopedia of Philosophy (Fall 2012 edn.).
5. J.S. Bell, Beables for quantum field theory. Phys. Rep. **137**, 49–54 (1986).
6. H. Everett, III., "Relative state" formulation of quantum mechanics. Rev. Mod. Phys. **29**(3), 454 (1957).
7. G.C. Ghirardi, A. Rimini, T. Weber, Unified dynamics for microscopic and macroscopic systems. Phys. Rev. D **34**(2), 470 (1986).
8. G.C. Ghirardi, P. Pearle, A. Rimini, Markov processes in Hilbert space and continuous spontaneous localization of systems of identical particles. Phys. Rev. A **42**(1), 78 (1990).
9. S. Weinberg, Testing quantum mechanics. Ann. Phys. **194**(2), 336–386 (1989).
10. N. Gisin, Weinberg's non-linear quantum mechanics and supraluminal communications. Phys. Lett. A **143**(1), 1–2 (1990).
11. A. Peres, Nonlinear variants of Schrödinger's equation violate the second law of thermodynamics. Phys. Rev. Lett. **63**(10), 1114 (1989).
12. W.F. Stinespring, Positive functions on C*-algebras. Proceedings of the American Mathematical Society **6**, 211–216 (1955).
13. D. Bohm, R. Schiller, J. Tiomno, A causal interpretation of the Pauli equation (A). Il Nuovo Cimento **1**, 48–66 (1955).

14. J.S. Bell, On the impossible pilot wave. Found. Phys. **12**(10), 989–999 (1982).
15. D. Bohm, A suggested interpretation of the quantum theory in terms of "hidden" variables II. Phys. Rev. **85**(2), 180 (1952).
16. S. Colin, A deterministic Bell model. Phys. Lett. A **317**(5), 349–358 (2003).
17. W. Struyve, H. Westman, A minimalist pilot-wave model for quantum electrodynamics. Proc. R. Soc. A: Math. Phys. Eng. Sci. **463**, 3115–3129 (2007).
18. R.W. Spekkens, Contextuality for preparations, transformations, and unsharp measurements. Phys. Rev. A **71**(5), 052108 (2005).
19. N. Harrigan, R.W. Spekkens, Einstein, incompleteness, and the epistemic view of quantum states. Found. Phys. **40**(2), 125–157 (2010).
20. T. Norsen, The theory of (exclusively) local beables. Found. Phys. **40**(12), 1858–1884 (2010).
21. P. Spirtes, C.N. Glymour, R. Scheines, Causation, Prediction, and Search, 2nd edn. (The MIT Press, 2001).
22. J. Pearl, Causality: Models, Reasoning, and Inference, 2nd edn. (Cambridge University Press, 2009).
23. J.S. Bell, The theory of local beables. Epistemo. Lett. **9**, 11–24 (1976).
24. J.S. Bell, Bertlmann's socks and the nature of reality. Le J. de Phys. Colloq. **42**(C2), 41–61 (1981).
25. M.S. Leifer, Quantum dynamics as an analog of conditional probability. Phys. Rev. A **74**(4), (2006).
26. M.S. Leifer, R.W. Spekkens, Towards a formulation of quantum theory as a causally neutral theory of Bayesian inference. Phys. Rev. A **88**(5), 052130 (2013).

Chapter 3
Recognising Top-Down Causation

George Ellis

The Theme

A key assumption underlying most present day physical thought is that causation in the hierarchy of complexity is bottom up all the way: particle physics underlies nuclear physics, nuclear physics underlies atomic physics, atomic physics underlies chemistry, and so on, and this is all that is taking place. Thus all the higher level subjects are in principle reducible to particle physics, which is therefore the only fundamental science; all the rest are derivative, even if we do not have the computing power to demonstrate this in detail. As famously claimed by Dirac, chemistry is just an application of quantum physics (see [60]). It is implied (or sometimes explicitly stated) that the same is true for biology and psychology.

Interlevel Causation

However there are many topics that one cannot understand by assuming this one-way flow of causation. Inter alia,

- The flourishing subject of social neuroscience makes clear how social influences act down on individual brain structure [13], giving the mechanism by which our minds get adapted to the society in which they are imbedded [5];
- the study of sensory systems shows conclusively that our senses do not work in a bottom up way with physical input from the environment uniquely determining what we experience; rather our expectations of what we should see play a key role [32];
- studies in physiology demonstrate that downward causation is key in physiological systems. For example it is needed to understand the functioning of the heart, where this form of causation can be represented as the influence of initial and boundary

G. Ellis(✉)
University of Cape Town, Cape Town, South Africa
e-mail: gfrellis@gmail.com

© Springer International Publishing Switzerland 2015
A. Aguirre et al. (eds.), *Questioning the Foundations of Physics*,
The Frontiers Collection, DOI 10.1007/978-3-319-13045-3_3

conditions on the solutions of the differential equations used to represent the lower level processes [51];

- epigenetic and developmental studies demonstrate that biological development is crucially influenced by the environment in which development takes place [34, 51];
- evolutionary theory makes clear that living beings are adapted to environmental niches, which means that environmental influences shape animal structure, function, and behavior [11].

In each case, the larger environment acts down to affect what happens to the components at lower levels of the hierarchy of structure. This does not occur by violating physical laws; on the contrary, it occurs through directing the effects of the laws of physics by setting constraints on lower level interactions.

Now many believe that insofar as this view is not just a trivial restatement of the obvious, it is wrong, because it implies denying the validity of the physics that underlies our material existence. This is however not the case; at no point do I deny the power and validity of the underlying physical causation. The crucial point is that even though physical laws always completely characterize the interactions at their own level of interaction (that between the physical components out of which complex entities arise), they do not by themselves determine unique outcomes either at the lower or higher levels. The specific outcomes that in fact occur are determined by the context in which those physical interactions take place; for example whether the electrons and protons considered are imbedded in a digital computer, a rock, a dog, a river, an elephant, an aircraft, or a trombone. Context has many dimensions: complexity arises out of suitable modular hierarchical structures, each layer influencing both those above and those below. Indeed that is why there are so many different subjects (chemistry, biochemistry, molecular biology, physiology, ecology, environmental science, evolutionary biology, ecology, psychology, anthropology, sociology, economics, politics, and so on) that characterize our complex existence. Only in the case of physical chemistry is there some chance of reducing the higher level emergent behaviour to "nothing but" the outcome of lower level causal interactions; and even there it does not actually work, inter alia because of the issue of the arrow of time (see sect. "The Arrow of Time").

What about physics itself? In this essay I make the case that top-down causation is also prevalent in physics [23], even though this is not often recognized as such. Thus my theme is [26],

> **Interlevel causation:** The assumption that all causation is bottom up is wrong, even in the case of physics. Both bottom up and top down effects occur in the hierarchy of complexity, and together enable higher level emergent behaviour involving true complexity to arise through the existence of inter-level feedback loops.

Some writers on this topic prefer to refer to contextual effects or whole-part constraints. These are perfectly acceptable terms, but I will make the case that the stronger term top-down causation is appropriate in many cases. As stated above, this is not an exercise in mysticism, or denial of physical causation; it is an important assertion about how causality, based essentially in physics at the bottom levels, works out in the real world.

Two Basic Issues

To set the scene, I give some definitions on which what follows is based.

Causation

The nature of causation is highly contested territory, although less so than before [53, 54]. I will take a pragmatic view:

Definition 1 (*Causal Effect*) If making a change in a quantity X results in a reliable demonstrable change in a quantity Y in a given context, then X has a causal effect on Y.

Example I press the key labelled A on my computer keyboard; the letter A appears on my computer screen.

Note: the effect may occur through intermediaries, e.g. X may cause C which in turn causes Y. It still remain true that (through this chain) X causes Y. What is at issue here is what mechanism enables X to influence Y. The overall outcome is unaffected by this issue.

Now there are of course a myriad of causal influences on any specific event: a network of causation is always in action. What we usually do is to have some specific context in mind where we keep almost all parameters and variables fixed, allowing just one or two remaining ones to vary; if they reliably cause some change in the dependent variable in that context, we then label them as the cause. For example in the case of a digital computer, we have

$$(Physics, \text{ computer structure, specific software, data}) \Rightarrow (output) \qquad (3.1)$$

Now in a particular run on a specific computer, the laws of physics do not change and the high level software loaded (e.g. Microsoft Word) will be fixed, so the above reduces to

$$(data) \underbrace{====\Rightarrow}_{MS\text{-}Word} (output) \qquad (3.2)$$

If however we load new high level software (e.g. now we run Photoshop) we will end up with a different relation than (3.2):

$$(data) \underbrace{====\Rightarrow}_{Photoshop} (output) \qquad (3.3)$$

Hence both the data and the software are causes of the output, in the relevant context. The laws of physics are also a cause, as is the existence of the Earth and the designer of the computer, but we usually don't bother to mention this.

Existence

Given this understanding of causation, it implies a view on ontology (existence) as follows: I assume that physical matter (comprised of electrons, protons, etc.) exists. Then the following criterion for existence makes sense

Definition 2 (*Existence*) If Y is a physical entity made up of ordinary matter, and X is some kind of entity that has a demonstrable causal effect on Y as per Definition 1, then we must acknowledge that X also exists (even if it is not made up of such matter).

This is clearly a sensible and testable criterion; in the example above, it leads to the conclusion that both the data and the relevant software exist. If we do not adopt this definition, we will have instances of uncaused changes in the world; I presume we wish to avoid that situation.

Hierarchy of Scales and Causation

The basic hierarchy of physical matter is indicated in Table 3.1,[1] indicating physical scales. It is also an indication of levels of bottom-up causation, with each lower level underlying processes at the one next above in functional terms; hence one can call it a hierarchy of causation. The computer hierarchy and life sciences hierarchy are rather different: those hierarchies are based on causation rather than scale (for the computer case, see section "Complex Structures: Digital Computers").

Now a core aspect of this hierarchy of emergent properties is that one needs different vocabularies and language at the different levels of description. The concepts that are useful at one level are simply inapplicable at other levels. The effective equations at the various levels are valid at a specific level, describing same-level (emergent) behaviour at that level. They are written in terms of the relevant variables at those levels. These emergent variables can sometimes be obtained by coarse graining of lower level states, but not always so (examples will be given below). One of the characteristics of truly complex systems is that higher level variables are not always just coarse grainings of lower level variables.

The key feature then is that the higher level dynamics is effectively decoupled from lower level laws and details of the lower level variables [60]: except for some highly structured systems, you don't have to know those details in order to understand higher level behaviour. Thus you don't have to know particle physics in order to be

---

[1] A fuller description is given here: http://www.mth.uct.ac.za/~ellis/cos0.html.

Table 3.1 The hierarchy of physical matter

Level	Domain	Scale (m)	Mass (kg)	Example
L17	Cosmology	10^{26}	10^{53}	Observable Universe
L16	Large Scale Structures	10^{23}	10^{47}	Great Attractor, Sloan Great wall
L15	Galaxy Clusters	10^{22}	10^{45}	Virgo cluster, Coma cluster
L14	Galaxies	10^{21}	10^{42}	M31, NGC 1300, M87
L13	Star clusters	10^{20}	10^{35}	Messier 92, Messier 69
L12	Stellar systems	10^{12}	10^{30}	Binaries, Solar system
L11	Stars	10^{10}	10^{30}	Sun, Proxima Centauri, Eta Carinae
L10	Planets	10^{9}	10^{24}	Earth, Mars, Jupiter
L9	Continents	10^{7}	10^{17}	Africa, Australia
L8	Land forms	10^{4}	10^{8}	Atlas mountains, Andes
L7	Macro objects	1	10	Rocks, chairs, computers
L6	Materials	10^{-2}	10^{-1}	Conductors, Insulators, semi-conductors
L5	Molecules/ chemistry	10^{-9}	10^{-25}	H_2O, SiO_2, $C_6H_{12}O_6$, $C_9H_{13}N_5O_{12}P_3$
L4	Atomic physics	10^{-10}	10^{26}	Hydrogen atom, Carbon atom
L3	Nuclear physics	10^{-14}	10^{27}	Neutron, Proton, Carbon nucleus
L2	Particle physics	10^{-15}	10^{33}	Quarks, electrons, gluons
L1	Quantum gravity	10^{-35}		Superstrings

either a motor mechanic or a zoologist. However you may need some knowledge of chemistry if you are a doctor.

A sensible view is that the entities at each classical level of the hierarchy (Table 3.1) are real [19, 51]. A chair is a real chair even though it is made of atoms, which in turn are real atoms even though they are made of a nucleus and electrons, and so on; and you too are real (else you could not read this paper), as is the computer on which you are reading it. Issues of ontology may be unclear at the quantum level [28, 35], but they are clear at the macro level.

In highly ordered structures, sometimes changes in some single micro state can have major deterministic outcomes at the macro level (which is of course the environment for the micro level); this cannot occur in systems without complex structure. Examples are,

1. A single error in microprogramming in a digital computer can bring the whole thing to a grinding halt;
2. A single swap of bases in a gene can lead to a change in DNA that results in predictable disease;
3. A single small poison pill can debilitate or kill an animal, as can damage to some very specific micro areas in the brain.

This important relation between micro structure and macro function is in contrast to statistical systems, where isolated micro changes have no effect at the macro level, and chaotic systems, where a micro change can indeed lead to a macro change, but it's unpredictable. Perhaps this dependable reliance on some specific lower level details is a characteristic of genuine complexity.

The sequel: I will now look in turn at digital computers (sect. "Complex Structures: Digital Computers"); life and the brain (sect. "Complex Structures: Life and the Brain"); astronomy and cosmology (sect. "Astronomy and Cosmology"); and physics (sect. "Contextual Effects in Physics"). The latter study will open the way to considering how there can be the needed causal slack for top-down effects to take place: how is there room at the bottom for all these effects, without overriding the lower level physics? (sect. "Room at the Bottom"). The essay ends with some comments on what this all implies for our understanding of causality (sect. "The Big Picture: The Nature of Causation").

Complex Structures: Digital Computers

Structured systems such as a computer constrain lower level interactions, and thereby paradoxically create new possibilities of complex behaviour. For example, the specific connections between p-type and n-type transistors can be used to create NOR, NAND, and NOT gates [45]; these can then be used to build up adders, decoders, flip-flops, and so on. It is the specific connections between them that channels causation and so enable the lower level entities to embody such logic; the physical structure constrains the movement of electrons so as to form a structured interaction network. The key physical element is that the structure breaks symmetry (cf: [2]), thereby enabling far more complex behaviour than is possible in isotropic structures such as a plasma, where electrons can go equally in any direction. This hardware structure takes the form of a modular physical hierarchy (networks, computers, logic boards, chips, gates, and so on [62]).

The Software Hierarchy

However hardware is only causally effective because of the software which animates it: by itself hardware can do nothing. Software is also modular and hierarchically structured [62], with the higher level logic of the program driving the lower level events. The software hierarchy for digital computers is shown in Table 3.2. All but the lowest level are virtual machines.

Entering data by key strokes is a macro activity, altering macro variables. This acts down (effect T1) to set in motion a great number of electrons at the micro physics level, which (effect D1) travel through transistors arranged as logic gates at the materials level; finally (effect B1) these cause specific patterns of light on

Table 3.2 The software
hierarchy in a digital
computer system

Levels	Software hierarchy
Level 7	Applications programs
Level 6	Problem oriented language level
Level 5	Assembly language level
Level 4	Operating system machine level
Level 3	Instruction set architecture level
Level 2	Microarchitecture level
Level 1	Digital logic level
Level 0	Device level

a computer screen at the macro level, readable as text. Thus we have a chain of top-down action T1 followed by lower level dynamical processes D1, followed by bottom up action B1, these actions composed together resulting in a same level effective macro action D2:

$$D2 = B1 \circ D1 \circ T1 \qquad (3.4)$$

This is how effective same-level dynamics D2 at the higher level emerges from the underlying lower level dynamics D1. This dynamics is particularly clear in the case of computer networking [40], where the sender and receiver are far apart. At the sender, causation flows downwards from the Application layer through the Transport, Network and Link layers to the Physical layer; that level transports binary code to the other computer through cable or wireless links; and then causation flows up the same set of layers at the receiver end, resulting in effective same-level transport of the desired message from source to destination. The same level effective action is not fictitious: it is a reliable determinable dynamic effect relating variables at the macro level of description at the sender and receiver. If this were not the case you would not be able to read this article, which you obtained by such a process over the internet. There was nothing fictitious about that action: it really happened. Emergent layers of causation are real [2, 19, 51, 60].

The result is that the user sees a same-level interaction take place, and is unaware of all the lower levels that made this possible. This is information hiding, which is crucial to all modular hierarchical systems [8]. The specific flow of electrons through physical gates at the physical level is determined by whether the high level software is a music playing program, word processor, spreadsheet, or whatever: a classic case of top-down causation (the lower level interactions would be different if different software were loaded, cf. (3.2) and (3.3)). Hence what in fact shapes the flow of electrons at the gate level is the logic of the algorithms implemented by the top level computer program [43, 45].

Key Issues

Four crucial points emerge.

A: Causal Efficacy of Non Physical entities: Both the program and the data are non-physical entities, indeed so is all software. A program is not a physical thing you can point to, but by Definition 2 it certainly exists. You can point to a CD or flashdrive where it is stored, but that is not the thing in itself: it is a medium in which it is stored. The program itself is an abstract entity, shaped by abstract logic. Is the software nothing but its realisation through a specific set of stored electronic states in the computer memory banks? No it is not because it is the precise pattern in those states that matters: a higher level relation that is not apparent at the scale of the electrons themselves. Its a relational thing (and if you get the relations between the symbols wrong, so you have a syntax error, it will all come to a grinding halt). This abstract nature of software is realised in the concept of virtual machines, which occur at every level in the computer hierarchy except the bottom one [62]. But this tower of virtual machines causes physical effects in the real world, for example when a computer controls a robot in an assembly line to create physical artefacts.

B: Logical relations rule at the higher levels: The dynamics at all levels is driven by the logic of the algorithms employed in the high level programs [41]. They decide what computations take place, and they have the power to change the world [43]. This abstract logic cannot be deduced from the laws of physics: they operate in a completely different realm. Furthermore the relevant higher level variables in those algorithms cannot be obtained by coarse graining any lower level physical states. They are not coarse-grained or emergent variables: they are assigned variables, with specific abstract properties that then mediate their behaviour.

C: Underlying physics allows arbitrary programs and data: Digital computers are universal computers. The underlying physics does not constrain the logic or type of computation possible, which Turing has shown is universal [14]. Physics does not constrain the data used, nor what can be computed (although it does constrain the speed at which this can be done). It enables the higher level actions rather than constraining them. The program logic dictates the course of things.

D: Multiple realisability at lower levels. The same high level logic can be implemented in many different ways: electronic (transistors), electrical (relays), hydraulic (valves), biological (interacting molecules) for example. The logic of the program can be realised by any of these underlying physical entities, which clearly demonstrates that it is not the lower level physics that is driving the causation. This multiple realisability is a key feature characterising top-down action [4]: when some high level logic is driving causation at lower levels, it does not matter how that logic is physically instantiated: it can be realised in many different ways. Thus the top-down map T1 in (3.5) is not unique: it can be realised both in different physical systems, and in different micro states of the same system.

Equivalence Classes

The last point means that we can consider as the essential variables in the hierarchy, the equivalence classes of lower level states that all that correspond to the same high level state [4]. When you control a top level variable, it may be implemented by any of the lower level states that correspond to the chosen high level state; which one occurs is immaterial, the high level dynamics is the same. You even can replace the lower level elements by others with the same functionality, the higher entity remains the same (a classic example: the cells in your body are now all different than they were 7 years ago; you are made up of different matter, but you are still the same person).

In digital computers, there are such equivalences all over the place:

- at the circuit level: one can use Boolean algebra to find equivalent circuits;
- at the implementation level: one can compile or interpret (giving completely different lower level functioning for same higher level outcome);
- at the hardware level, one can run the same high level software on different microprocessors;
- even more striking is the equivalence of hardware and software in much computing (there is a completely different nature of lower level entities for the same higher level outcomes).

In each case this indicates top-down effects are in action: the higher level function drives the lower level interactions, and does not care how it is implemented (information hiding is taking place).

As to the use of the computer for represent text, the keyboard letters are never exactly identical: yet the abstract letter "A" represented by them is still the letter "A" despite many possible variations. It is also the letter "A" if

- you change font (Times New Roman to Helvetica)
- you change to bold or italic
- you change size of the font
- you change colour of the font
- you change the medium from light on a computer screen to ink on paper

Such multiple realisability occurs at all levels in a text. One of the key problems in generating intelligent understanding is to assign all these different representations to the same abstract entity that they all represent. This way varied lower level representations of a higher level entity occur is characteristic of top-down causation [4]: what matters is the equivalence class of all these representations, which is the characteristic of the higher level entity, not which particular representation has been chosen (see the Appendix).

Implications

Hence although they are the ultimate in algorithmic causation as characterized so precisely by Turing [14], digital computers embody and demonstrate the causal efficacy of non-physical entities. The physics allows this; it does not control what takes place, rather it enables the higher level logic to be physically implemented. Computers exemplify the emergence of new kinds of causation out of the underlying physics, not implied by physics but rather by the logic of higher level possibilities as encoded in data structures and algorithms [8, 41, 43]. This leads to a different phenomenology at each of the levels of Table 3.2, described by effective laws for that level, and an appropriate language. A combination of bottom up causation and contextual affects (top-down influences) enables their complex functioning.

Complex Structures: Life and the Brain

Living systems are modular hierarchical system, for the same reasons as in the case of digital computers: this structuring enables complex behaviour *inter alia* it because it allows class structures with inheritance, information hiding, and abstraction [8]. The lower level interactions are constrained by recurrent network structures, thereby creating possibilities of truly complex behaviour. But the core dynamic that allows true complexity to emerge is adaptive selection. Biological systems are finely adapted to their physical, ecological, and social environments: and that cannot take place without topdown information flows from the environment to the structure and behaviour of organisms. This takes place on evolutionary, developmental, and functional timescales.

Microbiology

The rise of epigenetics has shown that the functioning of molecular biology is controlled by many epigenetic mechanisms that are sensitive to environmental effects, such as DNA methylation [34]. Consequently the view that biology is controlled bottom up by the actions of genes alone is fatally flawed.[2] Contextual effects are crucial in determining physical outcomes.

---

[2] For a comprehensive discussion, see the many links on Denis Noble's webpage at http://musicoflife.co.uk/.

Physiology

The molecular basis behind the physiology of an animal obeys the laws of physics and chemistry. But by themselves they do not create entities such as a sensory or circulatory system, nor determine their mode of functioning and resulting physical outcomes. When you study the physiology of the heart you find it cannot be understood except in terms of the interplay of bottom up and top down causation, which determines which specific molecular interactions take place where and at what time [50, 51]. Bottom up physics alone cannot explain how a heart comes into being, nor what its design is, nor its regular functioning.

The Brain

Top-down causation is prevalent at all levels in the brain: for example it is crucial to vision [32, 39] as well as the relation of the individual brain to society [13] and social decision making [63]. The hardware (the brain) can do nothing without the excitations that animate it: indeed this is the difference between life and death. The mind is not a physical entity, but it certainly is causally effective: proof is the existence of the computer on which you are reading this text. It could not exist if it had not been designed and manufactured according to someones plans, thereby proving the causal efficacy of thoughts, which like computer programs and data are not physical entities.

This is made possible firstly by the hierarchical structuring in the brain described in [31]. His "forward connections" are what I call bottom up, and his "backward connections" are what I call top-down (the difference in nomenclature is obviously immaterial). He makes quite clear that a mix of bottom up and top down causation is key as to how the brain works; backward connections mediate contextual effects and coordinate processing channels. For example, the visual hierarchy includes 10 levels of cortical processing; 14 levels if one includes the retina and lateral geniculate nucleus at the bottom as well as the entorhinal cortex and hippocampus at the top [27]. Secondly, it depends on context-dependent computation by recurrent dynamics in prefrontal cortex [46]. And thirdly, it happens by top-down reorganization of activity in the brain after learning tasks on developmental timescales [33, 61] and by environmental influence (of the effect of stress) on childrens physiological state by influencing telomere length in chromosomes [48]. On evolutionary timescales, basic social responses have been built into the brain through evolutionary processes even in animals as simple as C. elegans, where a hub-and-spoke circuit drives pheromone attraction and social behaviour [44].

These structural features and related cortical functioning are reflected in the way the brain functions at the higher functional levels. I will give two illustrations of to-down processes in psychology.

Example 1: Reading

How does reading work? Heres a remarkable thing.

- You can read this even through words are misspelt,
- and this through letters are wrong,
- And this through words missing.

How can it be we can make sense of garbled text in this way? One might think the brain would come to a grinding halt when confronted with such incomplete or grammatically incorrect text. But the brain does not work in a mechanistic way, first reading the letters, then assembling them into words, then assembling sentences. Instead our brains search for meaning all the time, predicting what should be seen and interpreting what we see based on our expectations in the current context.

Actually words by themselves may not make sense without their context. Consider:

- The horses ran across the plane,
- The plane landed rather fast,
- I used the plane to smooth the wood.

  – what 'plane means differs in each case, and is understood from the context. Even the nature of a word (noun or verb) can depend on context:

- Her wound hurt as she wound the clock

This example shows you cant reliably tell from spelling how to pronounce words in English, because not only the meaning, but even pronunciation depends on context.

The underlying key point is that we are all driven by a search for meaning: this is one of the most fundamental aspects of human nature, as profoundly recorded by Viktor Frankl in his book *Man's Search for Meaning* [30]. Understanding this helps us appreciate that reading is an ongoing holistic process: the brain predicts what should be seen, fills in what is missing, and interprets what is seen on the basis of what is already known and understood. And this is what happens when we learn to read, inspired by the search for understanding. One learns the rules of grammar and punctuation and spelling too of course; but such technical learning takes place as the process of meaning making unfolds. It is driven top down by our predictions on the basis of our understandings, based in meaning.

Example 2: Vision

Vision works in essentially the same way, as demonstrated by Dale Purves in his book *Brains: How They Seem to Work* [59]. The core of his argument is as follows [58]

> The evolution of biological systems that generate behaviorally useful visual percepts has inevitably been guided by many demands. Among these are: 1) the limited resolution of photoreceptor mosaics (thus the input signal is inherently noisy); 2) the limited number of

neurons available at higher processing levels (thus the information in retinal images must be abstracted in some way); and 3) the demands of metabolic efficiency (thus both wiring and signaling strategies are sharply constrained). The overarching obstacle in the evolution of vision, however, was recognized several centuries ago by George Berkeley, who pointed out that the information in images cannot be mapped unambiguously back onto real-world sources (Berkeley, 1975). In contemporary terms, information about the size, distance and orientation of objects in space are inevitably conflated in the retinal image. In consequence, the patterns of light in retinal stimuli cannot be related to their generative sources in the world by any logical operation on images as such. Nonetheless, to be successful, visually guided behavior must deal appropriately with the physical sources of light stimuli, a quandary referred to as the "inverse optics problem".

The resolution is top-down shaping of vision by the cortex, based in prediction of what we ought to see. Visual illusions are evidence that this is the way the visual system solves this problem [31, 59].

Adaptive Selection

As was mentioned above, adaptive selection is one of the most important types of top-down causation. It is top-down because the selection criteria are at a different conceptual level than the objects being selected: in causal terms, they represent a higher level of causation [22]. Darwinian selection is the special case when one has repeated adaptive selection with heredity and variation [56]. It is top-down because the result is crucially shaped by the environment, as demonstrated by numerous examples: for example a polar bear is white because the polar environment is white. Section "Complex Structures: Life and the Brain" of [47] emphasizes how downward causation is key to adaptive selection, and hence to evolution. An important aspect is that *multilevel* selection must occur in complex animals: environmental changes have no causal handle directly to genes but rather the chain of causation is via properties of the group, the individual, or physiological systems at the individual level, down to the effects of genes; that is, it is inherently a multilevel process [24].

The key point about adaptive selection (once off or repeated) is that it lets us locally go against the flow of entropy, and lets us build up and store useful information through the process of deleting what is irrelevant. Paul Davies and Sara Walker explain that this implies that evolutionary transitions are probably closely related to top-down causation [64]:

> Although it has been notoriously difficult to pin down precisely what it is that makes life so distinctive and remarkable, there is general agreement that its informational aspect is one key property, perhaps the key property. The unique informational narrative of living systems suggests that life may be characterized by context-dependent causal influences, and in particular, that top-down (or downward) causation – where higher-levels influence and constrain the dynamics of lower-levels in organizational hierarchies – may be a major contributor to the hierarchal structure of living systems. Here we propose that the origin of life may correspond to a physical transition associated with a fundamental shift in causal structure. The origin of life may therefore be characterized by a transition from bottom-up to top-down causation, mediated by a reversal in the dominant direction of the flow of

information from lower to higher levels of organization (bottom-up), to that from higher to lower levels of organization (top-down). Such a transition may be akin to a thermodynamic phase transition, with the crucial distinction that determining which phase (nonlife or life) a given system is in requires dynamical information and therefore can only be inferred by identifying causal relationships. We discuss one potential measure of such a transition, which is amenable to laboratory study, and how the proposed mechanism corresponds to the onset of the unique mode of (algorithmic) information processing characteristic of living systems.

However adaptive selection occurs far more widely than that; e.g. it occurs in state vector preparation [22], as I indicate below.

Astronomy and Cosmology

I now turn to the physical sciences: first astronomy and cosmology, and then physics itself.

Astronomy

In the context of astronomy/astrophysics, there is a growing literature on contextual effects such as suppression of star formation by powerful active galactic nuclei [52]. This is a top-down effect from galactic scale to stellar scale and thence to the scale of nuclear reactions. Such effects are often characterised as feedback effects.

Cosmology

In cosmology, there are three venerable applications of the idea of top-down effects: Olber's paradox, Machs Principle, and the Arrow of Time [7, 20] In each case it has been strongly suggested that boundary conditions on the Universe as a whole are the basic cause of crucial local effects (the dark night sky, the origin of inertia, and the local direction of time that characterizes increasing entropy). Machs Principle is not now much discussed, and I will not consider it further. More recent examples are nucleosynthesis and structure formation in the expanding universe, though they are not usually expressed in this way.

Nucleosynthesis

In the case of element formation in the early universe, macro-level variables (average cosmic densities) determine the expansion rate of the cosmos, which determines the temperature-time relation $T(t)$. The Einstein equations for a radiation dominated cosmology lead to a hot state evolving as

$$S(t) \propto t^{2/3}, \quad t = \left(\frac{c^2}{15.5\pi GaT^4}\right)^{1/2} \Rightarrow \frac{T}{10^{10}K} = \left[\frac{t}{1.92\,\text{sec}}\right]^2 \tag{3.5}$$

This is the context within which nucleosynthesis takes place, and in turn determines the rates of nuclear reactions (micro-level processes) and hence the outcome of nucleosynthesis, leading to macro variables such as the overall mass fraction of Helium and Lithium in the early Universe. Occurrence of this top-down effect is why we can use element abundance observations to constrain cosmological parameters [18].

Structure Formation

Another example is structure formation in the expanding universe, studied by examining the dynamics of perturbed cosmological models. Again macro-level variables occur as coefficients in the relevant equations, determining the growth of perturbations and hence leading to macro variables such as the power spectrum of structure in the universe. Occurrence of this top-down effect is the reason we can use observations of large scale structural features, such as power spectra and identification of baryon acoustic oscillations, to constrain cosmological parameters [18, 49].

Olbers Paradox

The calculations leading to understanding of the CBR spectrum [18, 49] are basically the present day version of the resolution of Olber's paradox (why the is the night sky not as bright as the surface of the Sun): one of the oldest calculations of global to local effects [7]. The essence of the resolution is that the universe is not infinite and static: it has expanded from a state with a low entropy to baryon ratio.

The Arrow of Time

Boltzmanns H-theorem wonderfully proves, in a purely bottom up way by coarse-graining the microphysics states, that entropy S always increases with time: $dS/dt > 0$. However the proof also works if one reverses the direction of time: define $T := -t$, exactly the same proof holds again, step by step, for this time T too, which proves that also $dS/dT > 0$ [55]. The same holds for Weinberg's proof of entropy increase ([66, p.150]), which is formulated in the language of quantum field theory and avoids approximations such as the Born approximation. His H-Theorem (3.6.20) will hold for both directions of time (just reverse the direction of time and relabel α to β: the derivation goes through as before) because the derivation depends only on unitarity. Unitary transformations however are time reversible; there is therefore nothing in the dynamics that can choose one time direction as against the other as far as any dynamical development is concerned, just as there is no intrinsic difference between the particles α and β.[3]

This is Loschmidt's paradox: because the microphysics considered in these two derivations is time reversible, whichever direction of time you choose, exactly the same proof shows that S cannot decrease in the opposite direction of time too. So you cannot determine the arrow of time from time reversible microphysics alone; it must come from somewhere else and the best candidate is special initial conditions at the start of the universe on a macro scale [1, 12, 55]. In fact this is obvious: if the micro physics is time symmetric, you cannot derive the macro arrow of time from it alone, because of the symmetry under $t \rightarrow -t$; it must derive from the cosmological context. A clear presentation of why some kind of non-local condition is necessary to resolve the arrow of time issue is given by [10]: bottom up effects alone are not able to resolve it.

Thus the time asymmetry can come not from the dynamic equations of the theory but from the boundary conditions. Consequently, one needs some global boundary condition to determine a consistent arrow of time in local physics: a top-down effect from the cosmological scale to everyday scales. This global coordination is plausibly provided by a macro-scale low entropy condition of the early universe [12, 55]. Some will claim this is "nothing but" the combined effect of all the particles and fields in the very early universe. Well yes and no: the effect is indeed due to all those particles, but it depends crucially on the specific relationship between them. Change that relationship from smooth to very clumpy: exactly the same particles will be there, but will have a very different effect. It's the spatial distribution that matters: the relationship between them is the specific cause of the local physical effect of existence of a unique arrow of time. You cannot describe that higher level relationship in terms of the variables and scales relevant at the lower levels in Table 3.1. And the outcome is crucially important: life would not be possible without a well-established local arrow of time.

Contextual Effects in Physics

Top-down causation happens wherever in addition to initial conditions, boundary conditions materially determine physical results; then environmental variables (a macro scale concept) act down to determine the values of physical fields locally. Such effects occur in both classical and quantum physics.

---

[3] This derivation does not refer to the time irreversibility of the weak interaction, which has no direct effect on everyday life: it cannot be the source of the arrow of time in physical chemistry and biology.

Classical Physics

Examples of top-down effects in classical physics are standing waves; hearing the shape of a drum [38]; outcomes of the reaction diffusion equation, for example in pattern formation in embryos. In each case the outcome is largely determined by the shape of the system boundary. Physicists often try to minimise such boundary effects through the idealisation of an isolated system; but no real system is isolated either in space or in time. Furthermore, that is in fact just another example of specific boundary conditions: asymptotic flatness or outgoing radiation conditions are just as much contextual constraints as periodic boundary conditions are.

There is nothing new in all this: its just that we don't usually talk about this as a top-down effect. It may be helpful to do so; so I give two other rather different examples.

The Ideal Gas Law

An example of top down action is given by the ideal gas law for gas constrained in a cylinder by a piston. The relation

$$PV = nRT \tag{3.6}$$

between the macroscopic variables pressure P, volume V, amount of gas n, and temperature T does not depend on detailed microscopic variables such as velocities, positions and masses of any specific molecules; indeed we don't know those values. The variables T and V at the macro level are the only handle we have on the state of a given body of gas n: we can't (except in unusual circumstances) manipulate the micro level variables directly. But we can change them by altering macro variables (e.g. by compressing the gas, so changing V), which then changes lower level states, speeding up the molecules so as to increase the pressure. This is the top down effect of the higher level variables on lower level states. The universal constant R is the link between the micro and macro states, because it relates the energy of micro states to the values observed at the bulk level.

Electromagnetism

The electromagnetic field is described by an anti-symmetric tensor F_{ab} made up of electric and magnetic field components. The micro level laws are (i) Maxwell's equations for the electromagnetic field, including the Gauss law, with particles such as electron and protons as sources; (ii) the Lorentz force law for the effect of electric and magnetic fields on charged particles such as protons and electrons. [29]. The lower level dynamics is the way the micro level electric and magnetic fields interact with each other according to Maxwell's equations, with the charged particles as sources, and these fields in turn exert forces on those particles via the Lorentz force law, that cause them to move if they are free to do so. Bottom up effects are the way that billions of electrons in motion at the micro level generate measurable electric currents at the macro level, such as in a generator, and the way that forces on billions of charged particles at the micro level can add up to generate macroscopic forces, such as in an electric motor.

Top down effects are for example the way that electric coils (macro entities, not describable in terms of micro level variables) constrain the motion of electrons in specific directions. This is why the micro fields they generate add up to macro level fields ([29, pp. 13-5 and 13-6]). Without the constraints exerted by the wires, no such macro fields would be generated. Similarly the constraints generated by the design of the electric motor ensure that the individual forces on electrons and protons are channelled so as to add up to measurable macro level forces. This constraining effect is top down action from the level of machine components to the level of protons and electrons. If one looks for example at the Feynman derivation of the magnetic field due to electrons flowing in a wire coil, the wire is represented as a structureless macro entity, even though it is made up of atoms and electrons. We just take this macro structure (the physical wire) for granted in all such derivations, without enquiring how it is made up of micro entities.

Thus the causal effectiveness of macro entities—the way the wire channels the flow of electrons along the direction of the wire—is taken for granted. Yes of course it is made up of atoms and electrons at the micro level, but that is irrelevant to it's function in the experiment, which role is due to the macro organisation embodied in this structure. These structural constraints act down to organise micro events, as is very clear in the case of the wire: its physical structure prevents electrons moving sideways out of the wire. This top-down aspect of what is going on is hidden because we take it for granted. It's just part of what we assume to be the case, so we don't notice it.

Quantum Physics

Top down effects occur in the context of quantum physics as well [22]. Here are some examples.

Band Structure and Quantum Cooperative Effects

The periodic crystal structure in a metal leads (via Blochs theorem) to lattice waves, and an electronic band structure depending on the particular solid involved, resulting in all the associated phenomena resulting from the band structure [67] such as electrical resistivity and optical absorption, as well as being the foundation of processes in solid-state devices such as transistors and solar cells. The entire machinery for describing the lattice periodicity refers to a scale much larger than that of the electron, and hence is not describable in terms appropriate to that scale.

Thus these effects all exist because of the macro level properties of the solid—the crystal structure—and hence represent top-down causation from that structure to the electron states. This can lead to existence of quasiparticles such as phonons (collective excitations in a periodic, elastic arrangement of atoms or molecules) that result from vibrations of the lattice structure. Because these are all based in top-down action, they are emergent phenomena in the sense that they simply would not exist if the macro-structure did not exist, and hence cannot be understood by a purely bottom-up analysis, as emphasized strongly by Laughlin [42].

Caldeira-Leggett Model

The Caldeira-Leggett model is a model for a system plus heat reservoir, used for the description of dissipation phenomena in solid state physics, as discussed in [9]. Here the Lagrangian of the composite system T consisting of the system S of interest and a heat reservoir B takes the form

$$L_T = L_S + L_B + L_I + L_{CT}, \tag{3.7}$$

where L_S is the Lagrangian for the system of interest, L_B that for the reservoir (a set of non-interacting harmonic oscillators), and L_I that for the interaction between them. The last term L_{CT} is a 'counter term', introduced to cancel an extra harmonic contribution that would come from the coupling to the environmental oscillators. This term represents a top-down effect from the environment to the system, because L_I completely represents the lower-level interactions between the system and the environment. The effect of the heat bath is more than the sum of its parts when $L_{CT} \neq 0$, because the summed effect of the parts on each other is given by L_I. The bottom up effects of lower level forces acting on the components of the system are completely described by L_S and L_I; they were carefully constructed that way. Similarly the bottom up dynamics of the environment is fully described by L_B. The term L_{CT} is thus not an outcome of the bottom up effects alone.

Superconductivity and Superfluidity

In superconductivity, the electrons—despite their repulsion for each other—form pairs ('Cooper pairs') which are the basic entities of the superconducting state. This happens by a cooperative process: the negatively charged electrons cause distortions of the lattice of positive ions in which they move, and the real attraction occurs between these distortions. It also can lead to superfluidity. These effects can only occur because of the specific nature of the lattice structure, which is thus the source of the existence of the Cooper pairs. The Nobel lecture by Laughlin [42] emphasizes the implications:

One of my favourite times in the academic year occurs in early spring when I give my class of extremely bright graduate students, who have mastered quantum mechanics but are otherwise unsuspecting and innocent, a take-home exam in which they are asked to deduce superfluidity from first principles. There is no doubt a very special place in hell being reserved for me at this very moment for this mean trick, for the task is impossible. Superfluidity, like the fractional Hall effect, is an emergent phenomenon - a low-energy collective effect of huge numbers of particles that cannot be deduced from the microscopic equations of motion in a rigorous way, and that disappears completely when the system is taken apart.... The world is full of things for which one's understanding, i.e. one's ability to predict what will happen in an experiment, is degraded by taking the system apart, including most delightfully the standard model of elementary particles itself.

The claim made here is that this dynamics is possible because of top-down causation.

Proton-Coupled Electron Transfer

Proton coupled electron transfer (PCET) describes the joint transfer of an electron and proton to or from a substrate [65], hence it can only occur in the presence of suitable substrate. This substrate is a higher level structure, i.e. at a larger scale than the electron and proton: what happens at the electron's level would not happen if the specific substrate structure were not there. Photosynthesis is an example of PCET, with the transfer of 24 e^- and 24 H^+ driven by at least 48 photons.

State Vector Preparation

State vector preparation is key to experimental set-ups, and is a non-unitary process because it can produce particles in a specific eigenstate from a stream of particles that are not in such a state. Indeed it acts rather like state vector reduction, being a transition that maps a mixed state to a pure state.

How can this non-unitary process happen in a way compatible with standard unitary quantum dynamics? The crucial feature is pointed out by Isham [37]: the outcome states are drawn from some collection E_i of initial states by being selected by some suitable apparatus, being chosen to have some specic spin state in the Stern-Gerlach experiment; the other states are discarded. This happens in two basic ways: separation and selection (as in the Stern Gerlach experiment), which is unitary up to the moment of selection when it is not, or selective absorption (as in the case of wire polarisers), which continuously absorbs energy. This top-down effect from the apparatus to the particles causes an effective non-unitary dynamics at the lower levels, which therefore cannot be described by the Schrödinger or Dirac equations.

In such situations, selection takes place from a (statistical) ensemble of initial states according to some higher level selection criterion, which is a form of top-down causation from the apparatus to the particles. This is a generic way one can create order out of a disordered set of states, and so generate useful information by throwing away what is meaningless. The apparatus is specifically designed to have this non-unitary effect on the lower level [22]. This is the essential process not only in state vector preparation but in all purification processes: which are the core of chemistry and chemical engineering. they are the foundation of technology.

Room at the Bottom

Given this evidence for top-down causation, the physicist asks, how can there be room at the bottom for top-down causation to take place? Isn't there over-determination because the lower level physics interactions already determine what will happen from the initial conditions?

There are various ways that top down causation can be effective, based in lower level physical operations:

Setting Constraints on Lower Level Interactions

Structural constraints break symmetries, and so for example create more general possibilities than are available to cellular automata. This has been explained above in the context of digital computers. It is also the case where effective potentials occur due to local matter structuring [23]. These are crucially dependent on the nature of the higher level structure (e.g. transistors connected together in integrated circuits in a computer, or networks of neurons connected by synapses in a brain) that cannot be described in terms of lower level variables.

It is the higher level patterns that are the essential causal variable in solid state physics by creating a specific band structure in solids (hence for example the search for materials that will permit high temperature superconductivity). Which specific lower level entities create them is irrelevant: you can move around specific protons and electrons while leaving the band structure unchanged. This multiple realisability is always the case with effective potentials.

Changing the Nature of the Constituent Entities

The billiard ball model of unchanging lower level entities underlying higher level structure is wrong. Hydrogen in a water molecule has completely different properties than when free; electrons bound in atom interact with radiation quite differently than when free; neutrons bound in nucleus have a half life of billions of years but they decay in 11 1/2 minutes when free. Assuming that the nature of an entity is characterised by the way it interacts with others, in each case the higher level context has changed the nature of the underlying entities (an effect that is commonplace in biology). In the case of string theory, the nature of fundamental particles depends on the string theory vacuum a non-local higher context for their existence. Their properties are not invariant, they depend on this vacuum.

Creating Constituent Entities

In many cases the lower level entities would not even exist without the higher level structure. For example, the very possibility of existence of phonons is a result of the physical structure of specific materials (section "Superconductivity and Superfluidity"). This structure is at a higher level of description than that of electrons. Top down effects clearly occur when lower level entities cannot exist outside their higher level context (again, a common effect in biology, where symbiosis is rife).

This is however foundational in quantum field theory, where even the existence of particles is context dependent;

> One of the lessons learner from the development of the subject [quantum field theory] has been the realisation that the particle concept does generally not have universal significance. Particles may register their presence on some detectors but not others, so there is an essential observer-dependent quality about them. One is still free to assert the presence of particles, but without specifying the state of motion of the detector, the concept is not very useful. ([6, p. 49]).

Deleting Lower Level Entities

There is not a fixed set of lower level entities when selection creates order out of disorder by deleting unwanted lower level entities or states: top-down action selects what the lower elements will be. This selective top-down process is what underlies state vector preparation in quantum physics [23, 37]. It is crucial in evolutionary biology and in the mind [22]. It is the way order is created for chaos and information is garnered from a jumble of incoming data.

Statistical Fluctuations and Quantum Uncertainty

Lower level physics is not determinate: random fluctuations and quantum indeterminism are both in evidence. What happens is not in the throes of iron determinism: random events take place at the micro level. These can get amplified to macro scales.

In complex systems, this unpredictable variability can result in an ensemble of lower level states from which a preferred outcome is selected according to higher level selection criteria. Thus top-down selection leading to increased complexity [25] is enabled by the randomness of lower level processes.

Overall, there are often not fixed lower level entities: their nature, or indeed their existence, is dependent on their higher level context.

The Big Picture: The Nature of Causation

The view put here, in agreement with [64], is that top-down effects is the key to the rise of genuine complexity (such as computers and human beings) in the universe.

Emergence

We need to explain genuine emergence, with all its complexity, and the reality of all the levels:

> We seek reality, but what is reality? The physiologists tell us that organisms are formed of cells; the chemists add that cells themselves are formed of atoms. Does this mean that these atoms or these cells constitute reality, or rather the sole reality? The way in which these cells are rearranged and from which results the unity of the individual, is not it also a reality much more interesting than that of the isolated elements, and should a naturalist who had never studied the elephant except by means of the microscope think himself sufficiently acquainted with that animal?—Henri Poincare [57].

Each higher level is real with its own autonomy. This is in accord with the view put by Denis Noble [51].

The Main Thesis

How does this come into being? The main thesis of this paper is

> **Hypothesis**: bottom up emergence by itself is strictly limited in terms of the complexity it can give rise to. Emergence of genuine complexity is characterised by a reversal of information flow from bottom up to top down.

The degree of complexity that can arise by bottom-up causation alone is strictly limited. Sand piles, the game of life, bird flocks, or any dynamics governed by a local rule [3] do not compare in complexity with a single cell or an animal body. Spontaneously broken symmetry is powerful [2], but not as powerful as symmetry breaking that is guided top-down by adaptive selection to create ordered structures (such as brains and computers).

Some kind of coordination of effects is needed for such complexity to emerge. David Deutsch has a classic comment on the topic in his book *The Fabric of Reality* [17]

For example, consider one particular copper atom at the tip of the nose of the statue of Sir Winston Churchill that stands in Parliament Square in London. Let me try to explain why that copper atom is there. It is because Churchill served as prime minister in the House of Commons nearby; and because his ideas and leadership contributed to the Allied victory in the Second World War; and because it is customary to honor such people by putting up statues of them; and because bronze, a traditional material for such statues, contains copper, an so on. Thus we explain a low-level physical observation– the presence of a copper atom at a particular location– through extremely high-level theories about emergent phenomena such as ideas, leadership, war and tradition. There is no reason why there should exist, even in principle, any lower-level explanation of the presence of that copper atom than the one I have just given.

Another example is particle collisions at the LHC at CERN: these are the result of the top down effect of abstract thoughts in the minds of experimenters to the particle physics level. Without these thoughts, there would be no such collisions.

The Overall View

A key point I make in my essay is that top down effects don't occur via some mysterious non-physical downward force, they occur by higher level physical relations *setting constraints* on lower level interactions, which not only can change the nature of lower level entities (as in the case of the chameleon particles that might be the nature of dark matter), they can even lead to the very existence of such entities (e.g. phonons or Cooper pairs). So it is indeed a two-way causal flow which enables abstract entities to be causally effective (as in the case of digital computers) but does not violate normal physics. It is a largely unrecognised aspect of normal physics.

There are some great discussions of the nature of emergent phenomena in physics [2, 36, 42, 60] but none of them specifically mention the issue of top down causation. This paper proposes that recognising this feature will make it easier to comprehend the physical effects underlying emergence of genuine complexity, and may lead to useful new developments, particularly to do with the foundational nature of quantum theory.

A bottom-up view is taken as an underlying principle of faith by somehard core reductionists, who simply ignore the contextual effects that in fact occur: for example claiming that biology is controlled bottom up by genes alone, thereby ignoring all the discoveries of epigenetics, which prove this false [34, 50]. But such reductionism is always a cheat, because it is always only partial. An example is that Francis Crick famously wrote [15]

You, your joys and your sorrows, your memories and your ambitions, your sense of personal identity and free will, are in fact no more than the behavior of a vast assembly of nerve cells and their associated molecules.

But nerve cells and molecules are made of electrons plus protons and neutrons, which are themselves made of quarks ...so why not

You, your joys and your sorrows, your memories and your ambitions, your sense of personal
identity and free will, are in fact no more than the behavior of a vast assembly of quarks and
electrons

And these themselves are possibly vibrations of superstrings. Why does he stop
where he does?—because that's the causal level he understands best!—he's not a
particle physicist. But if he assumes that the level of cells and molecules is real, it's
an arbitrary assumption unless *all* levels are real—which is my position. It's the only
one that makes sense.

What I am pointing out in my essay is that physics does not by itself determine
what happens in the real world. Physics per se cannot account for the existence of
either a teapot or a Jumbo jet airliner, for example. At no point have I in any ways
denied that the laws of physics and chemistry apply at the lower levels. Of course
they do. The point is that they do not by themselves determine what happens. That
is determined by top down causation, as is abundantly clear for example in the case
of the computer. Some physical processes are not emergent but are entailed in a
top-down way. For example there is no bottom up process by which the computer
memory states embodying a Quicksort algorithm can emerge from the action of the
underlying physics acting in a purely bottom-up way. Indeed the same is true of the
processes leading to creation of a teacup or a pair of spectacles (see [21]).

If you believe this is wrong, please advise me of a physical law or process that
unambiguously determines how a tea cup can be created in a purely bottom-up way.
You will not be able to do so—it does not exist.

Appendix: Equivalence Classes

The key idea is that of functional equivalence classes. Whenever you can identify
existence of such equivalence classes, that is an indication that top-down causation
is taking place [4]. Indeed this is essentially the ontological nature of the higher level
effective entity: a computer program is in its nature the same as the set of all possible
implementations of the set of logical operations it entails. These are what enter into
the higher level effective relations; they can be described in many different ways,
and implemented in many different ways; what remains the same in those variants
is the core nature of the entity itself.

An equivalence relation is a binary relation $\sim$ satisfying three properties:

1. For every element a in X, $a \sim a$ (reflexivity),
2. For every two elements a and b in X, if $a \sim b$, then $b \sim a$ (symmetry), and
3. For every three elements a, b, and c in X, if $a \sim b$ and $b \sim c$, then $a \sim c$
   (transitivity).

The *equivalence class* of an element a is denoted $[a]$ and may be defined as the set
of elements that are related to a by $\sim$.

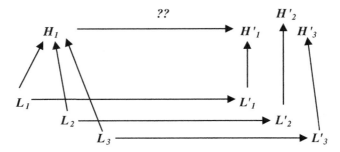

Fig. 3.1 The lower level dynamics does not lead to coherent higher level dynamics when the lower level dynamics acting on different lower level states corresponding to a single higher level state, give new lower level states corresponding to different higher level states

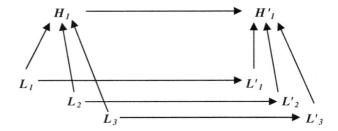

Fig. 3.2 The lower level dynamics leads to coherent higher level dynamics when the lower level dynamics acting on different lower level states corresponding to a single higher level state, give new lower level states corresponding to the same higher level state

The way this works is illustrated in (Figs. 3.1 and 3.2) higher level states H1 can be realised via various slower level states L1. This may or may not result in coherent higher level action arising out of the lower level dynamics.

When coherent dynamics emerges, the set of all lower states corresponding to a single higher level state form an equivalence class as far as the higher level actions are concerned. They are indeed the effective variables that matter, rather than the specific lower level state that instantiates the higher level one. This is why lower level equivalence classes are the key to understanding the dynamics (Figs. 3.1).

References

1. D.Z. Albert, *Time and Chance* (Harvard University Press, Cambridge, 2000)
2. P.W. Anderson, More is different. Science **177**, 393–396 (1972)
3. P. Arrighi, J. Grattange, The quantum game of life. Phys. World (2012), http://physicsworld.com/cws/article/indepth/2012/may/31/the-quantum-game-of-life
4. G. Auletta, G.F.R. Ellis, L. Jaeger, Top-down causation by information control: from a philosophical problem to a scientific research program. J. R. Soc. Interface **5**, 1159–1172 (2008)

5. P.L. Berger, T. Luckmann, *The Social Construction of Reality: A Treatise in the Sociology of Knowledge* (Penguin, London, 1991)
6. N.D. Birrell, P.C.W. Davies, *Quantum Fields in Curved Space* (Cambridge University Press, Cambridge, 1982)
7. H. Bondi, *Cosmology* (Cambridge University Press, Cambridge, 1960)
8. G. Booch, *Object-Oriented Analysis and Design with Applications* (Addison-Wesley Professional, 1993)
9. H.P. Breuer, F. Petruccione, *The Theory of Open Quantum Systems* (Clarendon Press, Oxford, 2006)
10. C. Callender, Thermodynamic asymmetry in time (2011), http://plato.stanford.edu/entries/time-thermo/
11. N.A. Campbell, J.B. Reece, *Biology* (Benjamin Cummings, San Francisco, 2005)
12. S. Carroll, *From Eternity to Here: The Quest for the Ultimate Arrow of Time* (Dutton, New York, 2010)
13. J.Y. Chiao, Cultural neuroscience. Prog. Brain Res. **178**, 287–304 (2009)
14. B.J. Copeland, *The Essential Turing* (Clarendon Press, Oxford, 2004)
15. F. Crick, *Astonishing Hypothesis: The Scientific Search for the Soul* (Scribner, New York, 1995)
16. J.P. Crutchfield, Between order and chaos. Nat. Phys. **8**, 17–24 (2011)
17. D. Deutsch, *The Fabric of Reality* (Penguin Books, London, 1998)
18. S. Dodelson, *Modern Cosmology* (Academic Press, New York, 2003)
19. A.S. Eddington, *The Nature of the Physical World* (Cambridge University Press, Cambridge, 1928)
20. G.F.R. Ellis, Cosmology and local physics. Int. J. Mod. Phys. **A17**, 2667–2672 (2002)
21. G.F.R. Ellis, Physics, complexity and causality. Nature **435**, 743 (2005)
22. G.F.R. Ellis, Top down causation and emergence: some comments on mechanisms. J. Roy Soc Interface Focus **2**, 126–140 (2012)
23. G.F.R. Ellis, On the limits of quantum theory: contextuality and the quantumclassical cut. Ann. Phys. **327**, 1890–1932 (2012)
24. G.F.R. Ellis, Multi-level selection in biology (2013), arXiv:1303.4001
25. G.F.R. Ellis, Necessity, purpose, and chance: the role of randomness and indeterminism in nature from complex macroscopic systems to relativistic cosmology (2014). Available at http://www.mth.uct.ac.za/~ellis/George_Ellis_Randomness.pdf
26. G.F.R. Ellis, D. Ellis, D. Noble, T. O'Connor (eds.), Top-down causation. Interface Focus **2**(1) (2012)
27. D.J. Felleman, D.C. Van Essen, Distributed hierarchical processing in the primate cerebral cortex. Cereb. Cortex **1**(1), 1–47 (1991)
28. R.P. Feynman, R.B. Leighton, M. Sands, *The Feynman Lectures on Physics: Volume III. Quantum Mechanics* (Addison-Wesley, Reading, 1965)
29. R.P. Feynman, R.B. Leighton, M. Sands, *Feynman Lectures in Physics Volume II: Electromagnetism* (Addison-Wesley, Reading, 1964)
30. V. Frankl, *Man's Search for Meaning* (Rider and Co., London, 2008)
31. K. Friston, A theory of cortical responses. Philos. Trans. R. Soc. (Lond.) B Biol. Sci. **36**, 360–815 (2005)
32. C. Frith, *Making Up the Mind: How the Brain Creates Our Mental World* (Blackwell, Malden, 2007)
33. C.D. Gilbert, M. Sigman, Brain states: top-down influences in sensory processing. Neuron **54**, 677–696 (2007)
34. S. Gilbert, D. Epel, *Ecological Developmental Biology* (Sinauer, Sunderland, 2009)
35. G. Greenstein, A.G. Zajonc, *The Quantum Challenge: Modern Research on the Foundations of Quantum Mechanics* (Jones and Bartlett, Sudbury, Mass, 2006)
36. S. Hartmann, Effective field theories, reductionism and scientific explanation. Stud. Hist. Philos. Sci. Part B **32**, 267–304 (2001)
37. C.J. Isham, *Lectures on Quantum Theory* (Imperial College Press, London, 1995)
38. M. Kac, Can one hear the shape of a drum? Am. Math. Mon. **73**, 1–23 (1966)

39. E.R. Kandel, *The Age of Insight* (Random House, New York, 2012)
40. J. Kurose, *Computer Networking: A Top-Down Approach* (Addison Wesley, Upper Saddle River, 2010)
41. R. Lafore, *Data Structures and Algorithms in Java* (SAMS, Indiana, 2002)
42. R.W. Laughlin, Fractional quantisation. Rev. Mod. Phys. **71**, 863–874 (1999)
43. J. MacCormick, *9 Algorithms that Changed the Future* (Princeton University Press, New Jersey, 2012)
44. E.Z. Macosko, N. Pokala, E.H. Feinberg, S.H. Chalasani, R.A. Butcher, J. Clardy, C.I. Bargmann, A hub-and-spoke circuit drives pheromone attraction and social behaviour in C. elegans. Nature **458**, 1171–1175 (2009)
45. M.M. Mano, C.R. Kine, *Logic and Computer Design Fundamentals* (Pearson, 2008)
46. V. Mante, D. Sussillo, K.V. Shenoy, Context-dependent computation by recurrent dynamics in prefrontal cortex. Nature **503**, 78–84 (2013)
47. M. Martnez, A. Moya, (2011) Natural selection and multi-level causation. Philos. Theory Biol. (2011), http://dx.doi.org/10.3998/ptb.6959004.0003.002. 3 May 2011
48. C. Mitchell, J. Hobcraft, S.S. McLanahan, S.R. Siegel, A. Berg, J. Brooks-Gunn, I. Garfinkel, D. Nottermand, Social disadvantage, genetic sensitivity, and childrens telomere length. Proc. Nati. Acad. Sci. **111**(16), 5944–5949 (2014)
49. S. Mukhanov, *Physical Foundations of Cosmology* (Cambridge University Press, Cambridge, 2005)
50. D. Noble, From the HodgkinHuxley axon to the virtual heart. J. Physiol. **580**, 15–22 (2007)
51. D. Noble, A theory of biological relativity: no privileged level of causation. Interface Focus **2**, 55–64 (2012)
52. M.J. Page et al., The suppression of star formation by active galactic nuclei. Nature **485**, 213–216 (2012)
53. J. Pearl, Causal diagrams for empirical research. Biometrika **82**, 669–688 (1995)
54. J. Pearl, *Causality: Models, Reasoning and Inference* (MIT Press, 2000)
55. R. Penrose, *Cycles of Time: An Extraordinary New View of the Universe* (Knopf, New York, 2011)
56. M. Pigliucci, An extended synthesis for evolutionary biology. Ann. N. Y. Acad. Sci. **1168**, 218228 (2009)
57. H. Poincare, The Value of Science. Originally published in 1913 (trans: G.B. Halstead), Cosimo Classics, 21(1913). ISBN: 978-1-60206-504-8
58. D. Purves, W. Wojtach, W.C. Howe, Visual ilusions. Scholarpedia (2007)
59. D. Purves, *Brains: How They Seem to Work* (FT Press Science, Upper Saddle River, 2010)
60. S. Schweber, Physics, community, and the crisis in physical theory. Phys. Today **46**, 34–40 (1993)
61. M. Sigman, H. Pan, Y. Yang, E. Stern, D. Silbersweig, C.D. Gilbert, Top-down reorganization of activity in the visual pathway after learning a shape identification task. Neuron **46**, 823–835 (2005)
62. A.S. Tanenbaum, *Structured Computer Organisation* (Prentice Hall, Englewood Cliffs, 1990)
63. R. van den Bos, J.W. Jolles, J.R. Homberg, Social modulation of decision-making: a cross-species review. Front. Hum. Neurosci. **7**, 301 (2013). www.frontiersin.org
64. S.I. Walker, L. Cisneros, P.C.W, Davies, Evolutionary transitions and top-down causation, in *Proceedings of Artificial Life XIII* (2012), http://xxx.lanl.gov/abs/1207.4808
65. D.R. Weinberg, C.J. Gagliardi, J.F. Hull, C.F. Murphy, C.A. Kent, B. Westlake, A. Paul, D.H. Ess, D.G. McCafferty, T.J. Meyer, Proton-coupled electron transfer. Chem. Rev. **107**, 5004–5064 (2007)
66. S. Weinberg, *The Quantum Theory of Fields. Volume 1: Foundations* (Cambridge University Press, Cambridge, 2005)
67. J.M. Ziman, *Principles of the Theory of Solids* (Cambridge University Press, Cambridge, 1979)

Chapter 4
On the Foundational Assumptions of Modern Physics

Benjamin F. Dribus

Abstract General relativity and the standard model of particle physics remain our most fundamental physical theories enjoying robust experimental confirmation. The foundational assumptions of physics changed rapidly during the early development of these theories, but the subsequent challenges of their refinement and the exploitation of their explanatory power turned attention away from foundational issues. Deep problems and anomalous observations remain unaddressed. New theories such as string theory seek to resolve these issues, but are presently untested. In this essay, I evaluate the foundational assumptions of modern physics and propose new physical principles. I reject the notion that spacetime is a manifold, the existence of static background structure in the universe, the symmetry interpretation of covariance, and a number of related assumptions. The central new principle I propose is the *causal metric hypothesis,* which characterizes the observed properties of the physical universe as manifestations of causal structure. More precisely, the *classical causal metric hypothesis* states that the metric properties of classical spacetime arise from a binary relation on a set, representing direct influences between pairs of events. Rafael Sorkin's maxim, "order plus number equals geometry" is a special case. The *quantum causal metric hypothesis* states that the phases associated with directed paths in causal configuration space, under Feynman's sum-over-histories approach to quantum theory, are determined by the causal structures of their constituent universes. The resulting approach to fundamental physics is called *quantum causal theory.*

Introduction

Relativity and Quantum Theory. Relativity and quantum theory emerged from mathematical and philosophical seeds in the works of Gauss, Riemann, Cayley, Hilbert, and others; were incorporated as physical theories by Einstein, Heisenberg, Schrödinger, Weyl, and their contemporaries; and matured as definitive predictive systems in the form of modern general relativity and the standard model of particle

B.F. Dribus (✉)
William Carey University, Hattiesburg, USA
e-mail: bdribus@wmcarey.edu; info@causalphysics.com

© Springer International Publishing Switzerland 2015
A. Aguirre et al. (eds.), *Questioning the Foundations of Physics,*
The Frontiers Collection, DOI 10.1007/978-3-319-13045-3_4

physics in the second half of the twentieth century. Among theories enjoying robust experimental confirmation, these two theories represent our deepest understanding of fundamental physics. The rapid alteration of foundational assumptions characterizing the early development of these theories later diminished as their fruit was harvested. Satisfactory unification of relativity and quantum theory proved to be an immense and forbidding challenge, resisting numerous optimistic early attempts, and an abundance of new experimental results amenable to description within the developing framework of quantum field theory decreased motivation for radical new departures.

Foundational Problems; New Theories. Recently the triumphs of quantum field theory have slowed, and unexplained phenomena such as dark matter and dark energy hint at new physics. In this environment, long-acknowledged foundational problems have gained new urgency. The fundamental structure of spacetime, the nature and significance of causality, the quantum-theoretic description of gravity, and unified understanding of physical law, have all attracted increased scrutiny. Untested new theories seek to address these issues, often incorporating new assumptions as alien to established physics as the assumptions of relativity and quantum theory were to the Newtonian paradigm. Among these new theories, string theory [1] abolishes point particles and introduces new dimensions, symmetries, and dualities; loop quantum gravity [2] undertakes the quantization of relativistic spacetime; noncommutative geometry [3] interprets spacetime as a noncommutative space; entropic gravity [4] attributes gravitation to the second law of thermodynamics; and causal set theory [5] discards manifold models of classical spacetime in favor of discrete partially ordered sets. While limited, this list represents a reasonable cross-section of the general approaches to new physics under active investigation.

Overview and Organization of This Essay. In this essay, I evaluate the foundational assumptions of modern physics and offer *speculative* new principles, partially overlapping some of the new theories mentioned above. These principles cannot, to my present knowledge, claim definitive experimental confirmation, but their consideration is reasonable alongside other untested theories. Among the assumptions I reject are the manifold structure of spacetime, the evolution of physical systems with respect to a universal time parameter, the existence of a static background structure serving as an "arena" for dynamical processes, the symmetry interpretation of covariance, the transitivity of the binary relation encoding causal structure, and the commutativity of spacetime. The central new principle I propose is the *causal metric hypothesis*, which characterizes the observed properties of the physical universe as manifestations of causal structure. For purposes of precision, it is convenient to formulate classical and quantum versions of the causal metric hypothesis. The classical version states that the properties of classical spacetime are manifestations of a binary relation on a set. Rafael Sorkin's maxim, "order plus number equals geometry," is a special case. The quantum version states that the phases associated with directed paths in causal configuration space are determined by the causal structures of their constituent universes. These ideas are explained in more detail below. The resulting approach to fundamental physics is called *quantum causal theory*.

This essay is organized as follows: in the section "Identifying the Foundational Assumptions," I identify and discuss the foundational assumptions of modern physics, focusing on assumptions enjoying wide recognition in the mainstream scientific community. I introduce three different classes of assumptions: general principles, formal postulates, and ancillary assumptions. I isolate six general principles of particular importance, and briefly cite a few others worthy of mention. I then discuss assumptions specific to relativity, nonrelativistic quantum theory, and quantum field theory. In the section "Vignette of Unexplained Phenomena," I briefly mention some empirical phenomena unexplained by these theories. In the section "Rejected Assumptions," I reject several existing assumptions, with motivation provided by the previous two sections. In the section "New Principles," I propose new physical principles, with particular focus on the *fundamental structure of spacetime*. In the section "Practical Considerations," I remark on the status of these assumptions and principles in light of the current state of experimental and theoretical physics, and suggest how the ideas presented in this essay might find their way into the laboratory.

Identifying the Foundational Assumptions

Three Classes of Assumptions. In the world of scientific thought, ideas from mathematics, philosophy, and the empirical realm converge in the form of *general physical principles*, which further crystallize into *formal postulates* of specific physical theories, while remaining colored and often distorted by *ancillary assumptions* involving issues of interpretation and biases from the prevailing intellectual environment. These principles, postulates, and ancilla are all *foundational assumptions* in the sense that basic science depends critically, and to some degree independently, on each. However, they also possess important distinguishing characteristics.

General physical principles represent attempts to capture deep physical truths that are often difficult to quantify. As a result, such principles often survive, via an evolutionary process of refinement and reinterpretation, through multiple scientific revolutions, while formal postulates and ancillary assumptions often die along with the specific theories built around them. For example, the general principle of *covariance*, which at its root involves an assertion of the observer-independence of physical law, has motivated a succession of mutually contradictory formal *invariance postulates*, such as Galilean invariance and Lorentz invariance, along the historical path from Newtonian physics through special relativity to general relativity and beyond. Parallel to these invariance postulates have followed a succession of mutually contradictory ancillary assumptions regarding the interpretation of time and related issues. Covariance itself, meanwhile, remains relevant even to nonmanifold models of spacetime.

Formal postulates of dubious aspect sometimes persist due to a lack of suitable alternatives, even when they contradict widely acknowledged general principles. For example, the general principle of *background independence* is usually taken for granted, at a philosophical level, in the modern physics community; yet the formal postulates underlying the standard model of particle physics, as well as many newer theories, fail to satisfy this principle. Ancillary assumptions can be particularly troublesome because of their tendency to escape serious scrutiny. Examples include the *luminiferous aether* in pre-relativistic physics, and some of the assumptions related to *Bell's inequalities* in the foundations of quantum theory. General principles and formal postulates are safer in this regard, since they attract the conscious focus of theorists.

Six General Physical Principles. Six crucial general principles of modern physics are *symmetry, conservation, covariance*, the *second law of thermodynamics, background independence*, and *causality*. These principles are intimately interrelated. Results such as *Noether's theorem* tie symmetries to conservation laws, and relativistic covariance is understood in terms of symmetry, at least locally. More generally, covariance may be interpreted in terms of *generalized order theory*. Both viewpoints involve isolating privileged information; either that fixed by a particular group action, or that contained in a distinguished suborder. Entropy, and thence the second law of thermodynamics, may also be expressed via partitioning of information: in ordinary statistical thermodynamics, entropy involves "microscopic refinements of macroscopic states;" while in discrete causal theory, it may be measured in terms of the cardinality of certain *Galois groups* of generalized order morphisms.

Background independence is usually understood as a statement about spacetime; that it is a dynamical entity subject to physical laws, such as Einstein's field equations, rather than a static object. Philosophically, background independence provides an example of the use of *parsimony* to achieve explanatory and predictive power; the less a theory assumes, the more it can potentially reveal. Background independence is one of the strengths of general relativity; relativistic spacetime geometry is determined via dynamics, not taken for granted. Improvement beyond relativity is conceivable. For example, Einstein's equations do not predict the *dimension* of spacetime; a theory providing a dynamical explanation of dimension would be superior in important ways. Causality is of central importance to physics, and to science in general, principally because *prediction* relies upon the discovery of causal relationships, together with the assumption of *reproducibility*. Classically, causality is often formalized as an *irreflexive, acyclic, transitive binary relation* on the set of spacetime events. It is related to covariance via order theory, to the second law of thermodynamics via the arrow of time, and to background independence via the general criteria of explanatory and predictive power. However, the deep meaning of causality, and its appropriate role in quantum theory, remain controversial.

Other General Principles. Other general principles deserving mention include *symmetry breaking*, physical versions of *superposition* including Feynman's *sum over histories, action principles, cluster decomposition* and other versions of *locality*, Einstein's *equivalence principle, scale-dependence* and *independence*, the *holographic principle, dualities* such as *S-duality*, and various principles involved in the interpretation of quantum theory. Untested modern theories rely on further principles, or refinements of principles already mentioned, whose importance is tied to their success. For example, Maldacena's *AdS/CFT correspondence* [6] is much more important if string theory is physically relevant than it would be otherwise. Pure mathematics, such as number theory, also offers general principles, and conjectured principles, with deep connections to physics. For example, *zeta functions*, and hence the *Riemann hypothesis*, are connected to quantum field theory via noncommutative geometry and the theory of *motives* [7]. The *Langlands program* is connected to physical symmetry and duality via *representation theory* and *conformal field theory*, and thence also to string theory [8].

Assumptions of Relativity and Quantum Theory. The following formal postulates and ancillary assumptions apply to general relativity and quantum theory, although some of them also survive in newer theories. General relativity postulates a four-dimensional pseudo-Riemannian manifold of Lorentz signature, interpreted as spacetime, whose curvature, interpreted as gravitation, is determined dynamically via interaction with matter and energy according to Einstein's field equations, and whose metric properties govern its causal structure. Singularities arise in the generic case, as noted by Penrose, Hawking and Ellis, and others.

Multiple approaches to nonrelativistic quantum theory exist. I will describe two, equivalent under suitable restrictions. The *Hilbert space approach* postulates complex Hilbert spaces whose elements represent probability amplitudes, self-adjoint operators whose eigenvalues represent the possible values of measurements, and time evolution according to Schrödinger's equation. In the simplest context, these amplitudes, operators, et cetera, represent the behavior of point particles. Feynman's *sum-over-histories approach* [9] postulates probability amplitudes given by complex sums over spaces of paths, interpreted as spacetime trajectories of point particles in the simplest context. In a path sum, each path contributes equally in magnitude, with phase determined by the *classical action*, given by integrating the *Lagrangian* along the path with respect to time. This version generalizes easily to relativistic and post-relativistic contexts.

Quantum field theory postulates operator fields that create and annihilate state vectors in complex Hilbert spaces. States corresponding to particular particle species are associated with particular representations of symmetry groups. The properties of Minkowski spacetime impose *external symmetries* encoded by the Poincaré group. *Internal symmetries*, such as those encoded by *gauge groups*, also play a critical role. The standard model of particle physics is expressed via the nonabelian *Yang-Mills* gauge theory, and includes particles, fields, and symmetry groups in remarkable accord with the observations of particle physicists over the last century.

Vignette of Unexplained Phenomena

Overview. Since the ascendancy of general relativity and the standard model, a variety of unexplained physical phenomena have been recognized. The large-scale dynamical anomalies attributed to *dark matter* and *dark energy*, the absence of a large *cosmological constant* arising from vacuum energy, and the apparent asymmetry between matter and antimatter in the observable universe, are a few of the most prominent examples. These phenomena suggest the promise of physical models that naturally incorporate *scale-dependence*, and that offer statistical or entropic explanations of small nonzero constants and inexact symmetries. Discrete order-theoretic and graph-theoretic models tend to perform well by these criteria.

Dark Matter. The *dark matter hypothesis* is based on the failure of astrophysical systems on the scale of galaxies to obey relativistic dynamics, assuming only the matter content detectable by non-gravitational means. In contrast, objects near the stellar scale seem to verify relativistic predictions remarkably well. The dark matter hypothesis has been compared unfavorably to the luminiferous aether, and various new dynamical laws have been proposed to account for observed behavior without invoking missing mass. However, this phenomenon does behave like ordinary matter in many respects, as observed in the collision of galaxies and in certain examples of *gravitational lensing*. If the dark matter hypothesis is valid, the matter involved seems unlikely to be accounted for by the standard model. Claims have been made of laboratory observations of new particles consistent with dark matter, but these are not broadly accepted at present.

Dark Energy. Dark energy is the entity invoked to explain the phenomenon interpreted as *acceleration of the expansion of the universe*. The cosmological constant appearing in the modified form of Einstein's equations is one possible type of dark energy. Predictions based on quantum field theory and the Planck scale yield a value for the cosmological constant roughly 120 orders of magnitude greater than observation implies. Interestingly, causal set theory suggests a fluctuating cosmological constant close to the observed value, based on a simple argument involving discreteness and the size of the Hubble radius in Planck units. Nonconstant models of dark energy, such as *quintessence*, have also been proposed, but any fluctuations in dark energy appear to occur on scales much larger than those of dark matter or ordinary matter and energy. Apparent anomalies in the motion of certain large galactic clusters, called *dark flow*, might reflect such fluctuations. Dark matter and dark energy extend the scale-dependence of phenomena already observed in conventional physics. Strong and weak interactions, electromagnetism, ordinary gravity, dark matter, and dark energy all dominate on different scales, each covering roughly equivalent ranges in a logarithmic sense. The extent of this scale-dependence was unknown during the development of relativity and quantum theory, and should command significant attention in the development of new models.

Matter-Antimatter Asymmetry. Our present understanding of antimatter comes almost entirely from quantum field theory, and it is reasonable to ask if matter-antimatter asymmetry in the observable universe might indicate a problem with quantum field theory itself, or at least with the standard model. Unexpected asymmetries have been successfully handled by quantum field theory in the past; the prototypical example is *CP violation*, which is itself related to the matter-antimatter problem. However, potential sources of matter-antimatter asymmetry in the standard model seem either too weak, or too strong, to account for observation. Interesting experimental issues regarding antimatter remain to be resolved. Until recently, little direct evidence existed to demonstrate that antimatter interacts *gravitationally* in the same way as matter, and it had even been suggested that local matter-antimatter asymmetry might result from a type of gravitational segregation. More conventionally, experiments designed to investigate matter-antimatter asymmetry have recently produced data suggesting rates for certain decay processes different than those predicted by the standard model. It seems too early to render judgment on the significance or meaning of these results, however.

Rejected Assumptions

Structural Assumptions; Metric Emergence. Some of the physical assumptions I reject in this essay are already widely doubted, but survive in old and new theories alike due to the unfamiliarity or intractability of their principal alternatives. Among these are the basic structural assumptions that spacetime is a real manifold, that physical systems evolve with respect to a universal time parameter, and that the universe possesses a static background structure serving as an immutable "arena" for dynamical processes. This last assumption is, of course, merely the negation of the general principle of background independence. General relativity includes the first of these assumptions, and the standard model includes all three. Since these assumptions are retained largely for operational reasons, their rejection is not very revolutionary. However, a successful theory abstaining from them would be revolutionary indeed. I reject them partly on general mathematical and philosophical grounds, and partly for the specific physical reason that they are incompatible with discrete quantum causal theory.

Another basic structural assumption I reject is that spacetime is commutative. This statement should be understood in the sense of Connes' noncommutative geometry [3]. Though this assumption is less-widely doubted in mainstream physics than those mentioned above, it has recently become the subject of justified scrutiny. A number of existing proposals about fundamental spacetime structure lead naturally to noncommutative spaces. For example, such spaces arise via the *deformation theory of Hopf algebras*, and in certain category-theoretic approaches to physics. Even "classical spaces" such as Minkowski spacetime may be "recognized as possessing noncommutative structures" in useful ways.

Along with these assumptions perish a number of corollaries. Spacetime dimension becomes an emergent property, no longer assumed to be constant, static, or an integer. Properties previously ascribed to a metric, in the sense of differential geometry, must either be discarded or assigned different structural origins. For example, given a geodesic between two events in relativistic spacetime, there exist many other "near-geodesics" between them; however, a nonmanifold model of spacetime might admit a unique "short" path between two events, with every other path being much "longer." Such reflections prompt reconsideration of the notions of distance and locality. Other metric properties could be similarly reexamined, but most important is to investigate what mechanisms supply the *appearance* of a metric at ordinary scales. This may be called the *problem of metric emergence.*

Assumptions About Causality. The answer I will propose to the problem of metric emergence involves reinterpreting the general physical principles of causality and covariance. This requires rejection of some common ancillary assumptions about these specific principles. First, I reject the assumption that the apparent metric properties of classical spacetime involve any information other than a set of events and a binary relation, the *causal relation*, encoding causal structure. This rejection amounts to a "negative version" of the *classical causal metric hypothesis*; the corresponding "positive version" is stated below. Theorems of Stephen Hawking [10] and David Malament [11] in the late 1970s hinted at this conclusion in a relativistic context, by demonstrating that "most" of the metric properties of relativistic spacetime may be recovered from its causal structure. Causal set theory already incorporates a version of this idea.

Second, I reject the assumption that the causal relation is transitive. This odd-seeming statement merely acknowledges the physical relevance of information about *direct* versus *indirect* causation. The usual transitive "causal order" may be recovered by closing the causal relation under transitivity. Third, I reject the assumption that the causal relation is acyclic. This rejection permits the existence of *causal cycles*, which already arise as *closed timelike curves* in certain solutions of general relativity. Causal cycles need raise no paradoxes; if they exist, they are properties of a binary relation, not "self-contradictory inhabitants" of a background structure.

Assumptions About Covariance. Turning to covariance, I reject the assumption that it is an instance of group symmetry, even locally; rather, it should be viewed in order-theoretic terms. For example, different frames of reference in relativity assign different time-orders to events separated by spacelike intervals; these orders correspond to different classes of refinements of the causal relation. This rejection is notable because progress in physics has historically involved invoking new symmetry principles, rather than rejecting existing ones. Since the time of Weyl, *group representation theory* has permeated theoretical physics as the mathematical expression of symmetry, and remains perhaps the most promising technical vehicle for short-term progress beyond the standard model. Over the long term, however, analogous constructs from order theory, and perhaps other notions more primitive than groups, will likely replace much of group representation theory in this role. Alternative approaches to covariance involving category theory and noncommutative geometry have already been proposed.

New Principles

Overview: Quantum Causal Theory. New principles I propose in this essay include the *causal metric hypothesis, iteration of structure* as a quantization principle, and *co-relative histories*. These principles, explained in more detail below, form the backbone of *quantum causal theory*, which is a general term I use to describe approaches to quantum spacetime and quantum gravity that take causal structure to be fundamental. Technical tools necessary to implement these ideas include a synthesis of *multicategory theory* and *categorification* in abstract algebra, involving "interchangeability of objects, morphisms, elements, and relations;" a refined version of *random graph dynamics*; and the theory of *semicategory algebras*. In particular, *path alge-*

bras encode the properties of both individual causal universes and their configuration spaces, while providing convenient methods of computation. Details of many of these ideas appear in my paper [12]. Here I focus only on the basic concepts.

Causal Metric Hypothesis. Foremost among the new principles I propose is the *causal metric hypothesis*. The philosophical content of this hypothesis is that *the observed properties of the physical universe are manifestations of causal structure.* To crystallize this idea into a precise, quantitative approach to physics, it is convenient to first state a classical version of the hypothesis, which serves as a precursor to the corresponding quantum version, just as classical notions form the building blocks of quantum theory in Feynman's sum-over-histories approach. The *classical causal metric hypothesis* may be stated as follows:

> The properties of classical spacetime arise from a binary relation $\prec$ on a set S, where elements of S represent spacetime events, and elements of $\prec$ represent direct influences; i.e., causal relations, between pairs of events.

Figure 4.1 illustrates the classical causal metric hypothesis, and demonstrates how it differs from the paradigm of general relativity. Figure 4.1a shows a region of relativistic spacetime, with distinguished events marked by nodes. In general relativity, the geometry of spacetime governs the scope of causal influence. For example, event *x may have been influenced by* all events in its geometric "past," shown in dark gray, and *may influence* all events in its geometric "future," shown in light gray. The classical causal metric hypothesis turns this picture on its head, taking "spacetime geometry" to be nothing more than a *way of describing actual influences*. Figure 4.1b shows a family of events, with direct influences indicated by edges running up the page. Under the classical causal metric hypothesis, the geometric "past" and "future" are *a posteriori* constructions. Rafael Sorkin's causal set maxim, "order plus number equals geometry," is a special case of the classical causal metric hypothesis.

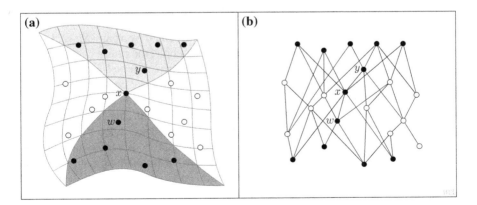

Fig. 4.1   a In general relativity, spacetime geometry governs the scope of causal influence; **b** under the classical causal metric hypothesis, "spacetime geometry" is merely a way of describing actual influences

The Causal Relation. The *binary relation* referenced in the classical causal metric hypothesis is a mathematical way of encoding direct influences between pairs of events, represented by edges in Fig. 4.1b. Such a relation, which I will call the *causal relation* in this context, may be viewed as a *generalized partial order*, with the word "order" indicating precedence and succession. For example, event x in Fig. 4.1b precedes event y; this is written $x \prec y$. In Sorkin's causal set theory, the causal relation is a partial order in the technical sense, but there are good reasons to generalize this picture; for example, by abstaining from transitivity and acyclicity, as already indicated above. However, the most interesting versions of causal theory I know of do impose "reasonable assumptions" on the causal relation; for example, *local finiteness*. More generally, assumptions about local structure are usually more reasonable to impose than their nonlocal counterparts, due to our ignorance of the global structure of the universe.

Recovery of Lorentzian manifold structure from the causal relation is necessary at some level of approximation, owing to the large-scale success of general relativity. The metric recovery theorems of Hawking and Malament, mentioned above, demonstrate that specifying appropriate *volume* data, as well as order data, is sufficient to recover continuum geometry. According to the classical causal metric hypothesis, this volume data should derive in some way from the pair $(S, \prec)$. The simplest dependence is the "trivial" one, in which a single unit of volume is assigned to each element of S, irrespective of $\prec$; this is the causal set approach, as encapsulated by Sorkin. However, the causal metric hypothesis allows for alternative methods of specifying volume data that depend on the causal relation $\prec$ in more complicated ways.

Iteration of Structure as a Quantization Principle. Feynman's *sum-over-histories* approach to quantum theory [9] is perhaps the most promising general approach under the causal metric hypothesis. Significant efforts have already been made to adapt this approach to causal set theory, although technical problems such as the *permeability of maximal antichains* complicate the picture. For this reason, and many others, it is preferable to work in *relation space*, as described in section 5 of my paper *On the Axioms of Causal Set Theory* [12]. Sums in this context involve paths in a "configuration space of classical universes," each represented by a pair $(S, \prec)$. I refer to such a space as a *causal configuration space*. For example, the causal configuration space of causal set theory is the space of all acyclic, transitive, interval-finite universes admitting an order embedding into the natural numbers. Causal configuration space inherits a *directed structure* induced by special morphisms between pairs of universes, called *transitions*. This directed structure may be viewed as a "higher-level analogue" of the directed structures on the individual universes $(S, \prec)$, encoded by the causal relations $\prec$. This emergence of higher-level directed structure on causal configuration space is a prototypical example of a recurring principle in quantum causal theory that I refer to as *iteration of structure*. In particular, *quantization* consists of passage from individual universes to causal configuration space. Mathematically, this may be viewed in terms of a generalized version of *categorification/decategorification*, in which structure is added or ignored by promoting elements or demoting objects.

Co-relative Histories; Kinematic Schemes. For technical reasons, transitions are *too specific* to be physically fundamental; they carry "gauge-like information."

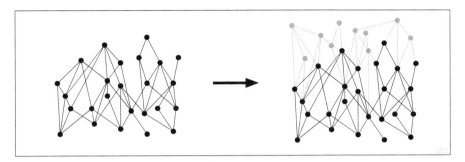

Fig. 4.2 A co-relative history. *Gray* indicates "new structure" in the target universe

Appropriate equivalence classes of transitions, which I call *co-relative histories*, are the physically significant building blocks of higher-level structure in causal configuration space, providing a refined version of iteration of structure. Figure 4.2 illustrates a co-relative history.

Co-relative histories replace the notion of *time evolution* in quantum causal theory. The target universe of a co-relative history may be viewed as a "later stage of development of its source universe." A suitable choice of co-relative histories, providing "evolutionary pathways for every possible universe," yields a special substructure of causal configuration space that I call a *kinematic scheme*.

Figure 4.3 modified from a similar figure in my paper [12], shows a portion of a kinematic scheme S that I refer to as the *positive sequential kinematic scheme*. The word "sequential" means that each co-relative history in S "adds a single element" to its source universe. The word "positive" means that the elements of each universe in S may be labeled by positive integers. The "generations" indicated by the large numbers 0, 1, 2, 3, 4 in Fig. 4.3 correspond to such labeling. The inset in Fig. 4.3 shows a portion of the abstract underlying directed graph corresponding to S; comparison of this graph to S itself illustrates iteration of structure. Each upward-directed path in S represents a "kinematic account" of the evolution of its terminal universe. The path terminating at the universe U in Fig. 4.3 is an example. Note that the "spacelike hypersurface;" i.e., maximal antichain, in S, represented by the three universes in double circles, is *permeated* by the path from $\oslash$ to U. This indicates that the relation space over S; i.e., the corresponding space of co-relative histories, provides a "superior viewpoint" in a structural sense. This is an example of the *relative viewpoint* advocated by Alexander Grothendieck, in which one studies *relationships between mathematical objects*, rather than studying each object individually. Gray indicates universes whose causal relations are intransitive. These universes distinguish S from the configuration spaces arising in causal set theory [13].

The theory of kinematic schemes provides a precise realization of the principle advocated by Robert Spekkens in his essay *The paradigm of kinematics and dynamics must yield to causal structure,* also appearing Chap. 2. Different kinematic schemes lead to different dynamical equations, all equally valid. For example, kinematic

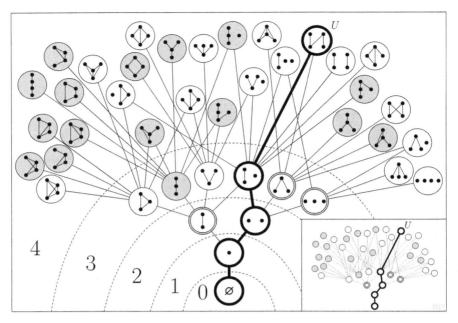

Fig. 4.3 Portion of the positive sequential kinematic scheme $\mathcal{S}$; *inset* shows the underlying directed structure; *large-font* numbers indicate generations; *double circles* represent a maximal antichain; *dark path* represents a permeating chain; *gray* indicates intransitive universes

schemes in which sources and targets differ by entire generations of elements govern discrete causal analogues of relativistic dynamics.

Dynamics; Quantum Causal Metric Hypothesis. The sum-over-histories approach to quantum theory, suitably adapted, assigns amplitudes to families of co-relative histories in a kinematic scheme. The sources of these co-relative histories are viewed as "initial universes," and the corresponding targets are viewed as "terminal universes." In ordinary quantum theory, such amplitudes are complex-valued, but the complex numbers cannot be taken for granted in the discrete causal context. Finite algebraic structures provide interesting alternatives. These amplitudes may be interpreted as encoding "probabilities" of reaching given families of terminal universes from given families of initial universes. They are computed by summing quantities called *phases* over paths between pairs of families of universes. The values of these phases are of great interest; they supply the specific physical content of the theory, just as choosing a Lagrangian supplies the physical content of a typical "conventional theory," via the corresponding action principle. The *quantum causal metric hypothesis* states that these phases "arise from causal structure" in an appropriate sense:

> The properties of quantum spacetime arise from a kinematic scheme $\mathcal{S}$. In particular, the phases associated with directed paths in $\mathcal{S}$, under the sum-over-histories approach to quantum theory, arise from the causal relations on the constituent universes of $\mathcal{S}$.

As mentioned above, technical advantages result from working in terms of the relation space over S; i.e., the corresponding space of co-relative histories.

Causal Schrödinger-type Equations. Given a suitable choice of phases, adapting and generalizing Feynman's reasoning to the quantum causal context enables the derivation of dynamical equations, which I refer to as *causal Schrödinger-type equations*. A special case of such an equation is

$$\psi^-_{R;\theta}(r) = \theta(r) \sum_{r^- \prec r} \psi^-_{R;\theta}(r^-),$$

where R is a subspace of the space of co-relative histories over a kinematic scheme, r^- and r are "consecutive" co-relative histories in R, $\prec$ is the binary relation on R induced by iteration of structure, θ is the *phase map*, and $\psi^-_{R;\theta}$ is the *past causal wave function*, defined by summing phases over the set of maximal irreducible directed paths in R terminating at r.

Practical Considerations

Current Status of Rejected Assumptions and New Principles. The rejected assumptions and new principles discussed in this essay occupy a variety of positions with respect to theory and experiment, some more precarious than others. Manifold structure of spacetime remains tenable, but the existence of a universal time parameter and static background structure have been doubtful ever since the first observations supporting general relativity. The idea that noncommutative geometry is essential to quantum spacetime is still conjectural. Consideration of the "negative version" of the causal metric hypothesis may be omitted in favor of the stronger "positive version," to which I return below. Intransitivity of the causal relation is obvious at large scales; for example, it is uncommon to be *directly* related to one's grandparents. At the fundamental scale, the issue may be treated technically by examining whether or not physical predictions depend on including intransitive universes in the sum over histories. A priori, the answer is yes, but special choices of phase maps might annul this. Regarding causal cycles, I know of no solid evidence of their existence; however, certain interesting interpretations of well-known phenomena do incorporate them. Inadequacy of the symmetry interpretation of covariance might be demonstrated only in conjunction with breakdown of manifold structure.

Turning to new principles, the causal metric hypothesis is most compelling in the discrete setting, due to the metric recovery theorems. There is at present no convincing experimental evidence of spacetime discreteness, but it is thus far infeasible to experimentally probe most regimes where such evidence might present itself. In this regard at least, the plausibility of the causal metric hypothesis must be judged indirectly at this time. The theory of co-relative histories can be neither "right" nor "wrong;" it represents a *viewpoint*, more useful in some contexts than others. The idea itself is quite general, but since "relationships" in category-like settings generally involve directed structure, the theory is most natural in the causal context. The same is true of iteration of structure; moreover, the conceptual utility of this idea seems greatest in the *discrete* causal setting. To demonstrate the contrast, *Einstein manifolds* possess directed structure, but configuration spaces of Einstein manifolds are generally nothing like Einstein manifolds themselves.

Recovery of Established Physics at Appropriate Scales. The parsimony of the new principles proposed in this essay renders recovery of established physics from these principles a substantial challenge, with a correspondingly great compensation if this challenge can be met. The metric emergence problem for flat Minkowski spacetime is the obvious first step toward both relativity and the standard model in this context, since along with it will emerge the usual algebraic notions regarding coordinate transformations and particle states. Note, however, that while the standard model adds particle states as *separate ingredients* to Minkowski spacetime, *both must emerge together* in the quantum causal context. Treating matter and energy as auxiliary data would defeat the purpose of the program by violating the causal metric hypothesis, as well as the principle of background independence. Based on our best guesses about the fundamental scale, the simplest "elementary particle" interactions currently accessible to observation might easily involve Avogadro's number of fundamental causal elements, or its square, or its cube. This is encouraging in the sense that such magnitudes allow for familiar mechanisms such as entropy, and novel ones such as graph-dynamical phase transitions, to produce sharp behavior and select for precise quantities. However, it is discouraging in the sense that interactions large enough to observe might be difficult to model.

Implications of Recent Observations. Last year, the Large Hadron Collider (LHC) at CERN detected a new particle with energy near 125 GeV and properties similar to the predicted properties of the standard model Higgs boson. Work is ongoing to analyze possible deviations from these predictions, but concern exists that the observed particle may match the standard model Higgs so precisely that the results will provide little or no help in pointing to new physics. Whether or not this is true, new high-energy particle physics may soon become technologically or economically infeasible in laboratory settings. This sharpens the need for creative ideas regarding the general problem of what experimental phenomena to search for and how to search for them. In the context of quantum causal theory, results one might look for experimentally include inexactness of symmetries, variation or small nonzero values of physical constants, and new kinds of scale-dependence. Quantities such as the emergent dimension of spacetime might vary with "energy density," though such effects might be extremely small.

Opportunities for observational physics exist beyond those afforded by traditional laboratory experiments, particularly in cosmological contexts. Shortly before publication of this volume, the BICEP experiment, which measures polarization in the cosmic microwave background, reported detection of so-called *B-modes of primordial gravitational waves*. This observation has been widely regarded as evidence in favor of the *inflationary hypothesis* in cosmology, which is based primarily on the apparent communication in the early universe of regions now widely separated. Inflation is thus rooted in *causal* considerations. In my paper [12], I propose a quantum-causal alternative to inflation, in which causal structure grew abruptly "sparser" in the early universe, due to a graph-dynamical phase transition. I am presently trying to connect this idea to experiment.

Connections to Quantum Information Theory. An intriguing possibility is that *quantum circuits* might provide relatively large-scale "windows" into fundamental-scale physics. Such circuits may be represented by small "causal universes" whose relations are weighted by *single-qubit unitary transformations*. In traditional quantum theory, important restrictions on such universes arise from results such as the *no-cloning theorem*. Such circuits are small at ordinary scales, but they are many orders of magnitude larger than the Planck scale. Only very simple quantum circuits have been constructed to date, but complex circuits may be built in the near future.

The behavior of quantum circuits might be related to fundamental-scale behavior in at least two different ways. First, and most optimistically, if spacetime possesses a sufficiently simple structure, appropriate quantum circuits might serve as *virtual fundamental-scale laboratories* easily accessible to future technology. Computations involving such circuits might then suggest unforeseen phenomena that could be detected independently at reasonable scales. Alternatively, breakdown of manifold structure at the fundamental scale might lead to detectable deviations from "ideal behavior" in quantum circuits. In particular, in the discrete context, the algebraic objects involved in standard quantum information theory, such as complex Lie groups, would require replacement by complicated discrete objects. Due to the information-theoretic sensitivity involved in the physical implementation of quantum circuits, quantum computing might provide an ideal setting in which to detect the deviations associated with such objects.

References

1. K. Becker, M. Becker, J. Schwarz, *String Theory and M-Theory: A Modern Introduction* (Cambridge University Press, Cambridge, 2007)
2. T. Theimann, *Modern Canonical Quantum General Relativity*. Cambridge Monographs on Mathematical Physics (Cambridge University Press, Cambridge, 2007)
3. A. Connes, *Noncommutative Geometry* (InterEditions, Paris, 1990). English version of Géométrie non commutative
4. E. Verlinde, On the origin of gravity and the laws of Newton. J. High Energy Phys. **2011**, 29 (2011)
5. R. Sorkin, Light, links and causal sets. J. Phys. Conf. Ser. **174**, 012018 (2009)
6. J. Maldacena, The large N limit of superconformal field theories and supergravity. Int. J. Theor. Phys. **38**(4), 1113–1133 (1999)
7. A. Connes, M. Marcolli, *Noncommutative Geometry, Quantum Fields and Motives* (Colloquium Publications, 2007)
8. E. Frenkel, *Lectures on the Langlands Program and Conformal Field Theory* (Springer, Berlin, 2005). (Based on lectures given by the author at the DARPA workshop "Langlands Program and Physics" at the Institute for Advanced Study, March 2004)
9. R.P. Feynman, Space-time approach to non-relativistic quantum mechanics. Rev. Mod. Phys. **20**, 367 (1948)
10. S.W. Hawking, A.R. King, P.J. McCarthy, A new topology for curved space-time which incorporates the causal, differential, and conformal structures. J. Math. Phys. **17**(2), 174–181 (1976)

11. D.B. Malament, The class of continuous timelike curves determines the topology of spacetime. J. Math. Phys. **18**(7), 1399–1404 (1977)
12. B.F. Dribus, On the axioms of causal set theory. Preprint. arXiv: http://arxiv-web3.library. cornell.edu/pdf/1311.2148v3.pdf
13. D. Rideout, R. Sorkin, Classical sequential growth dynamics for causal sets. Phys. Rev. **D61**(2), 024002 (2000)

Chapter 5
The Preferred System of Reference Reloaded

Israel Perez

Abstract According to Karl Popper assumptions are statements used to construct theories. During the construction of a theory whether the assumptions are either true or false turn out to be irrelevant in view of the fact that, actually, they gain their scientific value when the deductions derived from them suffice to explain observations. Science is enriched with assumptions of all kinds and physics is not exempted. Beyond doubt, some assumptions have been greatly beneficial for physics. They are usually embraced based on the kind of problems expected to be solved in a given moment of a science. Some have been quite useful and some others are discarded in a given moment and reconsidered in a later one. An illustrative example of this is the conception of light; first, according to Newton, as particle; then, according to Huygens, as wave; and then, again, according to Einstein, as particle. Likewise, once, according to Newton, a preferred system of reference (PSR) was assumed; then, according to Einstein, rejected; and then, here the assumption is reconsidered. It is claimed that the assumption that there is no PSR can be fundamentally wrong.

Introduction

One of the main objectives of present day physics is to formulate the so-called theory of everything (TOE), a unifying theory that will be capable of describing physical phenomena at all spatial and energy scales. For the past fifty years, legions of physicists have worked relentlessly on this problem without arriving at satisfactory results. This lack of success tells us two important things: first, that the problem is much more complex than originally thought; and, second, that we may have arrived at a dead end. Indeed, many researchers are realizing that theoretical physics is falling into a deep crisis. Such crisis is mirrored in an ever increasing number of

FQXi 2012 Contest, Which of our basic physical assumptions are wrong?

I. Perez (✉)
Department of Physics and Engineering Physics, University of Saskatchewan,
Saskatoon, SK, Canada
e-mail: cooguion@yahoo.com

© Springer International Publishing Switzerland 2015
A. Aguirre et al. (eds.), *Questioning the Foundations of Physics*,
The Frontiers Collection, DOI 10.1007/978-3-319-13045-3_5

publications proposing bold alternatives and yet no clear answers have been found. Actually, as more experimental evidence piles up, the puzzle aggravates. When we found ourselves in such situation, the most natural way to proceed is to revise the foundations and discard whatever that is blinding our sight. Here we shall review one of the most prominent principles in the history of physics that, if seriously reconsidered, can help to heal the current problems in physics, namely, the existence of a preferred system of reference (PSR).

Epistemological Background

Before we move on to our central topic, it would be appropriate to discuss the epistemology behind scientific theories. In this section we shall inform the reader that certain kind of 'scientific hypothesis' are natural components of physical theories and that they are legitimate insofar as they reinforce the theory. In doing this, we shall present a series of arguments that would be pivotal to understand why the PSR can still play a major role in the future of physics.

Assumptions: True, False or Useful?

For the benefit of this work, we shall see as synonyms the words: postulate, axiom, and principle. By these we understand a statement or proposition that serves as the foundation for a chain of reasoning in the construction of a theory. Generally, it is understood that this kind of propositions are true because they are so self-evident that no proof is demanded to demonstrate their veracity. For instance, the proposition '*A straight line segment can be drawn joining any two points*' is generally accepted as true. However, when giving a deeper thought, some propositions turn out to be uncertain. To illustrate this point, let us consider the following assertion: *Space and time are continuous*. If we now ask: is this statement true or false? Needless to say, the reply would be a shrug. In the face of the lack of certainty, the theorist can still proceed and argue that during the conception of a theory whether the statement is false or true turns out to be irrelevant provided that the predictions derived from such assumption reproduce the experimental evidence at hand. In other words, the assumption will gain its scientific value not from the preliminary judgement of the statement but from the experimental verification of the predictions which are derived from it. Only then, the theorist can vigorously contend that the assumption is not only true but also scientific. Thus, despite that the truthfulness or falsehood of a statement can be debatable, a theorist can *presume, for practical purposes*, that the proposition has the potential to be really true.

Theories, on the other hand, can also rest on the basis of 'false' statements as long as these are helpful to strengthen the theory. Examples of this sort can be found anywhere in physics. One typical case is the possibility of reversible processes in thermodynamics. After all, we all are aware that this notion was invented having in mind that, in practice, all processes are irreversible. Despite this, the principle has been highly beneficial for this branch of physics. Other commonly used 'false' assumptions are: point particles or rigid bodies. Therefore, at the end, for the theorist what is crucial is not the truthfulness or falsehood of the assumptions but their *usefulness* in solving particular problems.

Physical Theories

In a wide sense, we understand by physical theory a rational and logical construct composed of principles, concepts and definitions aim at explaining experimental observations. Theories are formulated by seeking *correlations and symmetries* among the observations. The summary of this search is then etched in the so-called laws of nature [1]. These are a set of statements written in mathematical language where the information of physical phenomena has been, so to speak, codified. According to Max Tegmark [2, 3], theories have two components: mathematical structures (equations) and baggage. Baggage are words that explain how the theories are connected to what we humans observe and intuitively understand, i.e., ontologies, 'physical' concepts and notions. And since physics is mainly written in mathematical language, Tegmark goes beyond and asserts that the universe is actually a complex mathematical structure. He remarks that as the theory becomes more fundamental, the level of abstraction increases up to the point that only mathematics would be capable of describing the universe and, therefore, the baggage would be gradually replaced by mathematics. According to him the TOE will have no baggage at all. At first sight, this position seems to be extremist but a deeper reflexion shows us that it may not be the case. To grasp the significance of this, first, one should ask what a mathematical structure is. The minimalistic view is that a mathematical structure is no other thing that a set of abstract objects connected by logical relations and operations. For our purposes, this definition suffices since the task of physics is to seek for 'physical' correlations. From this standpoint, one is then allowed to assert that the description of the universe can be reduced to a set of logical relations, i.e., physical laws. If we agree, this means that what can be said of the universe in terms of baggage can also be said with mathematics. Mathematics is, so to speak, a language in which our intuitive perceptions can be expressed more effectively.

Now, since physical theories use mathematical structures, their structure should be axiomatic. The axiomatization of physics allows us to apply the deductive method from which theorems and quantitative predictions are derived. Such predictions are usually tested in the light of experiments and when the predictions are corroborated, one says that the model has shown its mettle. On the contrary, if the model is incapable of reproducing the data, it should be discarded. In this sense, the job of a theoretical physicist is to single out the mathematical structures or models that fit the observations.

Physical Objects

By analogy with the case of physical assumptions, during the construction of the theory, whether physical objects really exist or not turns out to be irrelevant (because ontologies can have a metaphysical source). This is in view of the fact that the proposed concepts and objects will acquire their physical meaning once the model is faced with experimental evidence. This could be the case of strings, loops, taquions, axions, etc. In some other cases, the experimental observations mold the shape of the theory as well as the properties of its physical objects. For instance, the conception of electron was figured out from observations on electrolysis which suggested a minimum quantity of charge. Later, electrons were conceived as an intrinsic part of the atom and new physical properties such as spin were assigned. In brief, the notion of a 'physical' object strongly depends on the structure of the observations and the theoretical framework where the object is interpreted.

Assumptions and Principles in the History of Physics

Hidden Assumptions

Since ancient times people have built theories based on principles which were considered to be absolute truths, but as time went by some of them have been proven to be actually false. One particular case is the famous assumption that heavier objects fall faster than lighter ones. This principle was held as a physical law for hundreds of years but, as more theoretical and experimental evidence accumulated, Galileo showed that it could no longer be held. In some other cases, theories convey some hidden assumptions. For instance, classical mechanics is based on the three laws of motion and some definitions. In addition to these elements, the theory tacitly presupposes some unnoticed or disregarded assumptions such as: (a) measurements do not affect the physical system under study; (b) the speed of propagation of physical entities can have any velocity, even infinite; (c) physical quantities are continuous. Assumption (a) fails in the microscopic realm. This issue was rectified by quantum mechanics (QM) with the introduction of a powerful postulate: the uncertainty principle. The principle states, among other things, the probabilistic character of a measurement due to the fact that the measuring instrument considerably influences the response of the system under study. Assumption (b), on the other hand, finds its restriction within the context of relativity theory in which there is a maximum speed for the propagation of physical entities. And finally, assumption (c) also finds limitations in QM where some physical magnitudes are discrete. If we extrapolate this reasoning, we can figure out that surely our modern theories are still incomplete and may need a deep revision.

Some Physical Assumptions in the History of Physics

In view of our previous discussion, it is worth extracting from our theories a short list of some of the most typical assumptions. This will help us to be aware of how physics erects and sculpts its 'reality'. It is not the intention of this section to discuss either the truthfulness or the falsehood of each example; for it is evident that some can be true, false or uncertain. As we discussed above, what is of real value for physics is their usefulness in solving the problems that physics has at a given moment of its history. Some of the assumptions are:

- Time flows equally for all observers regardless of their state of motion or its position in a gravitational field.
- The earth/sun is the center of the universe.
- Rigid bodies, æther, vacuum, and fundamental particles (atoms) exist.
- There is no speed limit for the propagation of physical entities.
- The measuring process does not affect the system under study.
- Space, time, and all other physical magnitudes are continuous/discrete.
- Light is a wave/corpuscle.
- Space is static/dynamic; space is a condensed state of matter, space is a network of relationships.
- Energy, charge, torque, linear, and angular momenta are conserved.
- The laws of physics were created along with the creation of the universe and they do not evolve with time.
- All particles of a particular class are identical (e.g., electrons, quarks, positrons, etc.).
- The principle of: relativity, general covariance, uncertainty, equivalence, causality, exclusion, cosmology, etc. hold.
- There is an absolute system of reference.

Some of these assumptions have been definitely discarded, some are still in use, and some others have been reconsidered several times in several epochs. From these three cases, we would like to deal with the last one. One of the most famous cases is the reintroduction of the heliocentric model, first proposed by Aristarchus of Samos in the third century B.C. The model remained in the shadow for many centuries until it was revived by Copernicus et al. in the XV century. The history of science tells us that this model was by far more superior in describing the celestial mechanics than the Ptolemaic system. Another famous case is related to the nature of light. In 1905 Einstein reintroduced into physics the almost forgotten notion that light can be a particle, just as Newton had put forward more than two centuries ago. Armed with this idea, Einstein built a rational explanation of the photoelectric effect discovered in 1887 by Heinrich Hertz. These two examples teach us that no matter how old or controversial an assumption might be, its potential to solve problems justifies its reestablishment as a scientific hypothesis.

Unquestionably, some assumptions have caused a great impact more than others, not only to the structure of a given theory but also to the whole evolution of physics.

Due to their preeminent influence, this kind of proposals deserve both a special attention and a scrupulous assessment; for their arbitrary rejection could be detrimental for the progress of physics. In what follows, we shall discuss the last assumption from the list above. I shall argue that this is one of the principles that physics should revive if physics wishes to make considerable headway for years to come. To this end, I shall try to dissipate some of the misconceptions that have been appended to it for more than a century.

The Principle of Relativity Is Not at Variance with the Preferred System of Reference

Newton's Absolute Space

When Newton developed his laws of motion, he thought that they were valid in absolute space (AS). He contended that the water inside the famous bucket was rotating relative not to the bucket but to AS [4]. This experiment gave him confidence that any body possesses not only apparent (or relative) motion but also genuine motion and such motion can only be relative to space itself, or generally speaking, relative to a PSR. From this, it follows that if bodies move relative to AS, then this entity has to be something endowed with physical properties and our disquisition would reduce to identify them. For Newton, AS was an homogenous and isotropic background in which material bodies were embedded, some sort of rigid container mathematically represented by Euclidean space. What we all learn at school, on the other hand, is that this entity is not composed of a material substance, rather, it is total emptiness. It is not clear if Newton agreed with this view, but we have evidence that he actually thought that there was an ethereal and pervading material substance conveying gravitational interactions. In a letter sent to Bentley in 1692, Newton wrote:

> It is inconceivable that inanimate brute matter should, without the mediation of something else which is not material, operate upon and affect other matter, without mutual contact, as it must do if gravitation in the sense of Epicurus be essential and inherent in it. And this is one reason why I desired you would not ascribe 'innate gravity' to me. That gravity should be innate, inherent, and essential to matter, so that one body may act upon another at a distance, through a vacuum, without the mediation of anything else, by and through which their action and force may be conveyed from one to another, is to me so great an absurdity, that I believe no man who has in philosophical matters a competent faculty of thinking can ever fall into it.

Newton's theory of gravitation assumes that between two celestial bodies there is absolutely nothing mediating the interaction, instead, the alleged interaction occurs by gravitational fields acting at a distance. We note from this letter, however, that his theory does not reflect his actual view. His words seem to imply that he did not even believe in total emptiness, since in Newton's time by the word 'vacuum' people understood 'devoid of matter'. In spite of this, what is relevant for us is that he established in his theory that space was immovable. This assumption was precisely what the philosopher Ernst Mach disliked [5]. For he questioned the scientific utility of an entity that exists but is not affected by the matter it contains. Mach replied to the bucket experiment arguing that the water moved relative not to AS but to the stellar matter surrounding the bucket, because for him only relative motion was possible. Although Mach's argument is weighty, it is not clear what physical substance mediates the gravitational force. It was the insight of Einstein that shed light on the problem some years later. Starting in 1905, Einstein rejected the æther as the medium for the propagation of light and, by doing this, he left 'physical' space again absolutely empty, just as in Newton's theory. Einstein immediately realized this flaw and from 1907 to 1916 he embarked in a historical journey to try to 'materialize' Mach's ideas [6–9]. In the Einsteinian vision of the universe, space is mathematically represented by a pseudo-Riemannian manifold characterized by the metrical field $g_{\mu\nu}$ which contains the gravitational potentials. As a consequence, one has to conclude that the water in the bucket moves relative to the gravitational fields (GF). Einstein then finally replaced the material substance, conceived by both Newton and Maxwell, by the metric field [10]. Since then, the assumption that space can be made up of a material substance has been ruled out from physics (we will return to this topic below). Yet, physics has never ignored the power of intuition. In 1933 the Swiss astronomer Fritz Zwicky discovered some anomalies—now known as dark matter—in his studies of the Coma galaxy cluster. This evidence clearly suggests that there is something in space by far more complex than originally thought and that it could be indeed composed of an imponderable and invisible material substance: Newton's substance? The æther that Einstein rejected? Unfortunately, Fritz' discovery was ignored for about forty years until Vera Rubin et al. revived it in the 1970s. Still dark matter is one of the most puzzling problems in modern physics.

Invariance of Newton's Laws

Let us not digress from Newton's work and bear in mind henceforth the previous discussion. It is well established that Newton's laws are invariant with respect to Galilean transformations. This is in virtue of the Galilean Principle of Relativity (GPR)—In fact, in The Principia, Newton included this principle as corollary V and justified it with the aid of the second law. It states that all mechanical laws are the same in any inertial system of reference (ISR). But, what is the experimental meaning of this principle? It simply means that no mechanical experiment can tell whether an ISR is at rest or in motion relative to AS (this was well understood by Newton). The understanding of this statement is vital to make clear that *the GPR is not at variance with the existence of the PSR*. This being said, let us consider the following two key questions:

1. *Does the fact that the PSR cannot be experimentally detected mean that the PSR does not exist?*

2. *If the PSR cannot be experimentally detected, does the assumption become a meaningless assumption?*

To grasp the deep significance of these questions, let us contemplate the following situation borrowed from QM and that is beautifully discussed at length by Popper [11, pp. 211–217]. During the development of the atomic model, Niels Bohr imagined that electrons follow orbital paths around the nucleus. He assumed that electrons revolve with a given period T and from this the energy levels E_n and thus the emission spectrum of the hydrogen atom was computed. The key for the success of this approach is the assumption that electron orbits are quantized. However, if we take a closer look at the concept of 'path' and scrutinize the principles of QM, we find a serious difficulty. According to Heisenberg's principle, the experimental determination of two correlated observables A and B is limited by the uncertainty relation $\Delta A \Delta B \geq h/4\pi$, where h is Planck constant. In particular, if we assume the observables to be the momentum p_x and the position x, the principle tells us that: the higher the precision in the measurement of x, the higher the error in the measurement of p_x and vice-versa. This means that it is impossible to experimentally determine the particle's path (i.e., the simultaneous knowledge of both p_x and x) with a certainty exceeding the above expression. The reason rests in the fact that the measurement affects the pristine state of the particle. Hence, according to Heisenberg himself, it is meaningless to grant any *real* significance to the notion of 'path' [12]. As such, the path becomes an unobservable magnitude, i.e., a magnitude unaccessible to experimental verification and therefore it is not useful as a basis for theoretical predictions. The recognition of this, urges us to conclude that measurements cannot serve as a foundation to test physical reality, so to speak, there is no such a thing as physical reality since our instruments are not capable of revealing the true state of a system. In reality, the information we get from our measurements is the outcome of the interaction between the instrument and the system under study. If we conveyed these considerations to the extreme, we would arrive at dramatic conclusions: we would conclude that the factual character of physics is just a chimera. Fortunately, not everything is lost and here the probabilistic and statistical interpretation of QM comes to the rescue. Since we cannot access the precise state of a particle, at least we can tell with some degree of certainty the probable outcome of an experiment.

On the other hand, the formalism of QM through the Schödinger equation allows us to calculate with certainty the particle's path up to a moment before the measurement, i.e., the formalism assumes that there is a path. Evidently, this is at variance with Heisenberg's principle. So, does the path physically exist or not? The answer is not trivial, but being physics a factual science, we understand that the experimental data are essential to sustain the scientific credibility of a theory; for the data collected give us some information of the state of a system. But we have also learned that our measuring processes modify the absolute state. Hence, despite that the path cannot be exactly determined, it is scientifically legitimate to presuppose that the particle's path physically exists, just as the formalism assumes. Admittedly, the fact that the measurement destroys the knowledge of the actual path, does not imply that the physical notion of 'path' has no scientific value and, at the same time, does not encourage us to reject QM altogether on the basis that the theory is dealing with unobservables (in the words of Popper, metaphysical constructs).

One more example of this type is illustrative to reinforce the view that the lack of experimental evidence does not suffice to reject a hypothesis regardless of its apparently unobservable character. Consider the postulate that space and time are continuous. Here once more, we have no conclusive experimental evidence to thoroughly sustain this assumption. In spite of this, good reasons can be advanced for trusting our postulate; actually, our theories have indeed shown that it can be true. Thus, having in mind these two examples, the answer to the first key question is clearly in the negative, for if one accepts the existence of a non PSR, one cannot deny the existence of the PSR since the GPR assures the equality of the mechanical laws in all ISR. Then, the second question is immediately answered also in the negative. From here we conclude that the GPR should not be understood as the exclusion of only one ISR, but quite the contrary, as the inclusion of all of them. Evidently, the arbitrary rejection of the PSR can be, in the long term, detrimental for the advancement of physics because we would deprive our theories from elements indispensable for their logical consistency.

Invariance of the Laws of Physics

It is unquestionable that the PSR assumption in classical mechanics resulted highly beneficial for the progress of physics for more than 200 years. The extension of this assumption to electromagnetic phenomena was also very fruitful. It achieved its highest peak with the development of electrodynamics. By the mid 1860s Maxwell predicted that light was some kind of electromagnetic wave that travels through the æther. Some years later, in 1887–8, Hertz could generate Maxwell's waves, leaving no doubt that Maxwell was in the right way [13]. Nonetheless, the mere corroboration of electromagnetic radiation did not suffice to establish the existence of the medium. Maxwell was aware that in his equations the æther did not appear, that is, they conserve the same form whether there was æther or not. Yet, for him, the material substance was indispensable to avoid the action at a distance that dominated gravitational interactions. Besides, all known waves hitherto required a medium to propagate and it was natural to assume that light waves could not be the exception. By the middle of the 1870s, Maxwell's theory was still under construction and many experiments were in line waiting for a satisfactory explanation. In the following years, Maxwell expanded the scope of the theory but, unfortunately, he prematurely died in 1879. During the next decade, a new generation of physicists resumed Maxwell's work. For this reason, Oliver Heaviside, Oliver Lodge, and George FitzGerald were called *The Maxwellians*. These brilliant scientists shaped Maxwell's theory nearly as we know it today [14, 15]. But in spite of the great advances, they did not solve the æther issue, still, the equations had no explicit link to the æther. Fortunately, both Hertz and Paul Dirac (six decades later) also realized Maxwell's problem and promptly modified the equations [13, 16, 17]. With the aim of accounting for effects of charged bodies in motion relative to the æther, Hertz replaced the partial time derivatives by total (also known as convective, Lagrangian, material or substantial) time derivatives. At that time, his formulation did not attract much attention because some of the predictions were in disagreement with experiments on insulators. Incidentally, modern investigations have revealed that Hertz' formulation was not incorrect at all and that the observed discrepancies can be attributed to quantum effects [18]. Indeed, Dirac in 1951 also proposed a new electrodynamics and discussed that quantum principles play an important role for reviving the æther concept and when considering the topic seriously, the æther was crucial to build a satisfactory theory of the electron (now known as quantum electrodynamics).

The problem of the æther was not only theoretical but also experimental. It was imperative to show that the ubiquitous substance was not a mere idea. To prove its existence, physicists engaged in an epic hunt by the end of the XIX century. In 1887, Michelson and Morley carried out their famous interferometric experiment which, according to the thinking of that time, would tell them whether the PSR existed or not (below we dispel some misconceptions about these kind of experiments). As is well known, the results were negative, and by analogy with the experimental implications of the GPR, later, from 1900 to 1905, Larmor [19, 20], Lorentz [21, 22], and Poincaré [23] realized that no electromagnetic experiment can tell whether an ISR is at rest or in motion relative to the æther. Such discovery was called simply the *Principle of Relativity* (PR) and it is considered as a generalization of the GPR. Thus, in spite of its undetectability, Larmor, Lorentz, and Poincaré answered the above key questions in the negative, whilst Einstein held the opposite opinion; he was actually appealing to the principle of parsimony [6]. Since no experiment can tell whether an ISR is at 'real rest' or in 'real motion', Einstein declared that these statements are meaningless (cf. with Heisenberg's opinion above). For him, just as Mach, only relative motion is measurable and hence has real meaning. Nonetheless, if we strictly follow this line of thought, motion would adopt a fictitious character. These theoretical perplexities were apparently 'overlooked' by Einstein but exposed years later by H. Ives and G. Stilwell when they experimentally corroborated time dilation [24, 25]—We shall discuss in the following sections the importance of the PSR on this issue.

To comply with the PR, physicists were prompted to construct a new dynamics which is now known as *Relativistic Dynamics* [26]. Both Maxwell's laws and the new kinematical and dynamical laws are said to be Lorentz invariant. The new symmetry assures that not only the form of the laws of physics (LP) but also the values of their constants remain the same in all ISR. This inevitably leads us to ask again: *Is then the PR at variance with the existence of the PSR?* Certainly, the answer goes in the negative [16, 17] for no experiment forces us to reject the PSR [22, 23, 27]. By analogy with the GPR, the PR should be understood not as the discrimination of the PSR, but quite the opposite, as the inclusion of the PSR for the description of physical phenomena. Within this context, Lorentz invariance experimentally means that any experiment carried out in the PSR will lead to the same LP that can be found in any other ISR. The history of physics tells us however that modern theories have discarded it following Einstein's canon [6, 10]. But if one upholds the opinion that the PSR is not an issue of parsimony but of usefulness and logical consistency in the physics, one can claim that the assumption that there is no PSR is fundamentally wrong. Let us make some other considerations to support this claim.

Experimental and Theoretical Considerations in Favor of the PSR

Immediately after the development of the special theory of relativity (SR) a hot debate not only on the existence of the PSR but also on the constancy of the speed of light set in both on theoretical and experimental grounds. Even today many researchers in the fields of physics and the philosophy of physics have kept alive these topics from an epistemological perspective. Thanks to their perseverance, the good news is that substantial advances have been made in the last decades. Although not widely known, specially among the physics community, now it has been understood some key aspects that can be fundamental that could be fundamental for the future of physics.

Misinterpretation of Experiments: The Michelson-Morley Experiment

In the first place, experimental arguments against the PSR have been misleading since the advent of SR. Interferometric and non-interferometric experiments performed during the XIX and XX centuries have been considered as proofs that the æther does not exist. For example, it is common to find in textbooks statements such as: *if the æther existed the Michelson-Morley experiment (MMX) would have shown any variation in the fringe shift N of the interference pattern.* This is evidently misleading because, as we discussed above, in virtue of the PR, no electromagnetic experiment can tell about the existence of the PSR. This is quite clear from the PR and it would be worth dissipating the misconceptions around the experiment because most interpretations suffer from the same drawback.

The experiment had the purpose of measuring the absolute speed of the Earth relative to the æther—to illustrate my point, I will use the words 'vacuum' and 'æther' interchangeably. In Fig. 5.1 we show the MMX as seen from both systems of reference, the vacuum (S) and the Earth (S') that moves with speed $v < c$ relative to S. For the sake of illustration, the extension of the wave fronts for the four electromagnetic waves have been accentuated to visualize the Doppler effect and their propagation vectors are displayed as well. According to Michelson and Morley the determination of v would be obtained by measuring changes in the phase difference δ produced at the point P by the interference of two electromagnetic waves. Then, from the theory of interference as calculated in S, the phase difference is

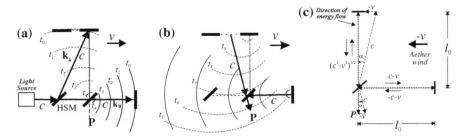

Fig. 5.1 The Michelson-Morley experiment. **a** Forward and **b** backward advance of light waves as seen from the observer at rest in the vacuum. *Arrows* represent the propagation vectors of the four wave fronts. *Solid arcs* are for longitudinal wave motion and *dashed arcs* for transversal wave motion. **c** As judged from Earth, the vacuum is passing by. Using Galilean transformations, the one-way speed of light becomes anisotropic

$$\delta = k(s_\parallel - s_\perp) = k\delta s = \omega\delta t, \tag{5.1}$$

here $\delta s = c\delta t$ is the difference in the optical path length (OPL) for the waves in the longitudinal and transversal journeys, respectively; and $\delta t = t_\parallel - t_\perp$ the corresponding time difference. In these expressions we have used the relations: $c = \omega/k = \nu\lambda$ for the one-way speed of light in vacuum, with $k = 2\pi/\lambda$ the wave number, λ the wavelength of light and $\omega = 2\pi\nu$ the angular frequency. From the preceding formulation it is evident that a fringe shift exists whenever $d\delta/dt \neq 0$. This is achieved in practice by varying the OPL or changing the speed of the beams of light. Recall also that the experiment was designed to revolve, so we have to consider an angular dependence θ in the phase representing the revolution of the plane of the interferometer with respect to the motion of the Earth (there is still another angle to be considered but to convey our idea we do not need to include it here [28–30]). Ignoring the length contraction effect, the expression for the phase in the system S is:

$$\delta = \frac{\omega}{c}\left\{l_0\gamma^2\left[(1 - \beta^2\sin^2\theta)^{1/2} - (1 - \beta^2\cos^2\theta)^{1/2}\right]\right\}, \tag{5.2}$$

here $\beta = v/c$, $\gamma = 1/\sqrt{1 - \beta^2}$, and l_0 is the length of the arms as measured at rest in S. This equation tells us that $\delta = \delta(\beta, \theta)$, but if the apparatus is not rotated we arrive, to a first approximation, at the traditional expression found in most textbooks: $\delta \approx (l_0\omega/c)\beta^2$. If we assume the Earth as an ISR then $d\delta/dt = 0$ and no fringe shift will be observed. So, the rotation of the apparatus is indispensable to observe a fringe shift. The maximum phase occurs after a $\pi/4$ rotation and N (the number of fringes) is calculated by the difference before (B) and after (A) rotation; then we have $N = \delta_A - \delta_B \approx (2l_0\omega/c)\beta^2$. However, when we consider length contraction in Eq. (5.2), we find that the OPL is the same for both light beams, so $\delta = 0$ and hence no fringe shift is observed regardless of the variations of θ and/or β. This justifies why the experiment failed to observe a positive result. Given this outcome we ask:

does this mean that there is no vacuum? To give a definite answer, let us now discuss the physics from the perspective of an observer in S'.

First, we emphasize that in Eq. (5.1) we have made used of the relation $c = \lambda \nu$ to express δ in terms of time. This change is possible because in the solutions of the wave equation, ω and k have a linear dispersion relation, i.e., the group and the phase velocities $V \equiv \partial \omega / \partial k = \omega / k = c$ are the same in all directions (isotropy of the one-way speed of light in vacuum). That the one-way speed of light is isotropic in at least the system S, does not follow from SR but it is a direct consequence of electrodynamics. Our problem is then to find out if this is also true in any other ISR in motion relative to the vacuum. Immediately, some closely related questions come to our mind: Where do the alleged anisotropy of the one-way speed of light find in most relativity textbooks come from? What is the physical basis for postulating that '*the speed of light is independent of the state of motion of the source or the observer*'? In the last statement it is implied that the velocity of light can depend on the velocity of the source or the observer. But what is the rational source that prompted physicists to conceive such possibility?

Before the discovery of the Lorentz transformations (LT) the only known transformations relating two ISR, moving with relative velocity $\mathbf{v}$, were the Galilean transformations:

$$\mathbf{r}' = \mathbf{r} - \mathbf{v}t; \qquad t' = t. \tag{5.3}$$

Note that the time relation expresses that time flows equally for all inertial observers, meaning that there is a unique rate of flow in all ISR. As a consequence of these transformations, physicists were induced to believe that the speed of light could acquire different numerical values in frames in motion relative to the æther and, in consequence, the wave numbers k's or the frequencies ω's for each of the light beams involved in an experiment would not take on, in general, the same values. This can be easily shown by applying these transformations to the phase of the wave function $\Psi = a \exp[2\pi i (\mathbf{n} \cdot \mathbf{r} - ct)/\lambda]$, where a is the amplitude and $\mathbf{n}$ a unit vector. After a straightforward calculation, the phase in the system S' becomes: $2\pi i [\mathbf{n} \cdot \mathbf{r}' - (c - \mathbf{n} \cdot \mathbf{v})]/\lambda$. Therefore, in S', the speed of light is $c - \mathbf{n} \cdot \mathbf{v}$, which is anisotropic. Evidently, the error in this prediction is the misapprehension that electrodynamics and Galilean relativity are compatible formulations. This is what Einstein spotted in his famous *Gedankenexperiment* about the race with light rays. While Maxwell's theory states that the one-way speed of light is a constant relative to the æther, the Galilean addition of velocities dictates that the speed of light must be velocity dependent.

Although we have already identified our 'naïve' mistake, let us further proceed with our analysis. As seen from S (refer back to Fig. 5.1c), the speed of energy flow (or energy flux given by the Poynting vector) for the longitudinal and oblique beams is c, therefore the velocity of the energy flow in the y-direction is $\sqrt{c^2 - v^2}$. According to the observer in S', the vacuum is passing by with velocity $\mathbf{v}' = -v\,\hat{\mathbf{x}}'$, and if we apply Galilean relativity to light propagation, the velocities of the energy flow for the four beams in the frame S' must take on the values:

$$\mathbf{c}'_{\parallel\pm} = \pm(c \mp v)\,\hat{\mathbf{x}}'; \quad \mathbf{c}'_{\perp\pm} = \pm\sqrt{c^2 - v^2}\,\hat{\mathbf{y}}', \tag{5.4}$$

where $\pm$ stands for forward and backward directions, respectively. Thus, for the parallel direction the wave fronts travel the OPL: $s'_1 = l'_0 = t'_1(c - v)$ and $s'_2 = l'_0 = t'_2(c + v)$; for the forward and backward journeys, respectively. Accordingly, the time spent in the longitudinal journey is $t'_\parallel = (2l'_0/c)\gamma^2$. The time for the transversal journey is calculated as follows. The transversal distance in one direction is $s'_3 = s'_4 = l'_0 = c'_\perp t'_\perp/2$. Solving for $t'_\perp$ and taking the difference $\delta t'$, we obtain to a first approximation: $\delta \approx \omega(2l'_0/c)\beta^2$. Since the Earth is in motion relative to the vacuum, then $t' = t\gamma^{-1}$ and $l'_0 = l_0\gamma^{-1}$. Taking into account these effects in our previous calculations, we also find that $\delta = 0$. Showing once more that the experiment cannot determine the Earth's velocity relative to the vacuum.

We must remark, that the absolute speed of light waves never changes regardless of the speed S' relative to S, because the light waves travel through the vacuum and its speed is determined by the properties of the medium. If we assume that the medium is static, isotropic, homogeneous and its temperature remains constant, we have no reason to believe that the speed of light would change. The alleged anisotropy of the speed of light is just a *fictitious* effect caused by the relative motion between the Earth and the vacuum. This automatically means that the speed of the waves is independent of the motion of the source or the observer (second postulate of SR). Thus, the null result of these kind of experiments does not prove that there is no medium. Some experimentalists that concluded that there is no medium, made the same 'mistake' of convoluting Galilean relativity and electrodynamics [31, pp. 518–524].

If we have made clear that no experiment of this kind rules out the PSR, we are faced again with the two key questions above and, therefore, the issue may become only a matter of usefulness and coherence in the logic of a theory. Einstein rejected the æther, first, because, from the theoretical viewpoint, SR could not make special distinctions among ISR; actually, for him the æther assumption was not wrong but appeared to be superfluous. And second, because, from the experimental viewpoint, there was no unambiguous evidence of its existence. Nevertheless, according to the discussion of the previous section, the first argument is weak, for if one follows such line of thought then Newton's AS would have been rejected as well from classical mechanics since the GPR guarantees that all ISR are equivalent. In Einstein's epoch, the second argument had a great weight, however, the discussion given above and the experimental evidence accumulated after the 1930s, strongly disagrees with Einstein's view. The experimental evidence we are referring to is this. Consider the following hypothetical situation. Imagine that before the discovery of relativity, particle accelerators had been already developed. And assume that the ALICE, ATLAS and CMS collaborations at the large hadron collider had released the news, well-known today, that the quantum vacuum is actually a perfect fluid [32]. If this fluid were assumed to be at rest and not significantly affected by the presence of material particles it would immediately be identified as the æther or AS; just in the same way as in 2012 many physicists sympathized with the discovered boson at the LHC and identified it as the Higgs boson despite that they did not know yet its other physical properties (spin, etc.). So, if by 1905, physicists had already discovered the presence of dark matter, the background radiation, the presence of a perfect fluid and the Casimir effect, would physicists, despite the success of relativity, have good reasons to discard the medium for light and thus the PSR? Indeed, the answer would be in the negative. The concept would be maintained because the experimental evidence would have suggested its presence.

On the Experimental 'Corroboration' of the Second Postulate of Special Relativity

In second place, research on experimental methods used for the measurement of the speed of any physical entity, shows that when the paths of the physical entities involved in the measurement form a closed circuit, what the experiment measures is an average speed, i.e., a harmonic mean of the speed or the so-called two-way speed. This implies that it is not feasible to measure the one-way speed of light [33]. Since the second postulate of SR tacitly states that there is a finite isotropic speed c, the studies reveal that this postulate has never been experimentally corroborated. Under such scenario, one can raise sharp objections against either SR or electrodynamics similar to those raised against the PSR. But despite that nature conspires to hide from us this crucial knowledge, there are enough reasons to hold the postulate. And, once more, this lack of experimental evidence does not impel us to reject it.

Misinterpretation of Newton's Theory

The structure of a physical theory is fixed and cannot be modified at will. A genuine scientific theory cannot, when faced with an irreconcilable fact, be rectified a little to make it agree. There are some colleagues (see for instance [34, p. 6]) who have claimed that Newtonian mechanics is just as relativistic as SR under the argument that Galilean transformations leave invariant Newton's laws. They even contend, led by the relativity school, that Galilean invariance demonstrates that there is no PSR. This is, of course, nonsense for AS is a principle of the theory and it cannot be arbitrarily eliminated. In fact, we shall see below that Minkowski space plays the role of absolute space in SR.

Relative Motion Leads to Quandaries

It is worth discussing briefly how quandaries arise in SR due to the idea that only relative motion is significant. Here I shall consider the case of time dilation which is usually confused with the so-called clock paradox. I do not treat the original clock paradox [6] since it has shown to be misleading [35]. Instead, I slightly modify the situation to identify where the perplexing part of SR is. The problem is related to the topic of relative motion versus absolute motion. Imagine three synchronized clocks placed in line at three equidistant points A, B, and C. Consider that clocks at A and C are moved simultaneously (that is, at $t_A = t_B = t_C = 0$) towards B with the same constant velocity v (and by symmetry, the same initial acceleration if you wish). Now we wonder whether time really dilates or not for clocks in motion. (i) According to SR, an observer at rest next to clock B will figure out that, since both clocks A and C are moving towards B at the same speed, they will arrive at B synchronized among each other but lagging behind clock B by the factor $\sqrt{1 - (v/c)^2}$. So far so good, but this is not the end of the story. Relative motion strictly dictates that the two clocks A and C are not only moving relative to each other at constant speed V, but also relative to the clock B at speed v. Since according to Einstein there is no PSR, this means that absolute motion is meaningless absolute motion is meaningless. On the basis of this theoretical restriction, it is equally legitimate to judge the situation from the standpoint of an observer in the ISR of clock A. (ii) From this perspective, clocks B and C are approaching clock A at speeds v and V, respectively. And by symmetry, when the three clocks meet, the observer at A will find that, both clocks B and C, will lag behind clock A in proportion to their relative velocities, $\sqrt{1 - (v/c)^2}$ and $\sqrt{1 - (V/c)^2}$, respectively. Moreover, since $V > v$, he will assert that clock C will be lagging behind clock B. (iii) With the same right and by the same argument, a third observer in the ISR of clock C will claim that when the three clocks meet, clocks B and A will be lagging behind clock C in proportion to their relative velocities, etc. Certainly, according to the view that only relative motion is meaningful, the tree options are equally legitimate, although it is obvious that if the experiment is performed the three options cannot be true. In view of these baffling conclusions, it is impossible to decide solely on the grounds of the principles of the theory itself, what would be the *actual* outcome of an experiment like this (a similar situation occurs with the stellar aberration). By 1937, Ives and Stilwell realized about these quandaries and discussed the topic at length [24, 25, 36]. They carried out a series of experiments to test time dilation and pointed out that the source of the problem is the omission of the æther. If we reintroduce the PSR in our picture, we will have a logical criterion to decide what will be the actual outcome of the experiment since, in this case, only absolute motion is meaningful (below we discuss how to distinguish absolute from relative motion). Even if we were not able to determine the real state of motion of an ISR, we can still theoretically assume that either the ISR of clock B is at rest or moving at speed w relative to the PSR. Under this scenario, we realize that the flow of time of clock B will remain constant at all times whereas the flow of time for the clocks A and C will be altered since they are absolutely moving (for detail calculations on this view see Ives and Stilwell works [24, 25, 36]). Therefore, from the absolute point of view, options (ii) and (iii) are naïve and can be discarded at once. We are left then with option (i). Whether this option is true or not would depend on the adopted clock synchronization convention, topic which is outside the scope of present work [33]. This example constitutes a logical justification to reconsider the PSR. Einstein rejected it because he considered it superfluous, now we see that parsimony leads to logical predicaments. What we learn here is that parsimony is not always the best choice; for if a theory A, assuming the PSR, explains the same amount of observations as another theory B, in which no PSR is assumed, one should chose theory A because it is free from perplexities. The theory A we refer to is not SR with a PSR, but Lorentz' æther theory [21] [not to be confused with FitzGerald-Lorentz hypothesis about length contraction].

The Law of Inertia and the Conservations Laws

When we work within the context of Newtonian mechanics, we are usually unaware how the law of inertia was defined. The law states that: *Every body persists in its state of rest or of uniform motion in a right line, unless it is compelled to change that state by forces impressed thereon.* But with respect to what system of reference is this true? To answer, let us imagine that we have two systems of reference, S and S', with S' moving along the x direction with velocity V and acceleration W relative to S. We now have a particle that moves also along the x direction with velocity v and acceleration w relative to S as well. According to Galilean relativity, the velocity v' and the acceleration w' in S' are given, respectively, by: $v' = v - V$, $w' = w - W$. If the particle moves by inertia relative to S, we have in S' that $w' = -W \neq 0$, that is, the observer in S' cannot figure out that the particle is moving by inertia. For this reason the law of inertia loses its meaning if we do not specify to what system of reference the law refers to. In consequence, in order for this law to be meaningful, we need to define a special system where the law of inertia holds. Following Newton, such system is AS and by virtue of the Galilean transformations, the law of inertia is also true in any other system moving uniformly relatively to AS. It follows that a system in which Newton's laws hold is an ISR.

The recognition of Newton's laws and AS invites us to accept their consequences since space (Euclidean) and time are isotropic and homogeneous. From these properties, as we know from Noether's first theorem, the conservation of momentum and energy follow. Because Minkowski space-time has these properties, the law of inertia and the conservation of energy-momentum hold also true in SR. In his article on cosmological considerations, [37] Einstein, led by Mach, manifested his disagreement with Newton enunciating the so-called 'relativity of inertia': *In a consistent theory of relativity there can be no inertia relatively to 'space', but only inertia of masses relatively to one another.* This statement is just another version of *Mach's principle.* To incorporate this principle in General Relativity (GR), the law of inertia and the sacred conservations laws had to be redefined. Lorentz, Klein, Einstein, et al. soon realized this 'inconvenience' and tried to amend it. The reason is quite evident: for Einstein, inertia is due to the presence of other masses and relativistic space is, in general, dynamic (non-Euclidean). In an attempt to save energy conservation, Einstein introduced the pseudo-tensor t_i^k and claimed that the total momentum and energy, $J_i = \int (T_{\mu\nu} + t_i^4) dV$ (with $T_{\mu\nu}$ the stress-energy tensor), of the closed system are, to a large extent, independent of the coordinate system [38, 39]. The contemporary version acknowledges, however, that the 'conserved quantities' are not in general conserved in GR and other diffeomorphism covariant theories of gravity [40, 41]. The problem consists in that in GR, gravitation is represented by the metric tensor $g_{\mu\nu}$ (that underlies the geometry of space) and the gravitational energy contained in the geometrical part cannot be, in practice, localized. In mathematical terms, this is implied in the divergence of the stress-energy tensor

$$T^{\mu\nu};_{\mu} = 0, \tag{5.5}$$

which expresses the exchange of energy-momentum between matter and the gravitational field. For asymptotically flat and stationary spacetimes at infinity (i.e., spaces that tend to Minkowski space), one can always find an energy conservation law by integration of Eq. (5.5). But, it is no longer possible for general spacetimes.

As for the law of inertia, Einstein worked out a cosmological model where he first considered the scenario of an open and expanding universe [37]. To solve the gravitational field equations, one needs to provide the boundary conditions. By a suitable choice of a reference system, the $g_{\mu\nu}$ in spatial infinity tends to Minkowski metric $\eta_{\mu\nu}$. He rejected this possibility because he first realized that the reference system would represent a PSR contradicting the '*spirit of relativity*'; and, secondly, because this choice would discriminate Mach's principle. He then opted for avoiding boundary conditions at infinity and considered the possibility of a finite and closed universe. For this purpose, he modified his field equations introducing the famous cosmological constant Λ (for details on the physical meaning of Λ, see A. Harvey et al. [42]). The modified equations read:

$$R_{\mu\nu} - \frac{1}{2}Rg_{\mu\nu} = T_{\mu\nu} + \Lambda g_{\mu\nu}, \tag{5.6}$$

where $R_{\mu\nu}$ is the Ricci tensor and R the scalar curvature. With this new term, he thought he had succeeded not only in 'satisfying' Mach's principle but also in removing the boundary conditions. His joy, however, did not last much because, in the same year, the Dutch astronomer de Sitter found a vacuum solution in which matter ($T_{\mu\nu} = 0$) was not necessary to define an ISR [43]. Three decades later, Pauli recognized that the problem was still open [39, p. 182] and Steven Weinberg expressed in his book of 1972 that the answer given by GR to this problem through the equivalence principle '*lies somewhere between that of Newton and Mach*' [44, pp. 86–88]. More recently, some physicists claim that the Lense-Thirring effect contemplates some effects of Mach's principle [45], although, most specialists agree that GR is neither completely Machian nor absolutely relativistic [46, p. 106], implying that, after all, both the PSR and Newton's law of inertia are still very alive. In addition to this failure, a closer inspection of the ontology of space in GR reveals peculiarities that require a careful examination.

Are Space and Vacuum the Same Physical Entities?

The understanding of the nature of the vacuum might be crucial to achieve the TOE because this problem is closely related to the energy density of the vacuum ρ_{vac} and the cosmological constant problem. The 'naïve' calculations for ρ_{vac} obtained from quantum field theory (QFT), yield a value that differs by about 120 orders of magnitude when compared to the value ρ_Λ obtained from cosmological observations where it is believed that GR is the correct theory. Up to now, physicists are still perplexed for such a big difference [43, 47, 48]. This can indicate that either GR, or QFT, or both are in need of serious revision. After many years of study, some researchers suspect (as I do) that the geometrical representation of space may not be the best choice for the future of physics [49]. In fact, this notion seems to be at variance with the notion of vacuum in QFT. In this theory the vacuum has a ground state energy different from zero, the so-called zero point energy. Nevertheless, it appears that GR posses a distinct understanding. Taking a closer look at Eq. (5.6), we notice the following peculiarities: The left hand side represents the geometry of space where the energy of the GF is included. On the right hand side, we find the $T_{\mu\nu}$ where we include 'matter' and the Λ-term that can be understood as a repulsive force due to the vacuum energy (known as dark energy). Since the latter can be put on the left hand side too, one can interpret it as gravitational energy rather than vacuum energy (this is still under debate). If we leave it on the right side, the vacuum is viewed, in its rest frame, as a perfect fluid with energy density ρ_{vac} and isotropic pressure p_{vac}, both related by $w = -P_{vac}/\rho_{vac}$. For this fluid the stress-energy tensor reads

$$T_{\mu\nu}^{vac} = (\rho_{vac} + p_{vac})u_\mu u_\nu + p_{vac}g_{\mu\nu}, \tag{5.7}$$

where u^μ is the fluid four-velocity. If we assume a motionless fluid, the first term in this expression is zero. Hence, ρ_{vac} is proportional to Λ and it is legitimate to assume that $\rho_{vac} = \rho_\Lambda = \Lambda/(8\pi G)$. Now, Einstein equations for 'empty space' with no vacuum energy ($\Lambda = 0$) read $R_{\mu\nu} = 0$. Solutions for these equations are Minkowski and Euclidean spaces. It is worth noting that Einstein considered Minkowski metric as a special case of metric with constant gravitational potentials Φ. Since $\Phi =$ constant, there is no GF in this space. Moreover, from the geodesic equation

$$\frac{d^2x^\mu}{ds^2} = -\Gamma^\mu_{\alpha\beta}\frac{dx^\alpha}{ds}\frac{dx^\beta}{ds}, \tag{5.8}$$

where $\Gamma^\mu_{\alpha\beta}$ is the Christofel symbol and s a scalar parameter of motion, it follows that a test particle moves in straight line (Newton's law of inertia). This rationale also applies to Euclidean space, implying also that, in this space, there is a constant gravitational potential and, therefore, zero GF. Nonetheless, it is difficult to reconcile ourselves with this interpretation given that there are no sources of gravitation and given that, in Newtonian mechanics, Euclidean space represents AS, i.e., total emptiness. We are thus tempted to think that if Euclidean space represents the PRS in Newton's theory, Minkowski space represents the PSR in SR. Naturally, for GR, Minkowski space is not a realistic space, although our analysis is exposing the substantival character of space in GR. Considering now that the vacuum has nonzero energy, Einstein's equations read $R_{\mu\nu} - \frac{1}{2}Rg_{\mu\nu} = \Lambda g_{\mu\nu}$. If $\Lambda > 0$, $\rho_{vac} > 0$ and $p_{vac} < 0$, one of the solutions is the *de Sitter flat space*. If, $\Lambda < 0$, $\rho_{vac} < 0$ and $p_{vac} > 0$ we have the *anti-de Sitter space*. In the former, space is open and expands; and in the latter, space is close and the expansion decelerates. So, tests particles will move apart in a de Sitter space; implying inertia without matter (in contradiction to Mach's principle). Finally, if $T_{\mu\nu} \neq 0$, the field equations are of the form (5.6) and one of the solutions is the well-known Friedmann-Walker-Robertson space which represents our 'real' expanding universe. In any case, we see that regardless of our considerations on the right hand side of Eq. (5.6), there is always space (except when all components of $g_{\mu\nu}$ are zero) which is subsequently filled, according to our considerations, with 'stuff'. In this sense, GR represents 'space' as a container (the pseudo-Riemannian manifold) that responds according to the energy-matter content. Is there any problem with this? Indeed, in GR we can have space seen as perfect emptiness, and space seen as an energetic perfect fluid (the vacuum). We see that, just as the Euclidean manifold plays the role of the background in Newtonian mechanics, the pseudo-Riemannian manifold along with the metric is playing the role of a substratum, since space can exist even if $T_{\mu\nu} = 0$ and $\Lambda = 0$.

The fact that in GR is possible to have space without sources of any kind seems to be in contradiction with the notion of vacuum as seen from QM and electrodynamics. Whilst geometrical spaces have no electromagnetic properties per se (compare to the Reissner-Nordström metric), the vacuum of electrodynamics has intrinsic finite electric permittivity ϵ and magnetic permeability μ. The assumption of the existence of a perfectly empty space is, just as the assumption of the existence of rigid bodies, a false but useful assumption. That the vacuum is an actual physical entity can be demonstrated even from the perspective of electrodynamics. To get the feeling of this, consider the following situation [50]. Suppose that a coil with n turns is energized and carries a current I. Accordingly, the magnetic induction of the coil is $B = \mu_0 nI + \mu_0 M$, where nI is the magnetic intensity and M is the magnetization induced in the coil. If we carry out an experiment where we keep the current constant and reduce the density of matter, B decreases. As we continue to eliminate 'all matter' then $M = 0$ and $B = \mu_0 nI$. This result experimentally demonstrates that the vacuum is a *paramagnetic medium* with magnetic permeability $\mu_0 = 4\pi 10^{-7}$ N/A^2. And because this property is exclusive of matter, the experiment tells us that the vacuum is not deprived of 'material substance' at all. In contrast, if physical space were totally empty, one would expect null electromagnetic properties.

On the other hand, the field of condensed matter has made important advances, particularly, in the field of Bose-Einstein condensates and superfluids. Giving the mathematical analogies of these systems with the quantum vacuum, some physicists have suggested that the vacuum can be a condensed state of matter [51]. One of the consequences of this approach is that perhaps the equivalence principle and some other symmetries such as Lorentz invariance and gauge invariance may not be fundamental at all but emergent from the low-energy sector of the quantum vacuum.

Indeed, F. Witenberg showed that assuming the vacuum as a kind of plasma composed of positive and negative massive particles interacting by the Planck force over a Planck length, one can derive QM and Lorentz invariance as asymptotic approximations for energies small compared to the Planck energy [52]. He finally concluded that Minkowski spacetime is superfluous for physics provided that Lorentz invariance is assumed as a dynamic symmetry, just as conceived in Lorentz' æther theory where length contraction and time dilation are explained in terms of atomic deformations and motions through the medium. Certainly, this would imply that electromagnetic fields, and no less particles, are states and excitations, respectively, of the vacuum. Following a similar line of thought, M. Urban et al. recently showed that the origin of the speed of light (and the permeability μ_0 and permittivity ϵ_0 constants) is the result of interaction of photons with fermions pairs spontaneously produced in the quantum vacuum. This implies, again, that the vacuum is the medium for light and that the speed of light is strictly defined relative to it. As for the law of inertia, B. Haisch et al. put forward a quantum mechanism to justify its origin. They showed that inertia can originate from the quantum vacuum without alluding to Mach's principle [53]. Admittedly, all this evidence strongly suggests that the vacuum is some sort of diluted material fluid. If we trust this view, the energy density ρ_{vac} found from QFT might be downright correct. Admitting the vacuum as a material substance capable of transmitting gravitation, just as Newton devised it, prompts us to deeply revise the geometrical interpretation of space and gravitation in GR.

General Relativity Is Not Fully Relativistic and the Speed of Light Is Not Constant

Absolute Motion Versus Relative Motion

In his celebrated scholium, [4] Newton taught us how to distinguish false motion from real motion; there he wrote:

> The causes by which true and relative motions are distinguished, one from the other, are the forces impressed upon bodies to generate motion. True motion is neither generated nor altered, but by some force impressed upon the body moved; but relative motion may be generated or altered without any force impressed upon the body...

Then, at the end, he topped off:

> But how we are to collect the true motions from their causes, effects, and apparent differences; and viceversa, how from the motions, either true or apparent, we may come to the knowledge of their causes and effects, shall be explained more at large in the following tract. For to this end it was that I composed it.

Newton's Principia was completely devoted to demonstrate that absolute motion exists and can be distinguished from relative one by forces. In Newton's theory, an observer in a non-inertial system (NIS), say the Earth, that rotates with angular velocity ω and translates with acceleration $\ddot{\mathbf{R}}$ relative to an ISR will observe a series of forces acting on a particle of mass m:

$$\mathbf{F}' = \mathbf{S} + m\mathbf{g} - m\ddot{\mathbf{R}} - m\dot{\omega} \times \mathbf{r} - m\omega \times (\omega \times \mathbf{r}) - 2m\omega \times \mathbf{v}, \qquad (5.9)$$

where $\mathbf{S}$ is the sum of external forces, $\mathbf{g}$ the gravitational acceleration and $\mathbf{v}$ the particle's velocity as measured on Earth. The fourth term appears in case ω is not constant, the fifth term is the centrifugal force and the last one is the Coriolis force. All these additional forces are known as *inertial, fictitious or pseudo forces*. The adjective 'fictitious' and the prefix 'pseudo' speak for themselves. In this theory, *these are not real forces* because their nature arise from relative motion. Not convinced, a relativist will claim that the Earth observer 'feels', i.e., measures these forces and, hence, they are real for him; consequently, the adjective 'fictitious' fades away. A Newtonian in turn will reply that if motion were purely relative, the Earth could be considered as static frame subjected to the pseudo-forces of Eq. (5.9) and, in consequence, the view that the world rotates around the Earth would be equally true. This line of reasoning will send us back to the idea of the Earth as the center of the universe and one would not be able to decide whether the Earth *really* rotates or not (similar to the time dilation quandary discussed in section "Relative Motion Leads to Quandaries"). For a Newtonian, the relativist view is, needless to say, naïve. For if an experiment could be conceived to measure the effects of the pseudo-forces, we would be demonstrating that AS exists. The Focault pendulum is a beautiful example that Newton was right. Exploiting the effects of the Coriolis force, the experiment not only gives geocentrism a *coup de grâce*, but also informs us that absolute rotation can be measured even if we were enclosed in a laboratory without observing the fixed stars. The experiment shows that the Earth's angular velocity relative to AS (Euclidean space) can be determined by just measuring the rotation of the oscillation plane as function of time. Likewise, the Michelson-Gale experiment shows clear evidence that, without looking at the sky, the Earth absolutely rotates relative to the vacuum [54]. This experiment not only measures ω but also teaches us that the vacuum is the medium for light. If we now judge these experiments from the standpoint of SR, the Earth revolves relative to a system either in motion or at rest relative to the Minkowskian background (physically speaking the vacuum). If ω is small, the calculations from both SR and Newton's theory agree. And what does GR have to say about this? For GR, as in the case of the Newton's bucket, the Earth rotates relative to its GF so that the fictitious forces become genuine GF (see the Kerr field and the Lense-Thirring effect). In the case of the Focualt experiment, GR includes tiny corrections that, in practice, cannot be distinguished from Newton's results. We thus see once more that $g_{\mu\nu}$ plays the role of background for the rotation of the Earth, by analogy with the Euclidean metric in Newtonian theory. But just as one cannot place a system of reference at absolute rest relative to AS, one cannot place a system of reference at rest relative to the GF. Thus, to determine ω astronomers use Eq. (5.4) and assume a special ISR, the so-called fixed-space system or International Celestial Reference System. Such system, evidently, is an ideal candidate for a PSR.

In the development of GR, Einstein sought to justify inertia, and therefore rotational motion, relative to the masses of the universe through both Mach's principle and the equivalence principle. We saw above that he did not succeed. Furthermore, Einstein did not succeed either in creating a fully relativistic theory. This means that not all systems of reference are equivalent. That this is the case can be seen from the principle of general covariance [44, pp. 91–93]. In 1917, Kretschmann [55] recognized in a critical study of GR, that the principle does not imply that the LP most be relativistic, but only that their form must be the same under general coordinate transformations. In fact, even Newton's laws can be written in covariant form. Thus, general covariance is not a PR but a principle that imposes restrictions between matter and geometry [56, 57]; for this reason, John Wheeler suggested that Einstein's theory should be called, instead of 'general' relativity, *Geometrodynamics*. Today, some physicists still look for a fully relativistic theory where Mach's principle could be embraced [46]. This, indeed, indicates that GR is not hermetic to accept a PSR, even going against its own spirit. Both astronomy and cosmology have always been in need of a special system of reference to assess the celestial dynamics and define a cosmic time. The cosmic microwave background radiation also strongly demands a special system. It seems to me that the PSR is valuable to satisfactorily account for physical phenomena at all scales [58–60].

Covariance and the Variation of the Speed of Light

Before we close this treatise, it is worth elucidating the fact that SR has actually only one postulate, i.e., the PR, since the second one is already tacitly included in electrodynamics. This postulate is valid insofar as one deals with ISR, but invariance no longer holds for NIS—or appealing to the equivalence principle for systems of reference in GF—. This means that the value of their constants and physical quantities may acquire different values in different NIS. As we showed in Eq. (5.9), the same occurs in Newton's theory. Covariance, by contrast, only demands that the form of the LP must remain the same. As early as 1911, Einstein was aware of this [7]. He knew, for instance, that the only cause that could change the path of light is by varying the speed of the different parts of a wave front. During the development of GR, he emphasized that the assumption of the constancy of the speed of light must be abandoned for NIS [7–9]. However, the principle of general covariance (also known as diffeomorphism covariance or general invariance) demands that the metric tensor $g_{\mu\nu}$ must change whereas all constants must remain the same under general coordinate transformations. Since then it is widely believed that the speed of light is a universal constant at any point of a GF. This could be true insofar as we understand space as GR does, but we have shown above that the vacuum can be seen as a diluted material fluid. Under this assumption, we can reinterpret the bending of light just as a simple refraction phenomenon. One can keep the vacuum static and assume it as a inhomogenous medium with degraded refraction index that vary as function of position in the GF. The gradient depends on the gravitational potential which, in turn, will make the speed of light function of position. Thus, within this context, the 'warping' of space can be physically understood as the change in the density of the medium [61]. Certainly, this will not account for the perihelion of Mercury or other gravitational phenomena, but it gives us a hint on how to build a unified theory and reinterpret gravitational effects.

Final Remarks

Throughout the course of this treatise I briefly reviewed the role played by the PSR in physics. In doing so, I presented a series of epistemological, experimental, and theoretical arguments to dispel the series of misconceptions around this central tenet, and, at the same time, I gave weighty reasons to champion its reintroduction into physics. I also pointed out that the geometrization of space may not be the most appropriate for the future of physics. Instead, the experimental evidence at hand suggests that space is a dynamical condensed state of matter. Due to the lack of space, I cannot discuss here the progress that has already been advanced based on these radical ideas and I prefer to leave it for a future contribution. The purpose of this work is to show that the PSR is not in conflict with physics and that the vacuum can be understood in a different way. Once we accept this, the next step is to unify the concepts of particle and wave using the notion of quasiparticles. In this sense, a field would become a state of the vacuum and a particle an excitation. The implications of this insight may impact physics at all scales leading to the TOE without invoking exotic assumptions (multiverses, extra dimensions, etc.). In my opinion, there are enough experimental and theoretical elements for a new revolution in physics. Thomas Kuhn taught us that a paradigm shift might be a thorny episode in the evolution of science [62]. The PSR assumption constitutes a paradigm shift that would request a drastic change in the way of understanding reality. Some 'established' facts such as the expansion of the universe and the big bang model may need to be revised in the light of this new paradigm. Inevitably, this will lead us at some point to the bucket problem. And just as Newton held, here it is claimed that the water moves relative to the vacuum, provided that we understand elementary particles as quasiparticles and the vacuum as a dynamical 'material' fluid.

Acknowledgments This research was supported by the Natural Sciences and Engineering Research Council of Canada, the Canada Research Program and CONACYT Mexico under grant 186142. The author is thankful to the University of Saskatchewan, Prof. Alex Moewes for his support in this project, and the FQXi organizers of the 2012 contest for opening the doors to new and fresh ideas fundamental for the progress of physics.

References

1. P. Weingartner, P. Mittelstaedt, *Laws of Nature* (Springer, Berlin, 2005)
2. M. Tegmark, Is the theory of everything merely the ultimate ensemble theory? Ann. Phys. **270**, 1 (1998)
3. M. Tegmark, The mathematical universe. Found. Phys. **38**, 101 (2008)
4. Sir Isaac Newton, *Newton's Principia, The Mathematical Principles of the Natural Philosophy*, 1st English edn. (Daniel Adee, New York, 1846)
5. E. Mach, *The Science of Mechanics*, 5th English edn. (The Open Court Publishing Company, La Salle, 1942)
6. A. Einstein, Zur elektrodynamik bewegter körper. Ann. Phys. **17**, 891 (1905)
7. A. Einstein, Über den einflutss der schwerkraft auf die ausbreitung des lichtes. Ann. Phys. **35**, 898 (1911)
8. A. Einstein, Die grundlage der allgemeinen relativitätstheorie. Ann. Phys. **49**, 769 (1916)
9. A. Einstein, Hamilton's principle and the general theory of relativity. S. B. Preuss. Akad. Wiss. (1916)
10. A. Einstein, Aether and the theory of relativity. Lectured delivered at Leyden (1920)
11. K. Popper, *The Logic of Scientific Discovery*, 1st edn. (Routledge Classics, London, 2009)
12. W. Heisenberg, The physical principles of the quantum theory. Z. Phys. (English Translation 1930, C. Eckart, F.C. Hoyt, **33** 879 (1925))
13. H. Hertz, *Electric Waves* (MacMillan, London, 1900)
14. J.G. O'Hara, W. Pricha, *Hertz and the Maxwellians* (Peter Peregrinus Ltd., London, 1987)
15. B.J. Hunt, *The Maxwellians* (Cornell University Press, Ithaca, 1991)
16. P.A.M. Dirac, Is there an æther? Nature Lett. Editor **168**, 906 (1951)
17. P.A.M. Dirac, Is there an æther? Nature Lett. Editor **169**, 702 (1952)
18. M.J. Pinheiro, Do Maxwell's equations need revision?—a methodological note (2005), arXiv:physics/0511103
19. J. Larmor, On a dynamical theory of the electric and luminiferous medium. Philos. Trans. R. Soc. **190**, 205–300 (1897)
20. J. Larmor, *Aether and Matter* (C. J. Clay and Sons, Cambridge, 1900)
21. H.A. Lorentz, Simplified theory of electrical and optical phenomena in moving systems. Zittingsverslag Akad. v. Wet **7**(507), 427 (1899)
22. H.A. Lorentz, Electromagnetic phenomena in a system moving with any velocity less than that of light. Proc., Amsterdam. **6**, 809–831 (1904)
23. H. Poincaré, Sur la dynamique de l'électron, on the dynamics of the electron. C. R. Acad. Sci **140**, 1504 (1905)
24. H.E. Ives, The aberration of clocks and the clock paradox. J. Opt. Soc. Am. **27**, 305 (1937)
25. H.E. Ives, G.R. Stilwell, An experimental study of the rate of a moving clock. J. Opt. Soc. Am. **28**, 215 (1938)
26. G. Granek, Poincaré's contributions to relativistic dynamics. Stud. Hist. Philos. Mod. Phys. **31**, 15 (2000)
27. J.S. Bell, *Speakable and Unspeakable in Quantum Mechanics*, 2nd edn. (Cambridge University Press, Cambridge, 2004)
28. G.A. Articolo, The Michelson-Morley experiment and the phase shift upon a 90° rotation. Am. J. Phys. **37**, 215 (1969)
29. P. Mazur, On the Michelson-Morley experiment. Am. J. Phys. **37**, 218 (1969)
30. M.M. Capria, F. Pambianco, On the Michelson-Morley experiment. Found. Phys. **24**, 885 (1994)
31. J.D. Jackson, *Classical Electrodynamics*, 3rd edn. (Wiley, New York, 1999)
32. L. James, I. Nagle, G. Bearden, W.A. Zajc, Quark-gluon plasma at the RHIC and the LHC: perfect fluid too perfect? New J. Phys. **13**, 075004 (2011)
33. I. Pérez, On the experimental determination of the one-way speed of light. Eur. J. Phys. **32**, 993–1005 (2011)

34. L.D. Landau, E.M. Lifshitz, *Mechanics, Volume 1 of Course of Theoretical Physics*, 3rd edn. (Butterworth-Heinemann, Oxford, 2000)
35. A. Grünbaum, The clock paradox in the special theory of relativity. Philos. Sci. **21**(3), 249–253 (1954)
36. H. Ives, G.R. Stilwell, An experimental study of the rate of a moving clock II. J. Opt. Soc. Am. **31**, 369 (1941)
37. A. Einstein, Kosmologische betrachtungen zur allgemeinen relativitätstheorie. S.B. Preuss, Akad. Wiss. 142–157, (1917)
38. A. Einstein, S.B. Preuss, Akad. Wiss. 448, (1918)
39. W. Pauli, *Theory of Relativity* (Dover Publications, New York, 1958)
40. Robert M. Wald, Andreas Zoupas, General definition of "conserved quantities" in general relativity and other theories of gravity. Phys. Rev. D **61**, 084027 (2000)
41. L.B. Szabados, Quasi-local energy-momentum and angular momentum in general relativity. Living Rev. Relativ. **12**, 4 (2009)
42. A. Harvey, E. Schucking, Einstein's mistake and the cosmological constant. Am. J. Phys. **68**, 723 (2000)
43. S. Weinberg, The cosmological constant problem. Rev. Mod. Phys. **61**, 1 (1989)
44. S. Weinberg, *Gravitation and Cosmology Principles and Applications of the General Theory of Relativity* (Wiley, New York, 1972)
45. H. Bondi, J. Samuel, The Lense-Thirring effect and Mach's principle. Phys. Lett. A **228**, 121–126 (1997)
46. J. Barbour, H. Pfister, *Mach's Principle: From Newton's Bucket to Quantum Gravity*, vol. 6 (Birkhäuser, Boston, 1995)
47. S. Carroll, The cosmological constant. Living Rev. Relativ. **4**, 1 (2001)
48. T. Padmanabhan, Cosmological constant—the weight of the vacuum. Phys. Rep. **380**, 235–320 (2003)
49. D. Dieks (ed.), *The Ontology of Spacetime*, vol. I & II (Elsevier, Amsterdam, 2006, 2008)
50. M. Urban, The quantum vacuum as the origin of the speed of light. Eur. J. Phys. D **67**, 58 (2013)
51. G. Volovik, *The Universe in a Helium Droplet* (Oxford University Press, New York, 2003)
52. F. Winterberg, Relativistic quantum mechanics as a consequence of the Planck mass plasma conjecture. Int. J. Theor. Phys. **46**, 3294–3311 (2007)
53. B. Haisch, A. Rueda, Y. Dobyns, Inertial mass and the quantum vacuum fields. Ann. Phys. **10**, 393–414 (2001)
54. A.A. Michelson, H.G. Gale, F. Pearson, The effect of the earth's rotation on the velocity of light part II. Astrophys. J. **61**, 140 (1925)
55. E. Kretschmann, Über den physikalischen Sinn der Relativitatspostulate. Ann. Phys. **53**, 575–614 (1917)
56. J.D. Norton, General covariance and the foundations of general relativity: eight decades of dispute. Rep. Prog. Phys. **56**, 791–858 (1993)
57. D. Dieks, Another look at general covariance and the equivalence of reference frames. Stud. Hist. Philos. Mod. Phys. **37**, 174–191 (2006)
58. E. Constanzo, M. Consoli, From classical to modern ether-drift experiments: the narrow window for a preferred frame. Phys. Lett. A **333**, 355–363 (2004)
59. T. Jacobson, D. Mattingly, Gravity with a dynamical preferred frame. Phys. Rev. D **64**, 024028 (2001)
60. V. Scarani, The speed of quantum information and the preferred frame: analysis of experimental data. Phys. Lett. A **276**, 1–7 (2000)
61. X.-H. Ye, L. Qiang, Inhomogeneous vacuum: an alternative interpretation to curved spacetime. Chin. Phys. Lett. **25**, 1571–1573 (2008)
62. T. Kuhn, *The Structure of Scientific Revolutions* (University of Chicago Press, Chicago, 1962)

Chapter 6
Right About Time?

Sean Gryb and Flavio Mercati

Abstract Have our fundamental theories got time right? Does size really matter? Or is physics all in the eyes of the beholder? In this essay, we question the origin of time and scale by reevaluating the nature of measurement. We then argue for a radical scenario, supported by a suggestive calculation, where the flow of time is inseparable from the measurement process. Our scenario breaks the bond of time and space and builds a new one: the marriage of time and scale.

Introduction

Near the end of the 19th century, physics appeared to be slowing down. The mechanics of Newton and others rested on solid ground, statistical mechanics explained the link between the microscopic and the macroscopic, Maxwell's equations unified electricity, magnetism, and light, and the steam engine had transformed society. But the blade of progress is double edged and, as more problems were sliced through, fewer legitimate fundamental issues remained. Physics, it seemed, was nearing an end.

Or was it? Among the few remaining unsolved issues were two experimental anomalies. As Lord Kelvin allegedly announced: "The beauty and clearness of the dynamical theory [...] is at present obscured by two clouds" [1]. One of these clouds was the ultra-violet catastrophe: an embarrassing prediction that hot objects like the

S. Gryb (✉)
Institute for Theoretical Physics, Utrecht University, Leuvenlaan 4,
3584 CE Utrecht, The Netherlands
e-mail: s.gryb@hef.ru.nl

S. Gryb
Institute for Mathematics, Astrophysics and Particle Physics,
Radboud University, Huygens Building, Heyendaalseweg 135, 6525 AJ
Nijmegen, The Netherlands

F. Mercati
Perimeter Institute for Theoretical Physics, 31 Caroline Street North,
Waterloo, AJ N2L 2Y5, Canada
e-mail: fmercati@perimeterinstitute.ca

© Springer International Publishing Switzerland 2015
A. Aguirre et al. (eds.), *Questioning the Foundations of Physics*,
The Frontiers Collection, DOI 10.1007/978-3-319-13045-3_6

sun should emit *infinite* energy. The other anomaly was an experiment by Michelson and Morley that measured the speed of light to be independent of how an observer was moving. Given the tremendous success of physics at that time, it would have been a safe bet that, soon, even these clouds would pass.

Never bet on a sure thing. The ultra-violet catastrophe led to the development of quantum mechanics and the Michelson–Morley experiment led to the development of relativity. These discoveries completely overturned our understanding of space, time, measurement, and the perception of reality. Physics was not over, it was just getting started.

Fast-forward a hundred years or so. Quantum mechanics and relativity rest on solid ground. The microchip and GPS have transformed society. These frameworks have led to an understanding that spans from the microscopic constituents of the nucleus to the large scale structure of the Universe. The corresponding models have become so widely accepted and successful that they have been dubbed *standard models* of particle physics and cosmology. Resultantly, the number of truly interesting questions appears to be slowly disappearing. In well over 30 years, there have been no experimental results in particle physics that cannot be explained within the basic framework laid out by the standard model of particle physics. With the ever increasing cost of particle physics' experiments, it seems that the data is drying up. But without input from experiment, how can physics proceed? It would appear that physics is, again, in danger of slowing down.

Or is it? Although the *number* of interesting fundamental questions appears to be decreasing, the *importance* of the remaining questions is growing. Consider two of the more disturbing experimental anomalies. The first is the *naturalness problem*, i.e., the presence of unnaturally large and small numbers in Nature. The most embarrassing of these numbers—and arguably the worst prediction of science—is the accelerated expansion of the Universe, which is some 120 orders of magnitude smaller than its natural value. The second is the *dark matter problem* that just under 85–90 % of the matter content of our Universe is of an exotic nature that we have not yet seen in the lab. It would seem that we actually understand very little of what is happening in our Universe!

The problem is not that we don't have enough data. The problem is that the data we do have does not seem to be amenable to explanation through incremental theoretical progress. The belief that physics is slowing down or, worse, that we are close to a final theory is just as as unimaginative now as it would have been before 1900. Our thesis here will be that the lesson to take from that period is that the way forward is to question the fundamental assumptions of our physical theories in a radical way. This is easier said than done: one must not throw out the baby with the bath water. What is needed is a careful examination of our physical principles in the context of real experimental facts to explain *more* data using *less* assumptions.

The purpose of this work is to point out three specific assumptions made by our physical theories that might be wrong. We will not offer a definite solution to these problems but suggest a new scenario, supported by a suggestive calculation, that puts these assumptions into a new light and unifies them. The three assumptions we will question are

1. Time and space are unified.
2. Scale is physical.
3. Physical laws are independent of the measurement process.

We will argue that these three assumptions inadvertently violate the same principle: the requirement that the laws of physics depend only on what is knowable through direct measurement. They fall into a unique category of assumptions that are challenged when we ask how to adapt the scientific method, developed for understanding processes in the lab, to the cosmological setting. In other words, how can we do science on the Universe *as a whole*?

We will not directly answer this question but, rather, suggest that this difficult issue may require a radical answer that questions the very origin of time. The flow of time, we will argue, may be fundamentally linked to the process of measurement. We will then support this argument with an intriguing calculation that recovers the black hole entropy law from a simple toy model. Before getting to this, let us explain the three questionable assumptions.

Three Questionable Assumptions

Many of our most basic physical assumptions are made in the first week of physics education. A good example is one of the first equations we are taught: the definition of velocity,

$$v = \frac{\Delta x}{\Delta t}. \tag{6.1}$$

It is perhaps a bit over-dramatic—but, at the same time, not inaccurate—to say that to give this equation a precise operational meaning has been an outstanding issue in physics for its entire history. This is because, to understand this equation, one has to have an operational definition of both x, t, and Δ. Great minds have pondered this question and their insights has led to scientific revolutions. This includes the development of Newtonian mechanics, relativity, and quantum mechanics.[1] Recently, the meaning of x and, in particular, t, have been the subject of a new debate whose origin is in a theory of quantum gravity. This brings us to our first questionable assumption.

Time and Space Are Unified

The theory of relativity changed our perception of time. As Minkowski put it in 1908 [2], "space by itself, and time by itself, are doomed to fade away into mere shadows,

---

[1] A lot to digest in the first week!

and only a kind of union of the two will preserve an independent reality". Nowhere is this more apparent than in the main equation physicists use to construct the solutions of general relativity (GR):

$$S_{\text{Einstein-Hilbert}} = \int d^4x \, (R + \mathcal{L}_{\text{matter}}) \sqrt{-g} \,. \tag{6.2}$$

Can you spot the t? It's hidden in the 4 of d^4x. But there are important structures hidden by this compact notation.

We will start by pointing out an invisible minus sign in Eq. (6.2). When calculating spacetime distances, one needs to use

$$x^2 + y^2 + z^2 - t^2, \tag{6.3}$$

which has a—in front of the t^2 instead of Pythagoras' +. The minus sign looks innocent but has important consequences for the solutions of Eq. (6.2). Importantly, the minus sign implies *causal structure*, which means that only events close enough to us so that light signals sent from these events can make it to us now can effect what is going on now. This, in turn, implies that generic solutions of GR can only be solved by specifying information at a particular *time* and then seeing how this information propagates into the future. Doing the converse, i.e., specifying information at a particular *place* and seeing how that information propagates to another place, is, in general, not consistent.[2] Thus, the minus sign already tells you that you have to use the theory in a way that treats time and space differently.

There are other ways to see how time and space are treated differently in gravity. In Julian Barbour's 2009 essay, *The Nature of Time* [3], he points out that Newton's "absolute" time is not "absolute" at all. Indeed, the Newtonian notion of *duration*— that is, how much time has ticked by between two distinct instants—can be *inferred* by the total change in the *spatial* separations of particles in the Universe. He derives the equation

$$\Delta t^2 \propto \sum_i \Delta d_i^2, \tag{6.4}$$

where the d_i are inter-particle separations in units where the masses of the particles are one. The factor of proportionality is important, but not for our argument. What is important is that changes in *time* can be inferred by changes in *distances* so that absolute duration is not an input of the classical theory. This equation can be generalized to gravity where it must be solved at every point in space. The implications for the quantum theory are severe: time completely drops out of the formalism.

Expert readers will recognize this as one of the facets of the *Problem of Time* [4]. The fact that there is no equivalent *Problem of Space* can be easily traced back to the points just made: time is singled out in gravity as the variable in terms of which

---

[2] Technically, the difference is in the elliptic versus hyperbolic nature of the evolution equations.

the evolution equations are solved. This in turn implies that local duration should be treated as an *inferred* quantity rather than something fundamental. Clearly, time and space are *not* treated on the same footing in the formalism of GR despite the rather misleading form of Eq. (6.2). Nevertheless, it is still true that the spacetime framework is incredibly useful and, as far as we know, correct. How can one reconcile this fact with the space-time asymmetry in the formalism itself? We will investigate this in Sect. "Time from Coarse Graining".

Scale Is Physical

Before even learning the definition of velocity, the novice physicist is typically introduced to an even more primary concept that usually makes up one's first physics lesson: *units*. Despite the rudimentary nature of units, they are probably the most inconsistently understood concept in all of physics. If you ask ten different physicists for the physical meaning of a unit, you will likely get ten different answers. To avoid confusion, most theoreticians set all dimensionful constants equal to 1. However, one can't predict anything until one has painfully reinserted these dimensionful quantities into the final result.

And yet, no one has *ever* directly observed a dimensionful quantity. This is because all measurements are comparisons. A 'meter' has no intrinsic operational meaning, only the ratio of two lengths does. One can define an object A to have a length of one meter and make a measurement that reveals that some other object B has twice the length of object A. Then, we can deduce that object B has a length of 2 meters. This, however, tells you nothing about the intrinsic absolute length of object A for if a demon doubled the intrinsic size of the Universe, the result of the experiment would be exactly the same. So, where do units come from?

Some units, like the unit of pressure, are the result of emergent physics. We understand how they are related to more "fundamental" units like meters and seconds. However, even our most fundamental theories of Nature have dimensionful quantities in them. The standard model of particle physics and classical GR require only a singe unit: *mass*. Scale or, more technically, *conformal invariance* is then broken by only two quantities with the units of mass. The first is the recently observed Higgs mass, which can be related to all the masses of the particles in the standard model. The second is the Plank mass, which sets the scale of quantum gravity. As already discussed, there is a naturalness problem associated with writing all other constants of nature as dimensionless quantities but this will not bother us to much here.

The presence of dimensionful quantities is an indication that our "fundamental" theories are not fundamental at all. Instead, scale independence should be a basic principle of a fundamental theory. As we will see in Sect. "Time from Coarse Graining", there is a formulation of gravity that is *nearly* scale invariant. We will try to address the "nearly" with the considerations of the next section.

Physical Laws Are Independent of the Measurement Process

There is one assumption that is so fundamental it doesn't even enter the physics curriculum. This is the assumption that the scientific method is generally applicable for describing everything in the Universe taken together. We know that the scientific method can be applied in the laboratory where external agents (i.e., scientists) carefully control the inputs of some subsystem of the Universe and observe the subsystem's response to these inputs. We don't know, however, whether it is possible to apply these techniques to the Universe as a whole. On the other hand, when it comes to quantum mechanics, we *do* know whether our formalism can be consistently applied to the Universe. The answer is 'NO'! The reasons are well understood—if not disappointingly under appreciated—and the problem even has a name: *the measurement problem*.

The measurement problem results from the fact that quantum mechanics is a framework more like statistical physics than classical mechanics. In statistical physics, one has *practical* limitations on one's knowledge of a system so one takes an educated guess at the results of a specific experiment by calculating a probability distribution for the outcome using one's current knowledge of the system. In quantum mechanics, one has *fundamental* limitations on one's knowledge of the system—essentially because of the uncertainty principle—so one can only make an educated guess at the outcome of a specific experiment by calculating a probability distribution for the outcome using one's current knowledge of the system. However, it would be strange to apply statistical mechanics to the whole Universe because the Universe itself is only given once. It is hard to imagine what an ensemble of Universes, for which one can calculate and give meaning to a probability distribution, would even mean.[3] The same is true in quantum mechanics, but the problem is worse. The framework itself is designed to give you a probability distribution for the outcome of some measurement but how does one even define a measurement when the observer itself is taken to be part of the system? The answer is not found in any interpretation of quantum mechanics, although the problem itself takes a different form in a given interpretation. The truth is that quantum mechanics requires some additional structure, which can be thought of as describing the observer, in order for it to make sense. In other words, quantum mechanics alone, without additional postulates, can never be a theory of the whole Universe.

As a consequence of this, any approach to quantum gravity that uses quantum mechanics unmodified—including all *major* approaches to quantum gravity—is not, and *can never be* a theory of the whole Universe. It could still be used for describing quantum gravity effects on isolated subsystems of the Universe, but that is not the

---

[3] This is one of the goals of the *Many Worlds* interpretation of quantum mechanics whose proponents believe that it is possible to make sense of such an ensemble using the standard axioms of classical probability theory (see [5] for a popular account). Whether it is sensible to apply these axioms to the Universe as a whole, however, is unclear. Furthermore, having to believe in an infinite number of unobservable parallel Universes is a big price to pay just to make sense of probabilities in quantum mechanics.

ambition of a full fledged quantum gravity theory. Given such a glaring foundational issue at the core of every major approach to quantum gravity, we believe that the attitude that we are nearing the end of physics is unjustified. The "shut-up and calculate" era is over. It is time for the quantum gravity community to return to these fundamental issues.

One approach is to change the ambitions of science. This is the safest and, in some ways, easiest option, but it would mean that science is inherently a restricted framework. The other possibility is to try to address the measurement problem directly. In the next section, we will give a radical proposal that embraces the role of the observer in our fundamental description of Nature. To understand how this comes about, we need one last ingredient: *renormalization*, or the art of averaging.

A Way Forward

The Art of Averaging

It is somewhat unfortunate that the great discoveries of the first half of the 20th century have overshadowed those of the second half of the century. One of these, the theory of *renormalization*, is potentially the uncelebrated triumph of twentieth century physics. Renormalization was born as rather ugly set of rules for removing some undesirable features of quantum field theories. From these humble beginnings, it has grown into one of the gems of physics. In its modern form due to Wilson [6], renormalization has become a powerful tool for understanding what happens in a general system when one lacks information about the details of its fine behavior. Renormalization's reach extends far beyond particle physics and explains, among other things, what happens during phase transitions. But, the theory of renormalization does even more: it helps us understand why physics is possible at all.

Imagine what it would be like if, to calculate everyday physics like the trajectory of Newton's apple, one would have to compute the motions of every quark, gluon, and electron in the apple and use quantum gravity to determine the trajectory. This would be completely impractical. Fortunately, one doesn't have to resort to this. High-school physics is sufficient to determine the motion of what is, fundamentally, an incredibly complicated system. This is possible because one can average, or *coarse grain*, over the detailed behavior of the microscopic components of the apple. Remarkably, the average motion is simple. This fact is the reason why Newtonian mechanics is expressible in terms of simple differential equations and why the standard model is made up of only a couple of interactions. In short, it is why physics is possible at all. The theory of renormalization provides a framework for understanding this.

The main idea behind renormalization is to be able to predict how the laws of physics will change when a coarse graining is performed. This is similar to what happens when one changes the magnification of a telescope. With a large magnification, one might be able to see the moons of Jupiter and some details of the structure of their atmospheres. But, if the magnification, or the *renormalization scale*, is steadily decreased, the resolution is no longer good enough to make out individual moons and the lens averages over these structures. The whole of Jupiter and its moons becomes a single dot. As we vary the renormalization scale, the laws of physics that govern the structures of the system change from the hydrodynamic laws governing atmospheres to Newton's law of gravity.

The theory of renormalization produces precise equations that say how the laws of physics will change, or *flow*, as we change the renormalization scale. In what follows, we will propose that flow under changes of scale may be related to the flow of time.

Time from Coarse Graining

We are now prepared to discuss an idea that puts our three questionable assumptions into a new light by highlighting a way in which they are connected. First, we point out that there is a way to *trade* a spacetime symmetry for conformal symmetry without altering the physical structures of GR. This approach, called *Shape Dynamics* (SD), was initially advocated by Barbour [7] and was developed in [8, 9]. Symmetry trading is allowed because symmetries don't affect the physical content of a theory. In SD, the irrelevance of duration in GR is traded for local scale invariance (we will come to the word "local" in a moment). This can be done without altering the physical predictions of the theory but at the cost of having to treat time and space on a different footing. In fact, the local scale invariance is only an invariance of *space*, so that local rods—not clocks—can be rescaled arbitrarily. Time, on the other hand, is treated differently. It is a global notion that depends on the total change in the Universe.

The equivalence between SD and GR is a rather remarkable thing. What can be proved is that a very large class of spacetimes that are solutions of GR can be reproduced by a framework that does not treat spacetime as fundamental. Instead, what is fundamental in SD is scale-invariant geometry. Recently [10], it has been discovered that some solutions of SD do not actually correspond to spacetimes as all, although they are still in agreement with experiment. These are solutions that describe certain kinds black holes in SD. In these solutions, there is no singularity where the curvature of spacetime becomes infinite. Rather, there is a traversable *worm hole* that connects the event horizon of a black hole to another region of space. This exciting discovery could pave the way to a completely different understanding of black holes.

Symmetry trading is the key to understanding how GR and SD are related. In 2 spatial dimensions, we know that this trading is possible because of an accidental mathematical relationship between the structure of conformal symmetry in 2 dimensions and the symmetries of 3 dimensional spacetime [11].[4] We are investigating

---

[4] Technically, this is the isomorphism between the conformal group in d spatial dimensions and the deSitter group in $d + 1$ dimensions.

whether this result will remain true in 3 spatial dimensions. If it does, it would mean that the spacetime picture and the conformal picture can coexist because of a mere mathematical accident.

We now come to a key point: in order for any time evolution to survive in SD, one cannot eliminate all of the scale. The *global* scale of the Universe cannot be traded since, then, no time would flow. Only a *redistribution* of scale from point to point is allowed (this is the significance of the word "local") but the overall size of the Universe cannot be traded. In other words, *global scale must remain for change to be possible.* How can we understand this global scale?

Consider a world with no scale and no time. In this world, only 3 dimensional Platonic shapes exist. This kind of world has a technical name, it is a *fixed point* of renormalization—"fixed" because such a world does not flow since the renormalization scale is meaningless. This cannot yet be our world because nothing happens in this world. Now, allow for something to happen and call this "something" a *measurement*. One thing we know about measurements is that they can never be perfect. We can only compare the smallest objects of our device to larger objects and coarse grain the rest. Try as we may, we can never fully resolve the Platonic shapes of the fixed point. Thus, coarse graining by real measurements produces flow away from the fixed point. But what about time? How can a measurement happen if no time has gone by? The scenario that we are suggesting is that the flow under the renormalization scale is exchangeable with the flow of time. Using the trading procedure of SD, the flow of time might be relatable to renormalization away from a theory of pure shape.

In this picture, time and measurement are inseparable. Like a diamond with many faces, scale and time are different reflections of a single entity. This scenario requires a radical revaluation of our notions of time, scale, and measurement.

To be sure, a lot of thought is still needed to turn this into a coherent picture. A couple of comments are in order. Firstly, some authors [12, 13] have investigated a similar scenario, called *holographic cosmology* using something called *gauge/gravity duality*. However, our approach suggests that one may not have to *assume* gauge/gravity duality for this scenario but, instead, can make use of symmetry trading in SD. Furthermore, our motivation and our method of implementation is more concrete. Secondly, in the context of scale-invariant particle "toy models", Barbour, Lostaglio, and one of the authors [14] have investigated a scenario where quantum effects 'ruin' the classical scale invariance. In these models, the quantum theory has an emergent scale, which can then be used as a clock that measures the quantum time evolution of the scale invariant shapes of the system. This simple model illustrates one way in which the radical scenario discussed here could implemented into a concrete theory. Finally, why should we expect that there is enough structure in a coarse graining of pure shapes to recover the rich structure of spacetime? A simple answer is the subject of the next section.

The Size that Matters

In this section, we perform a simple calculation suggesting that the coarse graining of shapes described in the last section could lead to gravity. This section is more technical than the others but this is necessary to set up our final result. Brave souls can find the details of the calculations in the "Technical Appendix".

We will consider a simple "toy model" that, remarkably, recovers a key feature of gravity. Before getting into the details of the model, we should quickly point out that this model should be taken more as an illustration of one way in which it is possible to define the notion of a coarse graining on shape space. The model should not be taken as a literal model for gravity or black holes, even though some of the results seem suggestive in this regard. Certainly much more work would be needed to flesh this out in a convincing way.

The model we will consider is a set of N free Newtonian point particles. To describe the calculation we will need to talk about two spaces: *Shape Space* and *Extended Configuration Space* (ECS). Shape Space is the space of all the shapes of the system. If $N = 3$, this is the space of all triangles. ECS is the space of all Cartesian coordinates of the particles. That is, the space of all ways you can put a shape into a Cartesian coordinate system. The ECS is larger than Shape Space because it has information about the position, orientation, and size of the shapes. Although this information is unphysical, it is convenient to work with it anyway because the math is simpler. This is called a *gauge theory*. We can work with gauge theories provided we remove, or *quotient*, out the unphysical information. To understand how this is done, examine Fig. (6.1) which shows schematically the relation between the ECS and Shape Space. Each point on Shape Space is a different shape of the system, like

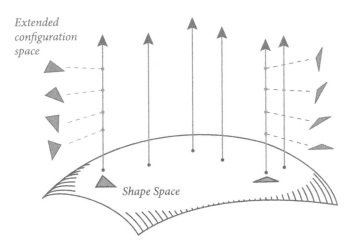

Fig. 6.1 Each point in Shape Space is a different shape (represented by *triangles*). These correspond to an equivalence class (represented by *arrows*) of points of the Extended Configuration Space describing the same shape with a different position, orientation, and size

a triangle. All the points along the arrows represent the same shape with a different position, orientation, or size. By picking a representative point along each arrow, we get a 1–to–1 correspondence between ECS and Shape Space. This is called *picking a gauge*. Mathematically, this is done by imposing constraints on the ECS. In our case, we need to specify a constraint that will select a triangle with a certain center of mass, orientation, and size. For technical reasons, we will assume that all particles are confined to a line so that we don't have to worry about orientation. To specify the size of the system, we can take the "length" of the system, R, on ECS. This is the *moment of inertia*. By fixing the center of mass and moment of inertia in ECS, we can work indirectly with Shape Space. The main advantage of doing this is that there is a natural notion of distance in ECS. This can be used to define the distance between two shapes, which is a key input of our calculations.

To describe the calculation, we need to specify a notion of *entropy* in Shape Space. Entropy can be thought of as the amount of information needed to specify a particular macroscopic state of the system. To make this precise, we can use the notion of distance on ECS to calculate a "volume" on Shape Space. This volume roughly corresponds to the number of shapes that satisfy a particular property describing the state. The more shapes that have this property, the more information is needed to specify the state. The entropy of that state is then related to its volume, Ω_m, divided by the total volume of Shape Space, Ω_{tot}. Explicitly,

$$S = -k_B \log \frac{\Omega_m}{\Omega_{\text{tot}}}, \tag{6.5}$$

where k_B is Boltzmann's constant.

We will be interested in states described by a subsystem of $n < N$ particles that have a certain center of mass x_0 and moment of inertia, r. To make sense of the volume, we need a familiar concept: coarse graining. We can approximate the volume of the state by chopping up the ECS into a grid of size ℓ. Physically, the coarse graining means that we have a measuring device with a finite resolution given by ℓ. Consider a state that is represented by some surface in ECS. This is illustrated in Fig. (6.2) by a line. The volume of the state is well approximated by counting the number of dark squares intersected by the line. In the "Technical Appendix", we calculate this volume explicitly. The result is

Fig. 6.2 *Left* Approximation of a line using a grid. *Right* Further approximation of the line as a strip of thickness equal to the grid spacing

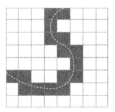 

$$\Omega_{\mathrm{m}} \propto \ell^2 \, r^{n-2} \left(R^2 - r^2 - \left(1 + \frac{m}{M-m} \right) \frac{m}{M} \, x_0^2 \right)^{\frac{N-n-2}{2}}, \qquad (6.6)$$

where M and R are the total mass and moment of inertia of the whole system and m is the mass of the subsystem. We can then compare this volume to the total volume of Shape Space, which goes like the volume of an $N-1$ dimensional sphere (the -1 is because of the center of mass gauge fixing). Thus,

$$\Omega_{\mathrm{tot}} \propto R^{N-1}. \qquad (6.7)$$

The resulting entropy is

$$S = \frac{1}{2} k_B \, \frac{N}{n} \left(\frac{r}{R} \right)^2 - k_B \, \log \frac{r}{R} + \cdots . \qquad (6.8)$$

Remarkably, the first term is exactly the entropy of a black hole calculated by Bekenstein and Hawking [15, 16]. More remarkably, the second term is exactly the first correction to the Bekenstein–Hawking result calculated in field theory [17, 18]. However, one should be careful not to interpret this result too literally. After all, we are considering only a very simplified case. A much more detailed analysis is necessary to draw any conclusions from this about real black holes. Note, however, that Erik Verlinde [19] discovered a way to interpret Newtonian gravity as an *entropic* force for systems whose entropy behaves in this way. It would appear that this simple model of a coarse graining of pure shapes has the right structure to reproduce Newtonian gravity.

Conclusions

We have questioned the basic assumptions that: (i) time and space should be treated on the same footing, (ii) scale should enter our fundamental theories of Nature, and (iii) the evolution of the Universe is independent of the measurement process. This has led us to a radical proposal: that time and scale emerge from a coarse graining of a theory of pure shape. The possibility that gravity could come out of this formalism was suggested by a simple toy model. The results of this model are non–trivial. The key result was that the entropy (6.8) scales like r^2, which, dimensionally, is an area. In three dimensions, this is the signature of *holography*. Thus, in this simple model, Shape Space is holographic. If this is a generic feature of Shape Space, it would be an important observation for quantum gravity.

Moreover, the toy model may shed light on the nature of the Plank length. In this model, the Plank length is the emergent length arising in ECS given by

$$L_{\mathrm{Planck}}^2 = G \, \hbar \propto \frac{R^2}{N} . \qquad (6.9)$$

This dimensionful quantity, however, is not observable in this model. What is physical, instead, it the dimensionless ratio r/R. This illustrates how a dimensionful quantity can emerge from a scale independent framework. Size doesn't matter—but a ratio of sizes does. The proof could be gravity.

Technical Appendix

The extended configuration space is $\mathbb{R}^N$: the space coordinates, r_i, $(i = 1, \ldots, N)$ of N particles in 1 dimension. To represent the reduced configuration space, or Shape Space, we can use a gauge fixing surface. To fix the translations, we can fix the center of mass to be at the origin of the coordinate system:

$$\sum_{i=1}^{N} m_i \, r_i = 0 \, . \quad (center \; of \; mass \; at \; the \; origin) \quad (6.10)$$

The equation above gives three constraints selecting three orthogonal planes through the origin whose orientation is determined by the masses m_i. A natural gauge-fixing for the generators of dilatations is to set the moment of inertia with respect to the center of mass to a constant[5] (the weak equation holds when the gauge-fixing (6.10) is applied):

$$\sum_{i<j} \frac{m_i m_j}{M^2} |r_i - r_j|^2 \approx \sum_{i=1}^{N} \frac{m_i}{M} |r_i|^2 = R^2 \, . \quad (fixed \; moment \; of \; inertia)$$

(6.11)

The last relation defines a sphere in $\mathbb{R}^N$ centered at the origin. Thus, Shape Space is the intersection of the $N - 1$-dimensional sphere (6.11) with the three orthogonal planes (6.10).

The flat Euclidean metric, $ds^2 = m_i \, \delta_{ij} \, \delta_{ab} \, dr_i^a \, dr_j^b$, is the natural metric on the extended configuration space Q. This metric induces the non-flat metric

$$ds_{\text{induced}}^2 = m_i \, \delta_{ij} \, \delta_{ab} \, dr_i^a \, dr_j^b \Big|_{Q_S} \, . \quad (6.12)$$

on Shape Space.

Description of a Macrostate in Shape Space

Consider an N-particle toy Universe with an n-particle subsystem, $n < N$. The particles in the subsystem have coordinates $x_i = r_i$, $(i = 1, \ldots, n)$, while the

---

[5] We are using here the notion of moment of inertia with respect to a point, which we rescaled by the total mass $M = \sum_i m_i$ to give it the dimensions of a squared length.

coordinates of all the other particles will be called $y_i = r_{n+i}$, $(i = 1, \ldots, N - n)$.
It is useful to define the coordinates of the center of mass of the subsystem and of
the rest of the Universe[6]:

$$x_0 = \sum_{i=1}^{n} \frac{m_i}{m} x_i, \qquad y_0 = \sum_{i=1}^{N-n} \frac{m_{n+i}}{M - m} y_i, \qquad m = \sum_{i=1}^{n} m_i, \qquad (6.13)$$

and the center-of-mass moment of inertia of the two subsystems

$$r = \sum_{i=1}^{n} \frac{m_i}{M} |x_i - x_0|^2, \qquad r' = \sum_{i=1}^{N-n} \frac{m_{n+i}}{M} |y_i - y_0|^2 . \qquad (6.14)$$

The relation between the moments of inertia of the total system and those of the two
subsystems is

$$R^2 = r^2 + (r')^2 + \left(1 + \frac{m}{M - m} \right) \frac{m}{M} x_0^2 . \qquad (6.15)$$

We define a macrostate as a state in which the moment of inertia of the subsystem,
r, and its center of mass, x_0, are constant. To calculate the Shape Space volume of
such a macrostate, we must integrate over all Shape Space coordinates x_i and y_i that
respect the conditions (6.13), (6.14), and (6.15) using the measure provided by the
induced metric (6.12). Let's make the following change of variables:

$$\tilde{x}_i = \sqrt{m_i} \, (x_i - x_0), \qquad \tilde{y}_i = \sqrt{m_{n+i}} \, (y_i - y_0) . \qquad (6.16)$$

Our equations become

$$\frac{1}{m} \sum_{i=1}^{n} \sqrt{m_i} \, \tilde{x}_i = 0, \qquad \frac{1}{M-m} \sum_{i=1}^{n} \sqrt{m_{n+i}} \, \tilde{y}_i = 0,$$

$$r = \frac{1}{M} \sum_{i=1}^{n} \tilde{x}_i^2, \quad r' = \frac{1}{M} \sum_{i=1}^{N-n} \tilde{y}_i^2, \quad R^2 = r^2 + (r')^2 + \left(1 + \frac{m}{M-m} \right) \frac{m}{M} x_0^2 .$$

$$(6.17)$$

In the new coordinates, the metric is the identity matrix (it loses the m_i factors
on the diagonal). The integral is over the direct product of an $(n - 2)$-dimensional
sphere of radius Mr and an $(N - n - 2)$-dimensional sphere of radius $Mr' = M\sqrt{R^2 - r^2 - \left(1 + \frac{m}{M-m} \right) \frac{m}{M} x_0^2}$ whose volume (calculated with a coarse-graining
of size ℓ) is:

---

[6] Notice that the two sets of coordinates must satisfy the relation $m \, x_0 + (M - m) y_0 = 0$ in order
to keep the total center of mass at the origin.

$$\Omega_{\mathrm{m}} = \ell^2 \frac{4\,\pi^{(N-n-1)/2}\pi^{(n-1)/2}}{\Gamma((N-n-1)/2)\Gamma((n-1)/2)} M^{N-4} r^{n-2}$$

$$\times \left(R^2 - r^2 - \left(1 + \frac{m}{M-m} \right) \frac{m}{M} x_0^2 \right)^{\frac{N-n-2}{2}}. \tag{6.18}$$

The total volume of Shape Space is that of an $(N-1)$-dimensional sphere of radius MR

$$\Omega_{\mathrm{tot}} = \frac{2\pi^{N/2}}{\Gamma(N/2)} M^{N-1} R^{N-1}. \tag{6.19}$$

Thus, the Shape Space volume per particle, in the limit $1 \ll n \ll N, r \ll r, m \ll M$ reduces to

$$\omega \propto \left(\frac{\ell}{r} \right)^{2/n} \frac{r}{R} \left(1 - \left(\frac{r}{R} \right)^2 - \left(1 + \frac{m}{M-m} \right) \frac{m}{M} \left(\frac{x_0}{R} \right)^2 \right)^{\frac{N}{2n}}, \tag{6.20}$$

and its logarithm has the expansion (remember that $x_0 < R$)

$$S = \frac{1}{2} k_B \frac{N}{n} \left(\frac{r}{R} \right)^2 - k_B \log \frac{r}{R} - \frac{2}{n} k_B \log \frac{\ell}{r} + \cdots . \tag{6.21}$$

Notice that the numerical factors change in the 3 dimensions. In that case, they are

$$S = \frac{3}{2} k_B \frac{N}{n} \left(\frac{r}{R} \right)^2 - 3 k_B \log \frac{r}{R} - \frac{4}{n} k_B \log \frac{\ell}{r} \cdots . \tag{6.22}$$

References

1. L. Kelvin, Nineteenth Century Clouds over the Dynamical Theory of Heat and Light, Philosophical Magazine, Sixth Series **2** 1–40 (1901). From a 1900, April 27, Royal Institution lecture
2. H. Minkowski, Space and time, *The principle of relativity: a collection of original memoirs on the special and general theory of relativity* (Dover, New York, 1952), pp. 75–91
3. J. Barbour, The Nature of Time. arXiv:0903.3489 [gr-qc]
4. C.J. Isham, Canonical quantum gravity and the problem of time. arXiv:gr-qc/9210011
5. D. Wallace, *The Emergent Multiverse: Quantum Theory according to the Everett Interpretation* (Oxford University Press, Oxford, 2012)
6. K. Wilson, J.B. Kogut, The Renormalization group and the epsilon expansion. Phys. Rept. **12**, 75–200 (1974)
7. J. Barbour, Dynamics of pure shape, relativity and the problem of time, in *Decoherence and Entropy in Complex Systems: Proceedings of the Conference DICE, Piombino 2002*, ed. by H.-T Elze. Springer Lecture Notes n Physics (2003)
8. H. Gomes, S. Gryb, T. Koslowski, Einstein gravity as a 3D conformally invariant theory. Class. Quant. Grav. **28**, 045005 (2011) arXiv:1010.2481 [gr-qc]

9. H. Gomes, S. Gryb, T. Koslowski, The link between general relativity and shape dynamics. Class. Quant. Grav. **29**, 075009 arXiv:1101.5974 [gr-qc]

10. H. Gomes, A birkhoff theorem for shape dynamics. Class. Quant. Grav. **31**, 085008 (2014) arXiv:1305.0310 [gr-qc]

11. S. Gryb, F. Mercati, 2 + 1 gravity as a conformal gauge theory in 2 dimensions (in preparation)

12. A. Strominger, Inflation and the dS/CFT correspondence, JHEP **0111** 049, (2001). arXiv:hep-th/0110087 [hep-th]

13. P. McFadden K. Skenderis, The holographic universe. J. Phys. Conf. Ser. **222**, 012007 (2010) arXiv:1001.2007 [hep-th]

14. J. Barbour, M. Lostaglio, F. Mercati, Scale anomaly as the origin of time. Gen. Rel. Grav. **45**, 911–938 (2013) arXiv:1301.6173 [gr-qc]

15. J.M. Bardeen, B. Carter, S.W. Hawking, The four laws of black hole mechanics. Commun. Math. Phys. **31**, 161–170 (1973)

16. J.D. Bekenstein, Black holes and entropy. Phys. Rev. **7**, 2333–2346 (1973)

17. K.S. Gupta, S. Sen, Further evidence for the conformal structure of a schwarzschild black hole in an algebraic approach. Phys. Lett. **B526**, 121–126 (2002) arXiv:hep-th/0112041

18. D. Birmingham, S. Sen, An exact black hole entropy bound. Phys. Rev. **D63**, 047501 (2001) arXiv:hep-th/0008051

19. E.P. Verlinde, On the origin of gravity and the laws of Newton. JHEP **04**, 029 (2011) arXiv:1001.0785 [hep-th]

Chapter 7
A Critical Look at the Standard Cosmological Picture

Daryl Janzen

Abstract The discovery that the Universe is accelerating in its expansion has brought the basic concept of cosmic expansion into question. An analysis of the evolution of this concept suggests that the paradigm that was finally settled into prior to that discovery was not the best option, as the observed acceleration lends empirical support to an alternative which could incidentally explain expansion in general. I suggest, then, that incomplete reasoning regarding the nature of cosmic time in the derivation of the standard model is the reason why the theory cannot coincide with this alternative concept. Therefore, through an investigation of the theoretical and empirical facts surrounding the nature of cosmic time, I argue that an enduring three-dimensional cosmic present must necessarily be assumed in relativistic cosmology—and in a stricter sense than it has been. Finally, I point to a related result which could offer a better explanation of the empirically constrained expansion rate.

Introduction

Many of our basic conceptions about the nature of physical reality inevitably turn out to have been false, as novel empirical evidence is obtained, or paradoxical implications stemming from those concepts are eventually realised. This was expressed well by Einstein, who wrote [1]

> What is essential, which is based solely on accidents of development?…Concepts that have proven useful in the order of things, easily attain such an authority over us that we forget their Earthly origins and accept them as unalterable facts….The path of scientific advance is often made impassable for a long time through such errors. It is therefore by no means an idle trifling, if we become practiced in analysing the long-familiar concepts, and show upon which circumstances their justification and applicability depend, as they have grown up, individually, from the facts of experience.

Or, as he put it some years later [2],

D. Janzen (✉)
Department of Physics and Engineering Physics,
University of Saskatchewan, 116 Science Place, Rm 163, Saskatoon, SK S7N 5E2, Canada
e-mail: daryl.janzen@usask.ca

© Springer International Publishing Switzerland 2015   103
A. Aguirre et al. (eds.), *Questioning the Foundations of Physics*,
The Frontiers Collection, DOI 10.1007/978-3-319-13045-3_7

> The belief in an external world independent of the percipient subject is the foundation of all science. But since our sense-perceptions inform us only indirectly of this external world, or Physical Reality, it is only by speculation that it can become comprehensible to us. From this it follows that our conceptions of Physical Reality can never be definitive; we must always be ready to alter them, to alter, that is, the axiomatic basis of physics, in order to take account of the facts of perception with the greatest possible logical completeness.

And so it is in the same spirit, that I shall argue against a number of concepts in the standard cosmological picture that have changed very little in the past century, by making note of original justifications upon which they were based, and weighing those against empirical data and theoretical developments that have been realised through the intervening years.

The essay will concentrate initially on the nature of cosmic expansion, which lacks an explanation in the standard cosmological model. Through a discussion of the early developments in cosmology, a familiarity with the pioneering conception of expansion, as being always driven by a cosmological constant Λ, will be developed, upon which basis it will be argued that the standard model—which cannot reconcile with this view—affords only a very limited description. Then, the nature of time in relativistic cosmology will be addressed, particularly with regard to the formulation of 'Weyl's postulate' of a cosmic rest-frame. The aim will therefore be towards a better explanation of cosmic expansion in general, along with the present acceleration that has recently become evident, by reconceiving the description of time in standard cosmology, as an approach to resolving this significant shortcoming of the big bang Friedman-Lemaître-Robertson-Walker (FLRW) models, and particularly the flat ΛCDM model that describes the data so well.

On Cosmic Expansion

The expansion of our Universe was first evidenced by redshift measurements of spiral nebulae, after the task of measuring their radial velocities was initiated in 1912 by Slipher; and shortly thereafter, de Sitter attempted the first relativistic interpretation of the observed shifts, noting that 'the frequency of light-vibrations diminishes with increasing distance from the origin of co-ordinates' due to the coefficient of the time-coordinate in his solution [3]. But the concept of an expanding Universe, filled with island galaxies that would all appear to be receding from any given location at rates increasing with distance, was yet to fully form.

For one thing, when de Sitter published his paper, he was able to quote only three reliable radial velocity measurements, which gave merely 2:1 odds in favour of his prediction. However, in 1923 Eddington produced an updated analysis of de Sitter space, and showed that the redshift de Sitter had predicted as a phenomenon of his statical geometry was in fact due to a cosmical repulsion brought in by the Λ-term, which would cause inertial particles to all recede exponentially from any one [4]. He used this result to support an argument for a truly expanding Universe, which would expand everywhere and at all times due to Λ. This, he supported with an updated list of redshifts from Slipher, which now gave 36:5 odds in favour of the expansion scenario.

That same year, Weyl published a third appendix to his *Raum, Zeit, Materie*, and an accompanying paper [5], where he calculated the redshift for the 'de Sitter cosmology',

$$ds^2 = -dt^2 + e^{2\sqrt{\frac{\Lambda}{3}}t}(dx^2 + dy^2 + dz^2), \tag{7.1}$$

the explicit form of which would only be found later, independently by Lemaître [6] and Robertson [7]. Weyl was as interested in the potential relevance of de Sitter's solution for an expanding cosmology as Eddington [5], and had indeed been confused when he received a postcard from Einstein later that year (Einstein Archives: [24–81.00]), stating,

> With reference to the cosmological problem, I am not of your opinion. Following de Sitter, we know that two sufficiently separate material points are accelerated from one another. If there is no quasi-static world, then away with the cosmological term.

Eight days after this was posted, Einstein's famous second note [8] on Friedman's paper, which he now referred to as 'correct and clarifying', arrived at *Zeitschrift für Physik*. Einstein evidently had in mind that the cosmic expansion can be described with Λ set to zero in Friedman's solution, and he might have thought Weyl would notice [8] and make the connection—but the latter evidently did not, as he wrote a dialogue the following year [9] in which the proponent of orthodox relativity[1] eventually states, 'If the cosmological term fails to help with leading through to Mach's principle, then I consider it to be generally useless, and am for the return to the elementary cosmology'—that being a particular foliation of Minkowski space, which, of the three cosmological models known to Weyl, was the only one with vanishing Λ.

At this point in the dialogue, the protagonist Paulus perseveres, citing the evidence for an expanding Universe, and therefore the de Sitter cosmology as the most likely of the three known alternatives. Weyl's excitement over its description is evident in Paulus' final statement: 'If I think about how, on the de Sitter hyperboloid the world lines of a star system with a common asymptote rise up from the infinite past (see Fig. 7.1), then I would like to say: the World is born from the eternal repose of "Father Æther"; but once disturbed by the "Spirit of Unrest" (*Hölderlin*), which is at home in the Agent of Matter, "in the breast of the Earth and Man", it will never come again to rest.' Indeed, as Eq. (7.1) indicates, and as illustrated in Fig. 7.1, the universe emerges from a single point at $t = -\infty$, even though slices of constant cosmic time are infinitely extended thereafter—and comoving geodesics *naturally* disperse throughout the course of cosmic time.

---

[1] The dialogue is set between Saints Peter and Paul, with the latter presenting Weyl's 'apostatical' and 'heretical' views against the 'Relativity Church'. The following statement, which seems to be loosely quoted from the postcard sent by Einstein, was made by Peter.

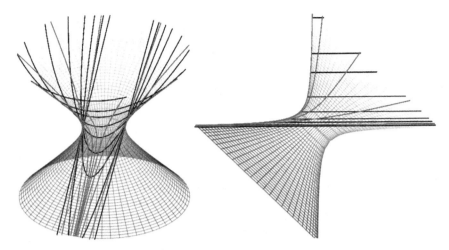

Fig. 7.1 Slices of constant time in the Lemaître-Robertson coordination of de Sitter space (*black lines*), along with comoving world lines (*red lines*), drawn on a two-dimensional slice of de Sitter space in three-dimensional Minkowski space

Thus, we have a sense of the concept of cosmic expansion that was common amongst the main thinkers in cosmology in the 1920s, who were considering the possibility of expansion driven by the cosmical repulsion in de Sitter space. Indeed, Hubble was aware of this concept, as he wrote of the 'de Sitter effect' when he published his confirmation of cosmic expansion in 1929 [10]; and de Sitter himself, in 1930, wrote of Λ as 'a measure of the inherent expanding force of the universe' [11]. Thus, along with the evidence that our Universe actually *does* expand, one had in-hand the description of a well-defined force to *always* drive that expansion.

It was therefore a huge blow to Eddington, e.g., when in 1932 Einstein and de Sitter [12] finally rejected that interpretation of cosmic expansion, in favour of a model that could afford no prior explanation for *why* the Universe *should* expand. As he put it [13],

> the theory recently suggested by Einstein and de Sitter, that in the beginning all the matter created was projected with a radial motion so as to disperse even faster than the present rate of dispersal of galaxies,[2] leaves me cold. One cannot deny the possibility, but it is difficult to see what mental satisfaction such a theory is supposed to afford.

To see why the big bang FLRW models with matter provide no explanation of expansion, for the reason stated by Eddington, we need only look at Friedman's equation,

$$\frac{\ddot{a}}{a} = \frac{\Lambda}{3} - \frac{\kappa}{2}\left(p + \frac{\rho}{3}\right), \qquad (7.2)$$

---

[2] They do not state this in words, but it is the meaning of their mathematical formulae. [Eddington's footnote].

which describes the dependence of the scale-factor, a, on Λ and the density, ρ, and pressure, p, of matter. Since $p + \rho/3$ goes like $1/a^4$ for radiation or $1/a^3$ for non-relativistic matter, the *decelerative* force due to finite matter-densities blows up exponentially as $a \rightarrow 0$, while the *accelerative* force due to Λ vanishes; so the 'inherent expanding force of the universe' only contributes to the expansion of space later on, when the relative contributions of matter and radiation have sufficiently weakened. Therefore, aside from Weyl's vacuous de Sitter cosmology, with its big bang singularity at $t = -\infty$, the big bang FLRW models can never *explain* the cosmic expansion they describe, which must be caused by the big bang singularity itself—i.e., where the theory blows up.

But since the cosmic microwave background radiation (CMBR) indicates that the Universe *did* begin in a hot dense state at a finite time in the past, the model Eddington had favoured instead (in which an unstable Einstein universe that existed since eternity would inevitably begin expanding purely due to Λ [14]) also can't be accepted.

The principal source of standard cosmology's great *explanatory deficit* is the fact that although the non-vacuous big bang FLRW models do *describe* expanding universes—and in particular the flat ΛCDM model describes the observed expansion of our Universe very well [15–26]—they afford no reason at all for *why* those universes *should* expand, since that could only be due to the initial singularity; i.e., as we follow the models back in time, looking for a possible cause of expansion, we eventually reach a point where the theory becomes undefined, and call that the cause of it all. In contrast, I've discussed two FLRW models, neither of which is empirically supported, which would otherwise better *explain* the expansion they describe, as the result of a force that is well-defined in theory.

The basic cause and nature of cosmic expansion, along with its recently-observed acceleration, are significant problems of the standard model; so, considering the evidence that the acceleration is best described by pure Λ [15–26], there is strong motivation to search for an alternative big bang model that would respect the pioneering concept of expansion, as a direct consequence of the 'de Sitter effect' in the modified Einstein field equations. It is therefore worth investigating the axiomatic basis of the Robertson-Walker (RW) line-element. As I will eventually argue that the problem lies in the basic assumptions pertaining to the description of cosmic time, I'll begin by discussing some issues related to the problem of accounting for a cosmic present.

The Cosmic Present

The problem of recognising a cosmic present is that, according to relativity theory, it should not be possible to assign one time-coordinate to the four-dimensional continuum of events that could be used to describe objective simultaneity, since two events that are described as simultaneous in one frame of reference will not be described as such by an observer in relative motion. However, as noted by Bondi [27],

The Newtonian concept of the uniform omnipresent even-flowing time was shown by special relativity to be devoid of physical meaning, but in 1923 H. Weyl suggested that the observed motions of the nebulae showed a regularity which could be interpreted as implying a certain geometrical property of the substratum .... This in turn implies that it is possible to introduce an omnipresent *cosmic time* which has the property of measuring *proper time* for every observer moving with the substratum. In other words, whereas special relativity shows that a set of arbitrarily moving observers could not find a common 'time', the substratum observers move in such a specialized way that such a public or cosmic time exists.

Although the existence of such a time concept seems in some ways to be opposed to the generality, which forms the very basis of the general theory of relativity, the development of relativistic cosmology is impossible without such an assumption.

In fact, as Einstein himself noted in 1917 [28],

The most important fact that we draw from experience as to the distribution of matter is that the relative velocities of the stars are very small as compared with the velocity of light. So I think that for the present we may base our reasoning upon the following approximative assumption. There is a system of reference relatively to which matter may be looked upon as being permanently at rest.

Thus, the assumption of a cosmic rest-frame—and a corresponding cosmic time— was justified in the derivation of Einstein's 'cylindrical' model.

While Einstein originally proposed this as an 'approximative assumption' that the empirical evidence seemed to support, the fact that he did restore absolute time when it came to the problem of describing the Universe on the largest scale was not lost on his peers. De Sitter was immediately critical of the absolute time variable in Einstein's model, noting that 'Such a fundamental difference between the time and the space-coordinates seems to be somewhat contradictory to the complete symmetry of the field-equations and the equations of motion' [29]. And a few years later, Eddington wrote that an objection to Einstein's theory may be urged, since [30] 'absolute space and time are restored for phenomena on a cosmical scale...Just as each limited observer has his own particular separation of space and time, so a being coexistive with the world might well have a special separation of space and time natural to him. It is the time for this being that is here dignified by the title "absolute."' Therefore, he concluded, 'Some may be inclined to challenge the right of the Einstein theory...to be called a relativity theory. Perhaps it has not all the characteristics which have at one time or another been associated with that name...'

Indeed, although the assumption of an absolute time in relativistic cosmology is definitely not in the spirit of relativity, the theory isn't fundamentally incompatible with such a definition. Furthermore, it is significant that despite such early criticisms, Einstein never wavered in assuming an absolute time when he came to consider the cosmological problem [12, 31, 32], i.e. as he always favoured the Friedman solutions (with $\Lambda = 0$), which begin by postulating the same.

So, we have two opposing descriptions of relativistic time—both of which are principally due to Einstein himself!—and what I'll now argue is that developments both in cosmology and in our understanding of relativity theory which have taken place in the past century demand the latter—that there is one absolute cosmic time relative to which every observer's proper time will measure, as space-time will be perceived differently due to their absolute motion through the cosmic present that must be uniquely and objectively defined—rather than the former implication of Einstein's 1905 theory of relativity [33].

In the case of special relativity, a description in which space-time emerges as a clearly defined absolute cosmic present endures, can be realised by considering four-dimensional Minkowski space as a background structure, and a three-dimensional universe that actually flows equably though it—with the past space-time continuum emerging as a purely ideal set of previous occurrences in the universe. Then, if we begin in the cosmic rest-frame, in which fundamental observers' world lines will be traced out orthogonal to the cosmic hyperplane, photons can be described as particles that move through that surface at the same rate as cosmic time, thus tracing out invariant null-lines in space-time. In this way, the evolution of separate bodies, all existing in one three-dimensional space, forms a graduating four-dimensional map.

The causal and inertial structures of special relativity are thus reconciled by describing the world lines of all observers in uniform motion through the cosmic present as their proper time axes, and rotating their proper spatial axes accordingly, so that light will be described as moving at the same rate in either direction of proper 'space'. And then, so that the speed of photons along invariant null-lines will actually be the same magnitude in all inertial frames, both the proper space and time axes in these local frames must also be scaled hyperbolically relative to each other.

This description of the emergence of space-time in a special relativistic universe can be illustrated in the following way. Consider a barograph, consisting of a pen, attached to a barometer, and a sheet of paper that scrolls under the pen by clockwork. The apparatus may be oriented so that the paper scrolls downwards, with changes in barometric pressure causing the pen to move purely horizontally. We restrict the speed of the pen's horizontal motion only so that it must always be less than the rate at which the paper scrolls underneath it. The trace of the barometric pressure therefore represents the world line of an arbitrarily moving observer in special relativistic space-time, with instantaneous velocity described in this frame by the ratio of its speed through the horizontal cosmic present and the graph paper's vertical speed, with 'speed' measured in either case relative to the ticking of the clockwork mechanism, which therefore cancels in the ratio.

Now, in order to illustrate the relativity of simultaneity, we detach the pen (call it $\mathscr{A}$) from the barometer so that it remains at rest absolutely, and add another pen, $\mathscr{B}$, to the apparatus, at the exact same height, which moves horizontally at a constant rate that's less than the constant rate that the paper scrolls along; therefore, with *absolute velocity* less than the absolute speed limit. Furthermore, we make $\mathscr{A}$ and $\mathscr{B}$ 'observers', by enabling them to send and receive signals that transmit horizontally at the same rate (in clockwork time) as absolute time rolls on (in clockwork time), thus tracing out lines on the graph paper with unit speed.

As this system evolves, the two 'timelike observers' can send these 'photons' back and forth while a special relativistic space-time diagram is traced out. If we'd rather plot the map of events in coordinates that give the relevant description from $\mathscr{B}$'s perspective, we use the Lorentz transformation equations corresponding to the description of the map as Minkowski space-time: a spacelike line is drawn, tilted

from the horizontal towards $\mathscr{B}$'s world line by the appropriate angle, and the events
along that surface are described as synchronous in that frame, even though they take
place sequentially in real time. In particular, at the evolving present, $\mathscr{B}$'s proper
spatial axis extends, in one direction, onto the empty sheet of graph paper in which
events have not yet occurred, and, in the other direction, into the past space-time
continuum of events that have already been traced onto the paper—while the real
present hyperplane, where truly simultaneous events are occurring, is tilted with
respect to that axis of relative synchronicity.

The main difference between this interpretation of special relativity and Einstein's
original one, is that 'simultaneity' and 'synchronicity' have objectively different
meanings for us, which coincide only in the absolute rest frame—whereas Einstein
established an 'operational' concept of simultaneity, so that it would be synonymous
with synchronicity, in section 1, part 1 of his first relativity paper [33]. Einstein's
definition of simultaneity is a basic assumption that's really no less arbitrary than
Newton's definitions of absolute space, time, and motion; and, as I'll argue, the
evidence from cosmology now stands against Einstein's wrong assumption, as it is
really more in line with Newton's.

The distinction between simultaneity and synchronicity in this different interpretation of relativity, can be understood more clearly through our barograph example,
by adding two more 'observers', $\mathscr{C}$ and $\mathscr{C}'$, which remain at rest relative to $\mathscr{B}$, with
$\mathscr{C}$ positioned along the same hyperplane as $\mathscr{A}$ and $\mathscr{B}$, and $\mathscr{C}'$ positioned precisely
at the intersection of $\mathscr{C}$'s world line (so that the world lines of $\mathscr{C}$ and $\mathscr{C}'$ exactly
coincide, as they are traced out on the space-time graph) and $\mathscr{B}$'s proper spatial
axis (therefore, on a different hyperplane than $\mathscr{A}$, $\mathscr{B}$, and $\mathscr{C}$); thus, $\mathscr{C}'$ shall not be
causally connected to $\mathscr{A}$, $\mathscr{B}$, and $\mathscr{C}$, since *by definition* information can only transmit
along the cosmic hyperplane; see Fig. 7.2.

The significant point that is clearly illustrated through the addition of $\mathscr{C}$ and $\mathscr{C}'$,
is that although in the proper coordinate system of $\mathscr{B}$ (or $\mathscr{C}$ or $\mathscr{C}'$), $\mathscr{C}'$ appears to
exist synchronously and at rest relative to $\mathscr{B}$, $\mathscr{C}$—which in contrast appears to exist
in $\mathscr{B}$'s (spacelike separated) past or future (depending on the direction of absolute
motion; in Fig. 7.2, $\mathscr{C}$ appears to exist in $\mathscr{B}$'s relative past)—is really the causally
connected neighbour that remains relatively at rest, with which it should be able to
synchronise its clock in the usual way; i.e., the synchronisation of $\mathscr{B}$'s and $\mathscr{C}$'s clocks

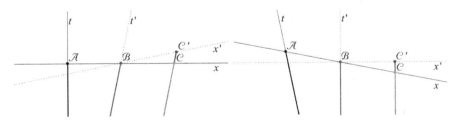

Fig. 7.2 Snapshots, in two proper reference frames, of an emergent space-time. Although the proper
times of $\mathscr{C}'$ and $\mathscr{B}$ appear to coincide, $\mathscr{C}'$ is disconnected from the causally coherent set, $\{\mathscr{A}, \mathscr{B}, \mathscr{C}\}$

will be *wrong* because *simultaneous noumena will not be perceived as synchronous phenomena in any but the cosmic rest-frame*.

According to this description, we should have to relinquish the concept that there can be no priviliged observers, as well as Einstein's light-postulate in its original form. With regard to the latter, consider that photons will still be described as travelling at a constant speed in all directions of all reference frames, due to the invariance of null-lines. But this won't actually be true, since an observer moving through the universe will keep pace better with a photon in their direction of motion, and will remain closer to that photon at all later times, on the cosmic hyperplane. Therefore, although light actually won't recede as quickly through the universe in the direction of absolute motion, it can always be described as such in the proper coordinate frame because it travels along invariant null-lines.

And with regard to the former concept, it is useful to note Galileo's argument that, to a person riding in the cabin of a moving ship, everything inside the cabin should occur just as if the ship were at rest. It was crucial for Galileo to make this point by *isolating* the inertial system from its relatively moving surroundings—as the point would have been less clear, e.g., if he had argued that when riding in the back of a wagon one can toss a ball straight in the air and have it fall back to the same point within the wagon. However, if one should argue that there *really* can't be privileged observers in the Universe, due to the relativity of inertia, one must go beyond this local-inertial effect—viz. the relativity of inertia—and consider the frame with respect to its cosmic surroundings—in which case the argument can't be justified.

For consider a neutrino, created in a star shortly after the Big Bang: in the neutrino's proper frame, only minutes may have elapsed since it left the star, throughout which time the galaxies would have formed, etc., all moving past it in roughly the same direction, at nearly the speed of light. Clearly the most reasonable interpretation, however, is that the neutrino has *really* been travelling through the Universe for the past 13.8 billion years—and this description may be given, with the cosmic present uniquely and objectively defined, in all frames including the neutrino's.

Furthermore, if we would assume that there are no privileged observers, it should be noted that the consequence of describing simultaneity and synchronicity as one and the same thing in all frames is a block universe [34]—a temporally singular 'absolute world' [35] in which 'the distinction between past, present, and future has only the significance of a stubborn illusion' [36]; i.e., 'The objective world simply *is*, it does not *happen*. Only to the gaze of my consciousness, crawling upward along the life line of my body, does a section of this world come to life as a fleeting image in space which continuously changes in time' [37]; 'There is no dynamics within space-time itself: nothing ever moves therein; nothing happens; nothing changes. ...one does not think of particles "moving through" space-time, or as "following along" their world lines. Rather, particles are just "in" space-time, once and for all, and the world line represents, all at once, the complete life history of the particle' [38].

And so I've argued against the simultaneity of synchronicity,—a reasonably intuitive concept held in common between the theories of both Newton and Einstein. But is there any *sensible* justification for the concept that the space in which events *really* take place simultaneously *must* be orthogonal to the proper time-axis of an inertial observer? When our theories are interpreted in this way, is that because one can, e.g., sit down on the floor with legs out in front, raise their right arm out to the side and their left arm up in the air, *and then stick out their tongue in the direction in which time is flowing*, for them as much as it is for their entire surroundings? Of course not. This is no more justified for someone who thus defines a right-handed coordinate system while sitting on solid ground, than it is for a person in the cabin of a ship—whether that is floating on water or flying through space. Therefore, intuition justifies only existence in space that endures with the ticking of everyone's watch—and relativity theory *demands* that this cannot be both coherently defined and synchronous with every inertial observer!

Now, although it may be argued that the alternative assumption of cosmic time is unobservable metaphysics, and therefore unscientific, that simply isn't true—for cosmology does provide strong empirical evidence of an absolute rest-frame in our Universe, as follows. As Einstein noted already in 1917 [28], there appears to be a frame relative to which the bodies of our Universe are at rest, on average. Now, Einstein had no idea of the scope of the Universe at that time, but already by 1923 Weyl realised the significance of this point, which has indeed stood the test of time, when he wrote that [5] 'Both the papers by de Sitter [3] and Eddington [4] lack this assumption on the "state of rest" of stars—by the way the only possible one compatible with the homogeneity of space and time. Without such an assumption nothing can be known about the redshift, of course.' For it is true, even in de Sitter space, that a cosmic time must be assumed in order to calculate redshifts; e.g., for particles in the comoving Lemaître-Robertson frame illustrated in Fig. 7.1 and described by Eq. (7.1), the redshift will be different from that in the frame of comoving particles in the three-sphere which contracts to a finite radius and subsequently expands (as illustrated by the gridlines of the de Sitter hyperboloid in Fig. 7.1) according to

$$ds^2 = -dT^2 + \frac{3}{\Lambda} \cosh^2\left(\sqrt{\frac{\Lambda}{3}}T\right) d\Omega_3{}^2, \tag{7.3}$$

where $d\Omega_3$ describes the three-sphere. The existence of more than one formally distinct RW cosmological model in one and the same space-time thus illustrates the importance of defining a cosmic time.

Since 1923, a number of novel observations have strengthened the evidence for a cosmic present, such as Hubble's confirmation of cosmic expansion, the detailed measurement of the expansion rate that has lately been afforded through type Ia supernovae observations, and the discovery of the CMBR, which gives a detailed signature of the cosmic rest-frame relative to which we are in fact moving, according to the common interpretation of its dipole anisotropy. Thus, the assumption of a cosmic present is now very well justified by empirical evidence.

Implications for Cosmology

Although many points should be considered in connection to the description of an absolute cosmic present, such as concepts of time travel, free will, and a causally coherent local description of gravitational collapse in the Universe—notwithstanding space-time curvature in general,—the one consequence that I will note pertains to cosmology and a better explanation of cosmic expansion.

To start, note that in deriving the general line-element for the background geometry of FLRW cosmology, Robertson required four basic assumptions [39]: i. a congruence of geodesics, ii. hypersurface orthogonality, iii. homogeneity, and iv. isotropy. i. and ii. are required to satisfy Weyl's postulate of a causal coherence amongst world lines in the entire Universe, by which every single event in the bundle of fundamental world lines is associated with a well-defined three-dimensional set of others with which it 'really' occurs simultaneously. However, it seems that ii. is therefore mostly required to satisfy the concept that synchronous events in a given inertial frame should have occurred simultaneously, against which I've argued above.

In special relativity, if we allow the fundamental world lines to *set* the cosmic rest-frame, then the cosmic hyperplane should be orthogonal—but that shouldn't be the case in general. Indeed, as I've shown in my Ph.D thesis [40], in the cosmological Schwarzschild-de Sitter (SdS) solution,

$$ds^2 = -\frac{r}{\frac{\Lambda}{3}r^3 + 2M - r}dr^2 + \frac{\frac{\Lambda}{3}r^3 + 2M - r}{r}dt^2 + r^2d\Omega^2, \qquad (7.4)$$

for which $\Lambda M^2 > 1/9, r > 0$ is *timelike*, and t is forever *spacelike*, the r-coordinate should well describe the cosmic time *and* factor of expansion in a universe in which, in the coordinates carried by fundamental observers, the cosmic present would not be synchronous, and r would evolve in proper time τ as

$$r(\tau) \propto \sinh^{2/3}[(\sqrt{3\Lambda}/2)\tau], \qquad (7.5)$$

which is *incidentally* also the flat ΛCDM scale-factor of the standard model that has been empirically constrained this past decade [15–26]; see "Appendix: Concerning Schwarzschild-de Sitter as a Cosmological Solution" for a derivation of Eq. (7.5) beginning from Eq. (7.4), and a discussion of the result's connection to cosmology. This is the rate of expansion that *all* observers would measure, if distant galaxies were themselves all roughly at rest with respect to fundamental world lines. But in contrast to FLRW theory, this universe actually has to expand—at all $r > 0$—as a result of the 'de Sitter effect'; i.e., if such a universe did come to exist at any infinitesimal time, it would *necessarily* expand—and in exactly the manner that we observe—which may be the closest to an explanation of that as we can achieve.

It is, of course, important to stress that this intriguing result is utterly meaningless if simultaneity should rather be defined as synchronicity in a given frame of reference. In that case, as Lemaître noted [41], the solution describes flat spatial slices extending from $r = 0$ to ∞, with particles continuously ejected from the origin. It is therefore only by reconceiving the relativistic concepts of time and simultaneity that SdS can be legitimated as a coherent cosmological model with a common origin—and one with the very factor of expansion that we've measured—which really *should* expand, according to the view of expansion as being always driven by Λ.

Acknowledgments Thanks to Craig Callender for reviewing an earlier draft and providing thoughtful feedback that greatly improved this essay. Thanks also to the many participants who commented on and discussed this paper throughout the contest, and FQXi for organising an excellent contest and providing criteria that helped shape the presentation of this argument.

Appendix: Concerning Schwarzschild-de Sitter as a Cosmological Solution

During the Essay Contest discussions, the critical remarks on this essay that were most important for me, and were by far the most probing, were those offered by George Ellis. Professor Ellis' criticism of the final section indicated, first of all, that the brief mention I made there of a result from my Ph.D thesis was too underdeveloped to pique much interest in it—and in fact that, stated as it was there, briefly and out of context of the explicit analysis leading from Eq. (7.4) to (7.5), the point was too easily missed. He wrote that the model is 'of course spatially inhomogeneous,' when the spatial slices *are* actually homogeneous, but rather are anisotropic; and when I pointed out to him that this is so because, in the cosmological form of the SdS solution $r > 0$ is *timelike* and t is forever *spacelike*, he replied that 'the coordinate notation is very misleading'.

So, one purpose of this appendix is to provide the intermediate calculation between Eqs. (7.4) and (7.5), that had to be left out of the original essay due to space limitations—and, in developing a familiarity with the common notation, through the little calculation, to ensure that no confusion remains in regard to the use of r as a timelike variable and t as a spacelike one. For the notation is necessary both in order to be consistent with every other treatment of the SdS metric to date, and because, regardless of whether r is timelike or spacelike in Eq. (7.4), it really does make sense to denote the coordinate with an 'r' because the space-time is isotropic (i.e. 'radially' symmetric) in that direction.

With these 'bookkeeping' items out of the way (after roughly the first four pages), the appendix moves on to address Professor Ellis' two more substantial criticisms, i.e. regarding the spatial anisotropy and the fact that the model has no dynamic matter in it; for, as he noted, the model 'is interesting geometrically, but it needs supplementation by a dynamic matter and radiation description in order to relate to our cosmic history'. These important points were discussed in the contest forum, but were difficult to adequately address in that setting, so the problem is given more proper treatment in the remaining pages of this appendix once the necessary mathematical results are in-hand. Specifically, in the course of developing a physical picture in which the SdS metric provides the description of a universe that *would* appear isotropic to fundamental observers who measure the same rate of expansion that we do (viz. as given by Eq. (7.5)), we will come to a possible, consistent resolution to the problem of accounting for dynamic matter, which leads to a critical examination of the consistency and justification of some of the most cherished assumptions of modern physics, thus further questioning its foundations.

We begin by writing down the equations of motion of 'radial' geodesics in the SdS geometry, using them to derive a description of the SdS cosmology that would be appropriate to use from the perspective of fundamental observers who evolve as they do, beginning from a common origin at $r = 0$, always essentially *because of* the induced field potential. It will be proved incidentally that the observed cosmological redshifts, in this homogeneous universe which is *not* orthogonal to the bundle of fundamental geodesics—and is therefore precluded by the a priori assumptions of standard FLRW cosmology—must evolve through the course of cosmic time, as a function of the proper time of fundamental observers, with the precise form of the flat ΛCDM scale-factor—i.e., with exactly the form that has been significantly constrained through observations of type Ia supernovae, baryon acoustic oscillations, and CMBR anisotropies [15–26].

Since the Lagrangian,

$$L = -\frac{r - 2M - \frac{\Lambda}{3}r^3}{r}\left(\frac{dt}{d\tau}\right)^2 + \frac{r}{r - 2M - \frac{\Lambda}{3}r^3}\left(\frac{dr}{d\tau}\right)^2 = -1, \qquad (7.6)$$

for timelike (r, t)-geodesics with proper time τ in the SdS geometry is independent of t, the Euler-Lagrange equations indicate that

$$E \equiv -\frac{1}{2}\frac{\partial L}{\partial(dt/d\tau)} = \frac{r - 2M - \frac{\Lambda}{3}r^3}{r}\left(\frac{dt}{d\tau}\right) \qquad (7.7)$$

is conserved $(-2dE/d\tau \equiv dL/dt = 0)$. Substituting Eq. (7.7) into (7.6), then, we find the corresponding equation of motion in r:

$$\left(\frac{dr}{d\tau}\right)^2 = E^2 - \frac{r - 2M - \frac{\Lambda}{3}r^3}{r}. \qquad (7.8)$$

While the value of E may be arbitrary, we want a value that distinguishes a particular set of geodesics as those describing particles that are 'fundamentally at rest'—i.e., we'll distinguish a preferred fundamental rest frame by choosing a particular value of E that meets certain physical requirements. In order to determine which value to use, we first note that where r *is* spacelike, Eq. (7.8) describes the specific (i.e., per unit rest-mass) kinetic energy of a test-particle, as the difference between its (conserved) *specific energy* and the gravitational field's *effective potential*,

$$V_{\text{eff}}(r) \equiv \frac{r - 2M - \frac{\Lambda}{3}r^3}{r}. \tag{7.9}$$

Then, a reasonable definition sets the 'fundamental frame' as the one in which the movement of particles in r and t is essentially *caused* by the non-trivial field potential—i.e., so that $dr/d\tau = 0$ just where the gravitational potential is identically trivial ($V_{\text{eff}} \equiv 1$), and the line-element, Eq. (7.4), reduces to that of Minkowski space. From Eq. (7.8), this amounts to setting $E^2 = 1$; therefore, a value $E^2 > 1$ corresponds to a particle that would not come to rest at $r = -\sqrt[3]{6M/\Lambda}$, where $V_{\text{eff}} \equiv 1$, but has momentum in r beyond that which would be imparted purely by the field.

As a check that the value $E^2 = 1$ is consistent with our aims, we can consider its physical meaning another way. First of all, note that where $V_{\text{eff}} \equiv 1$, at $r = -\sqrt[3]{6M/\Lambda}$, r is spacelike and t is timelike regardless of the values of M and Λ; therefore, it is consistent in any case to say that a particle with $E^2 > 1$ has non-vanishing spatial momentum there. Indeed, from Eq. (7.7), we find that $t = \tau$ at $r = -\sqrt[3]{6M/\Lambda}$ if, and only if, $E = 1$—so the sign of E should in fact be positive for a particle whose proper time increases with increasing t in the absence of gravity.

Furthermore, note that when $\Lambda = 0$, Eq. (7.8) reduces to

$$\left(\frac{dr}{d\tau}\right)^2 = E^2 - 1 + \frac{2M}{r}. \tag{7.10}$$

As such, Misner, Thorne and Wheeler describe E as a particle's 'specific energy at infinity', where the effective potential is trivial [42]. It is relevant to note their statement (on p. 658), that the conservation of 4-momentum 'allows and forces one to take over the [term] $E =$ "energy at infinity"…, valid for orbits that do reach to infinity, for an orbit that does not reach to infinity.' More generally, we should describe E for arbitrary M and Λ as the 'energy at vanishing field potential' even when $r = -\sqrt[3]{6M/\Lambda}$ is a negative mathematical abstraction that lies beyond a singularity at $r = 0$. In particular, we take $E = 1$ to be the *specific energy of a test-particle that is at rest with respect to the vanishing of the potential*. It's simply a matter of algebraic consistency.

Thus, we have $E = 1$ as the specific energy of particles that would come to rest in the absence of a gravitational field, which are therefore guided purely through the effective field potential. We therefore use the geodesics with $E = 1$ to define a preferred rest frame in the SdS geometries, and say that any particle whose world line is a geodesic with $E \neq 1$ is one that has uniform momentum relative to the fundamental rest frame.

We can now write the SdS line-element, Eq. (7.4), in the proper frame of a bundle of these fundamental geodesics, which evolve through t and r all with the same proper time, τ, and occupy constant positions in 'space'. Since Λ must be positive in order to satisfy the requirement, $\Lambda M^2 > 1/9$, for $r > 0$ to be timelike—i.e. the requirement for the SdS line-element to be cosmological rather than a local solution— it is more convenient to work with scale-invariant parameters $r \to r' = \sqrt{\frac{\Lambda}{3}}r$,

$t \to t' = \sqrt{\frac{\Lambda}{3}}t$, $\tau \to \tau' = \sqrt{\frac{\Lambda}{3}}\tau$, $M \to M' = \sqrt{\frac{\Lambda}{3}}M$, etc., normalising all dimensional quantities by the cosmic length-scale $\sqrt{3/\Lambda}$ (see, e.g., § 66 in [4] or § 66 in [43] for interesting discussions of this length parameter). This normalisation ultimately amounts to striking out the factor $\Lambda/3$ from the r^3-term in the line-element

$$\left(\text{since } \frac{\sqrt{\frac{\Lambda}{3}}r - 2\sqrt{\frac{\Lambda}{3}}M - \left(\sqrt{\frac{\Lambda}{3}}r\right)^3}{\sqrt{\frac{\Lambda}{3}}r} \to \frac{r - 2M - r^3}{r} \right), \text{ or e.g. writing the flat } \Lambda\text{CDM scale-}$$

factor, Eq. (7.5), as

$$r(\tau) \propto \sinh^{2/3}(3\tau/2), \tag{7.11}$$

and the corresponding Hubble parameter as

$$H \equiv \dot{r}/r = \coth(3\tau/2), \tag{7.12}$$

which exponentially approaches $H = 1$ on timescales $\tau \sim 2/3$.

The evolution of each geodesic through scale-invariant t and r is then given, through Eqs. (7.7) and (7.8) with $E = 1$, as

$$\partial_\tau t \equiv \frac{\partial t}{\partial \tau} = \frac{r}{r - 2M - r^3}, \tag{7.13}$$

$$(\partial_\tau r)^2 \equiv \left(\frac{\partial r}{\partial \tau}\right)^2 = \frac{2M + r^3}{r}. \tag{7.14}$$

Eq. (7.14) can be solved using $\int (u^2 + a)^{-1/2}\,du = \ln\left(u + \sqrt{u^2 + a}\right)$ after substituting $u^2 = 2M + r^3$. Taking the positive root (so τ increases with r), we have,

$$\tau = \int_{r(0)}^{r(\tau)} \sqrt{\frac{r}{2M + r^3}}\,dr = \frac{2}{3}\ln\left(\sqrt{2M + r^3} + r^{3/2}\right)\Big|_{r(0)}^{r(\tau)}, \tag{7.15}$$

where the lower limit on τ has been arbitrarily set to 0. Thus, in this frame we can express r as a function of each observer's proper time τ and an orthogonal (i.e. synchronous, with constant $\tau = 0$) spatial coordinate, $r(0)$, which may be arbitrarily rescaled without altering the description in any significant way.

Then, as long as M is nonzero, a convenient set of coordinates from which to proceed results from rescaling the spatial coordinate as[3]

---

[3] Note that this transformation is not valid when $M = 0$, which we are anyhow not interested in. An equivalent transformation in that case is found by setting $r(0) \equiv e^\chi$, whence $r(\tau, \chi) = e^{\tau + \chi}$, and Eqs. (7.21), (7.24), and (7.27), yield the line-element, $ds^2 = -d\tau^2 + r^2 \left(d\chi^2 + d\theta^2 + \sin^2\theta d\phi^2\right)$.

$$r(0) \equiv (2M)^{1/3} \sinh^{2/3} \left(\frac{3}{2} \chi \right); \; M \neq 0, \tag{7.16}$$

from which we find, after some rearranging of Eq. (7.15),[4]

$$r(\tau, \chi) = (2M)^{1/3} \sinh^{2/3} \left(\frac{3}{2} [\tau + \chi] \right), \tag{7.17}$$

which immediately shows the usefulness of rescaling the $r(0)$ as in Eq. (7.16), since it allows Eq. (7.15) to be solved explicitly for $r(\tau, \chi)$. As such, we immediately have the useful result (cf. Eq. (7.8)),

$$\partial_\chi r \equiv \frac{\partial r}{\partial \chi} = \partial_\tau r = \sqrt{\frac{2M + r^3}{r}}. \tag{7.18}$$

The transformation, $t(\tau, \chi)$, may then be calculated from

$$\frac{dt}{dr} = \frac{\partial_\tau t}{\partial_\tau r} = \frac{r}{r - 2M - r^3} \sqrt{\frac{r}{2M + r^3}}. \tag{7.19}$$

Then, to solve for $t(\tau, \chi)$, we can gauge the lower limits of the integrals over t and r, at $\tau = 0$, by requiring that their difference, defined by

$$t(\tau, \chi) = \int^{r(\tau, \chi)} \frac{r}{r - 2M - r^3} \sqrt{\frac{r}{2M + r^3}} dr - F(\chi), \tag{7.20}$$

sets

$$0 = g_{\chi\tau} = g_{tt} \partial_\tau t \partial_\chi t + g_{rr} \partial_\tau r \partial_\chi r. \tag{7.21}$$

(Thus, χ will be orthogonal to τ.) This calculation is straightforward[5]:

$$0 = -\frac{r}{r - 2M - r^3} + \partial_\chi F(\chi) + \frac{r}{r - 2M - r^3} \frac{2M + r^3}{r} \tag{7.22}$$

$$= \partial_\chi F(\chi) - 1, \tag{7.23}$$

---

[4] Note that the two identities, $e^x = \sinh(x) + \cosh(x)$ and $\text{arsinh}(x) = \ln\left(x + \sqrt{x^2 + 1}\right)$, are useful here. Eq. (7.17), along with our eventual line-element, Eq. (7.28), was originally found by Lemaître [41], although his solution to Eq. (7.14) (with dimensionality restored),

$$r = (6M/\Lambda)^{1/3} \sinh^{2/3} \left[3\sqrt{\Lambda}(t - t_0)/2 \right],$$

is too large in its argument by a factor of $\sqrt{3}$.

[5] Note that we don't actually have to solve the integral in Eq. (7.20), since only partial derivatives of t are needed here and below.

so that $F(\chi) = \chi$.

Now, it is a simple matter to work out the remaining metric components as follows: our choice of proper reference frame immediately requires

$$g_{\tau\tau} = g_{tt}(\partial_\tau t)^2 + g_{rr}(\partial_\tau r)^2 = -1, \tag{7.24}$$

according to the Lagrangian, Eq. (7.6); and by direct calculation, we find

$$g_{\chi\chi} = g_{tt}(\partial_\chi t)^2 + g_{rr}(\partial_\chi r)^2 \tag{7.25}$$

$$= -\frac{r - 2M - r^3}{r}\left(\frac{2M + r^3}{r - 2M - r^3}\right)^2 + \frac{r}{r - 2M - r^3}\frac{2M + r^3}{r} \tag{7.26}$$

$$= (\partial_\chi r)^2. \tag{7.27}$$

But this result is independent of any arbitrary rescaling of χ; for if we replaced $\chi = f(\xi)$ in Eq. (7.16), we would then find the metric to transform as $g_{\xi\xi} = g_{\chi\chi}(d\chi/d\xi)^2 = (\partial_\xi r)^2$, the other components remaining the same.

Therefore, the SdS metric in the proper frame of an observer who is cosmically 'at rest', in which the spatial coordinates are required, according to an appropriate definition of $F(\chi)$, to be orthogonal to τ,[6] can generally be written,

$$ds^2 = -d\tau^2 + (\partial_\chi r)^2 d\chi^2 + r^2\left(d\theta^2 + \sin^2\theta d\phi^2\right). \tag{7.28}$$

This proves Lemaître's result from 1949 [41],—that slices $d\tau = 0$ $\left(\Rightarrow (\partial_\chi r)^2 d\chi^2 = (dr/d\chi)^2 d\chi^2 = dr^2\right)$ are Euclidean, with line-element,

$$d\sigma^2 = dr^2 + r^2\left(d\theta^2 + \sin^2\theta d\phi^2\right). \tag{7.29}$$

However, in the course of our derivation we have also found that Lemaître's *physical* interpretation—that the 'geometry is Euclidean on the expanding set of particles which are ejected from the point singularity at the origin'—is wrong.

It is *wrong* to interpret this solution as describing the evolution of synchronous 'space' which always extends from $r = 0$ to $r = +\infty$ along lines of constant τ, being truncated at the $r = 0$ singularity at $\chi = -\tau$ from which particles are continuously ejected as τ increases. But this is exactly the interpretation one is apt to make, who is accustomed to thinking of synchronous spacelike hypersurfaces as 'space' that exists 'out there', regardless of the space-time geometry or the particular coordinate system used describe it.

As we noted from the outset, the 'radial' geodesics that we have now described by the lines $\chi = const.$, along which particles all measure their own proper time to increase with τ, describe the world lines of particles that are all fundamentally

---

[6] Note that the 'radially' symmetric part of Eq. (7.4) is already orthogonal to τ.

at rest—i.e., at rest with respect to the vanishing of the effective field potential. Therefore, these particles should not all emerge from the origin at different times, and then somehow evolve together as a coherent set; but *by Weyl's principle* they should all emerge from a *common origin*, and evolve through the field that varies *isotropically in r, together for all time.* In that case, space will be homogeneous, since the constant cosmic time ($dr = 0$) slices of the metric can be written independent of spatial coordinates; so every fundamental observer can arbitrarily set its spatial position as $\chi = 0$ and therefore its origin in time as $\tau = 0$.

The spaces of constant cosmic time should therefore be those slices for which $r(\tau + \chi) = const.$—i.e., we set $\bar{\tau} = \tau + \chi$ as the proper measure of cosmic time in the fundamental rest frame of the universe defined by this coherent bundle of geodesics, so that Eq. (7.17) becomes

$$r(\bar{\tau}) = (2M)^{1/3} \sinh^{2/3}(3\bar{\tau}/2). \tag{7.30}$$

The spacelike slices of constant $\bar{\tau}$ are at 45° angles in the (τ, χ)-plane, and are therefore definitely not synchronous with respect to the fundamental geodesics. However, given this definition of cosmic time, the redshift of light that was emitted at $\bar{\tau}_e$ and is observed now, at $\bar{\tau}_0$, should be

$$1 + z = \frac{r(\bar{\tau}_0)}{r(\bar{\tau}_e)}, \tag{7.31}$$

where $r(\bar{\tau})$ has exactly the form of the flat ΛCDM scale-factor (cf. Eq. (7.11)), which is exactly the form of expansion in our Universe that has been increasingly constrained over the last fifteen years [15–26].

Now, in order to properly theoretically interpret this result for the observed redshift in our SdS cosmology, it should be considered in relation to FLRW cosmology—and particularly the theory's basic assumptions. As noted in Sect. "Implications for Cosmology", the kinematical assumptions used to constrain the form of the line-element are: i. a congruence of geodesics, ii. hypersurface orthogonality, iii. homogeneity, and iv. isotropy. Assumptions i. and ii. have a lot to do with how one defines 'simultaneity', which I have discussed both in the context of special relativity in Sect. "The Cosmic Present", and now in the context of the SdS cosmology, in which simultaneous events that occur in the course of cosmic time are *not* synchronous *even in the fundamental rest frame.* As the discussion should indicate, the definition of 'simultaneity' is somewhat arbitrary—and it is an *assumption* in any case—and should be made with the physics in mind. Einstein obviously had the physics in mind when he proposed using an operational definition of simultaneity [33]; but it has since been realised that even special relativity, given this definition, comes to mean that time can't pass, etc., as noted in Sect. "The Cosmic Present".

Special relativity *should* therefore be taken as an advance on Newton's bucket argument, indicating that not only should acceleration be absolute, as Newton showed (see, e.g., [44] for a recent discussion of Newton's argument), but velocity should be as well, since time obviously passes—which it can't do, according to special relativity, if motion isn't *always* absolute. Usually, however, the opposite is done, and people who have been unwilling or unable to update the subjective and arbitrary definitions of simultaneity, etc., from those laid down by Einstein in 1905, have simply concluded that Physical Reality has to be a four-dimensional Block in which time doesn't pass, and the apparent passage of time is a stubborn illusion; see, e.g., Sect. 5 in [45] for a popular account of this, in addition to Refs. [34–38]. The discussion in Sect. "The Cosmic Present" shows how to move forward with a realistic, physical, and most importantly a *relativistic* description of objective temporal passage, which can be done only when 'simultaneity' is not equated with 'synchronicity' a priori; and another useful thought-experiment along those lines, which shows how perfectly acceptable it is to assume objective temporal passage in spite of relativistic effects, is presented in my more recent FQXi essay [46].

In contrast to the hardcore relativists who would give up temporal passage in favour of an operational definition of simultaneity, Einstein was the first relativist to renege on truly relative simultaneity when he assumed an absolute time in constructing his cosmological model [28]; and despite immediately being chastised by de Sitter over this [29], he never did balk in making the same assumption whenever he considered the cosmological problem [12, 31, 32]—as did just about every other cosmologist who followed, with very few notable exceptions (e.g., de Sitter [3, 29] was one, as there was no absolute time implicit in his model).

But whenever the assumption of absolute time has been made in cosmology, it has been made together with special relativity's baggage, as the slices of true simultaneity have been assumed to be synchronous in the fundamental rest frame. Now we see that, not only is the operational definition *wrong* in the case of special relativity (since it comes to require that time does not pass, which is realistically unacceptable), but here we've considered a general relativistic example in which equating 'simultaneity' and 'synchronicity' makes even less sense in terms of a reasonable physical interpretation of the mathematical description, since the interpretation is *causally incoherent*—i.e. Lemaître's interpretation, that the line-element Eq. (7.28) should describe an 'expanding set of particles which are ejected from the point singularity at the origin' represents *abominable physical insight*. The main argument of this essay was therefore, that while assumption i. of FLRW cosmology is justified from the point of view that relative temporal passage should be coherent, assumption ii. is not, and this unjustified special relativistic baggage should be shed by cosmologists—and really by all relativists, as it leads to further wrong interpretations of the physics.

Assumption iii. hardly requires discussion. It is a mathematical statement of the cosmological principle—that no observer holds a special place, but the Universe should look the same from every location—and is therefore as fundamental an assumption as the principle of relativity. Furthermore, our SdS universe *is* homogeneous, so there is no problem.

The final assumption, however, is a concern. The isotopy of our Universe is an empirical fact—it looks the same to us in every direction, and the evidence is that it must have done since its beginning. In contrast, the spatial slices of the cosmological SdS solution are *not* isotropic: they are a 2-sphere with extrinsic radius of curvature

r, multiplied by another dimension that scales differently with r. Furthermore, by Eq. (7.19) we know that all these fundamental world lines move uniformly through this third spatial dimension, t, as r increases.

The SdS cosmology therefore describes a universe that should be conceived as follows: First of all suppressing one spatial dimension, the universe can be thought of as a 2-sphere that expands from a point, with all fundamental observers forever motionless along the surface; then, the third spatial dimension should be thought of as a line at each point on the sphere, through which fundamental observers travel uniformly in the course of cosmic time.

Since it is a general relativistic solution, the distinction between curvature along that third dimension of space and motion through it is not well-defined. However, a possibility presents itself through an analogy with the *local* form of the SdS solution. As with all physically meaningful solutions of the Einstein field equations, this one begins with a physical concept from which a general line-element is written; then the line-element gets sent through the field equations and certain restrictions on functions of the field-variables emerge, allowing us to constrain the general form to something more specific that satisfies the requisite second-order differential equations. This is, e.g., also how FLRW cosmology is done—i.e., first the RW line-element satisfying four basic physical/geometrical requirements is written down, and then it is sent through the field equations to determine equations that restrict the form of the scale-factor's evolution, under a further assumption that finite matter densities in the universe should influence the expansion rate. The local SdS solution, too, is derived as the vacuum field that forever has spacelike radial symmetry about some central gravitating body, and the field equations are solved to restrict the form of the metric coefficients in the assumed coordinate system. But then, as Eddington noted [4],

> We reach the same result if we attempt to define symmetry by the propagation of light, so that the cone $ds = 0$ is taken as the standard of symmetry. It is clear that if the locus $ds = 0$ has complete symmetry about an axis (taken as the axis of t) ds^2 must be expressable by [the radially isotropic line-element with general functions for the metric coefficients].

Therefore, the local SdS metric corresponds to the situation in which light propagates isotropically, and its path in space-time is described by the null lines of a Lorentzian metric. Prior to algebraic abstraction (i.e. the assumption of a Lorentzian metric and a particular coordinate system), the geometrical picture is already set; and it is upon that basic geometrical set-up that the algebraic properties of the general relativistic field are imposed.

This construction of the local SdS solution through physical considerations of light-propagation can be used analogously in constructing a geometrical picture upon which the cosmological SdS solution can be based; however, a some more remarks are necessary before coming to that. First of all, as our discussion of the local SdS and FLRW solutions indicates, in general much of the *physics* enters into the mathematical description already in defining the basic geometrical picture and the corresponding line-element, which broadly sets-up the physical situation of interest. Only then is the basic physical picture further constrained by requiring that it satisfy the specific properties imposed by Einstein's field law. In fact, when it comes to the cosmological problem, and we begin *as always* by assuming what will be our *actual* space, and how it will roughly evolve as *actual* cosmic time passes—i.e. by *first* assuming prior *kinematical* definitions of absolute space and time, and then constructing an appropriate cosmological line-element, which is finally constrained through the *dynamical* field law—there is really a lot of room to make it as we like it.

But now we have a particular line-element in mind (viz the SdS cosmological solution), and we can use it in guiding our basic kinematical definitions. In particular, we have the description of a universe that *is* a two-dimensional sphere that expands as a well-defined function of the proper time of fundamental observers who all remain absolutely at rest, multiplied by another dimension through which those same observers *are* moving at a rate that varies through the course of cosmic time. According to the equivalence principle, it may be that the gravitational field is non-trivial along that particular direction of space, and therefore guides these fundamental observers along—or it could be that this direction is uniform as well, and that the fundamental observers are moving along it, and therefore describe it relatively differently.

What is interesting about this latter possibility, is that there would be a fundamental background metric describing the evolution of this uniform space, and that the metric used by fundamental observers to describe the evolution of space-time in their proper frames shouldn't necessarily *have* to be the same fundamental metric transformed to an appropriate coordinate system. The metric itself might be defined differently from the background metric, for other physical reasons. In this case, an affine connection defining those world lines as fundamental geodesics may not be *compatible* with the more basic metric, and could be taken as the covariant derivative of a different one. The picture starts to resemble teleparallelism much more closely than it does general relativity; but since the two theories are equivalent, and we have recognised that in any case the kinematical definitions bust be made first—i.e. since we must set-up the kinematical definitions in the first place, according to the physical situation we want to describe, before ensuring that the resulting line-element satisfies the field equations—we'll press on in this vein.

Let us suppose a situation where there is actually no gravitational mass at all, but fundamental inertial observers—the constituent *dynamical matter* of our system— are *really* moving uniformly through a universe that fundamentally *is* isotropic and homogeneous, and expands through the course of cosmic time. The fundamental metric for this universe should satisfy even the RW line-element's orthogonality assumption, although the slices of constant cosmic time would *not* be synchronous in the rest frame of the fundamental observers. Since space, in the two-dimensional slice of the SdS cosmology through which fundamental observers are *not* moving, really *is* spherical, the obvious choice is an expanding 3-sphere, with line-element

$$ds^2 = -dT^2 + R(T)^2 d\Omega_3{}^2, \qquad (7.32)$$

where the radius $R(T)$ varies, according to the vacuum field equations, as

$$R(T) = C_1 e^{\sqrt{\frac{\Lambda}{3}}T} + C_2 e^{-\sqrt{\frac{\Lambda}{3}}T}; \quad C_1 C_2 = \frac{3}{4\Lambda}. \tag{7.33}$$

In particular, because a teleparallel theory would require a parallelisable manifold, we note that this is true if

$$C_1 = C_2 = \frac{1}{2}\sqrt{\frac{3}{\Lambda}}, \tag{7.34}$$

and the de Sitter metric is recovered (cf. Eq. (7.3)). While there may be concern because in this case $R(0) = \sqrt{3/\Lambda} > 0$ is a minimum of contracting and expanding space, I will argue below that this may actually be an advantage.

This foliation of de Sitter space is particularly promising for a couple of reasons: i. the bundle of fundamental geodesics in Eq. (7.3) are world lines of massless particles, i.e. the ones at $r = 0$ for all t in Eq. (7.4) with $M = 0$; and ii. unlike e.g. the 2-sphere (or spheres of just about every dimension), the 3-sphere *is* parallelisable, so it is possible to define an objective direction of motion for the dynamic matter.

Now we are finally prepared to make use, by analogy, of Eddington's remark on the derivation of the local SdS metric in terms of light propagation along null lines. In contrast, we are now beginning from the description of a universe in which particles that *are not* moving through space are massless, but we want to write down a different Lorentzian metric to describe the situation from the perspective of particles that *are* all moving through it at a certain rate, who define null lines as the paths of relatively moving masselss particles—so, we will write down a new metric to use from the perspective of particles that all move along null lines pointing in one direction of de Sitter space, describing the relatively moving paths of massless particles that actually remain motionless in the 3-sphere, as null lines instead. This new line-element can be written,

$$\mathrm{d}s^2 = -A(r, t)\mathrm{d}r^2 + B(r, t)\mathrm{d}t^2 + r^2\mathrm{d}\Omega^2, \tag{7.35}$$

where r points in the *timelike* direction of the universe's increasing radius, and t describes the dimension of space through which the fundamental particles are moving. Solving Einstein's field equations proves the Jebsen-Birkhoff theorem—that A and B are independent of the spacelike variable t—and leads to the cosmological SdS solution, Eq. (7.4), as the abstract description of this physical picture.

Thus, we have come full circle to a statement of the line-element that we started with. Our analysis began with a proof that in this homogeneous universe, redshifts should evolve with exactly the form that they do in a flat ΛCDM universe; and in the last few pages we have aimed at describing a physical situation in which this line-element would apply in the proper reference frame of dynamical matter, and the observed large scale structure would be isotropic. And indeed, in this universe, in which the spatial anisotropy in the line-element is an artifact of the motion of fundamental observers through homogeneous and isotropic expanding space *and* their definition of space-time's null lines, this would be so—for as long as these fundamental observers are uniformly distributed in space that really *is* isotropic and homogeneous, all snapshots of constant cosmic time (and therefore the development, looking back in time with increasing radius all the way to the cosmological horizon) should appear isotropic, since these uniformly distributed observers would always be at rest *relative to one another*.

Having succeeded in showing how the SdS metric can be used to describe a homogeneous universe in which the distribution of dynamical matter appears isotropic from each point, and which would be measured to expand at exactly the rate described by the flat ΛCDM scale-factor—i.e. with exactly the form that has been observationally constrained [15–26], but which has created a number of problems because, from a basic theoretical standpoint, many aspects of the model are not what we expect— we should conclude with a brief discussion of potential implications emerging from the hypothesis that the SdS cosmology accounts for the fundamental background structure in our Universe.

The most obvious point to note is that the hypothesis would challenge what Misner, Thorne, and Wheeler called 'Einstein's explanation for "gravitation"'—that '*Space acts on matter, telling it how to move. In turn, matter reacts back on space, telling it how to curve*' [42]. For according to the SdS cosmology, the structure of the absolute background should remain unaffected by its matter-content, and only the geometry of space-time—the four-dimensional set of events that develops as things happen in the course of cosmic time—will depend on the locally-inertial frame of reference in which it is described.

When I mentioned to Professor Ellis in the essay contest discussions, that I think the results I have now presented in this appendix may provide some cause to seriously reconsider the assumption that the expansion rate of the Universe should be influenced by its matter-content—a fundamental assumption of standard cosmology, based on 'Einstein's explanation for "gravitation"', which Professor Loeb also challenged in his submission to the contest—his response was, 'Well its not only of cosmology its gravitational theory. It describes solar system dynamics, structure formation, black holes and their interactions, and gravitational waves. The assumption is that the gravitational dynamics that holds on small scales also holds on large scales. It's worked so far.' And indeed, it has worked so far—but that is not a good reason to deny consideration of alternate hypotheses. In fact, as noted by Einstein in the quotation that began this essay, 'Concepts that have proven useful in the order of things, easily attain such an authority over us that we forget their Earthly origins and accept them as unalterable facts...The path of scientific advance is often made impassable for a long time through such errors.' From Einstein's point of view, such a position as 'It's worked so far' is precisely what becomes the greatest barrier to scientific advance.

After making this point, Einstein went on to argue that it is really when we challenge ourselves to rework the basic concepts we have of Nature that fundamental advances are made, adding [1],

This type of analysis appears to the scholars, whose gaze is directed more at the particulars, most superfluous, splayed, and at times even ridiculous. The situation changes, however, when one of the habitually used concepts should be replaced by a sharper one, because the development of the science in question demanded it. Then, those who are faced with the fact that their own concepts do not proceed cleanly raise energetic protest and complain of revolutionary threats to their most sacred possessions.

In this same spirit, we should note that not only our lack of a technical reason for why the Universe should ever have come to expand—i.e. the great explanatory deficit of modern cosmology described in Sect. "The Cosmic Present"—but also both the cosmological constant problem [47] and the horizon problem [48] are significant problems under the assumption that cosmic expansion is determined by the Universe's matter-content. The former of these two is the problem that the vacuum does not appear to gravitate: an optimistic estimate, that would account for only the electron's contribution to the vacuum energy, still puts the theoretical value 10^{30} larger than the dark energy component that cosmologists have measured experimentally [49]. The latter problem is that we should not expect the observable part of a general relativistic big bang universe to be isotropic, since almost all of what can now be seen would not have been in causal contact prior to structure formation—and indeed, the light reaching us now from antipodal points of our cosmic event horizon has come from regions that *still* remain out of causal contact with each other, since they are only just becoming visible to *us*, at the half-way point between them.

While there is no accepted solution to the cosmological constant problem, the horizon problem is supposed to be resolved by the inflation scenario [48]—an epoch of exponential cosmic expansion, proposed to have taken place almost instantly after the Big Bang, which would have carried off our cosmic horizon much faster than the speed of light, leaving it at such a great distance that it is only now coming back into view. Recently, provisional detection of B-mode polarisation in the CMBR [50] that is consistent with the theory of gravitational waves from inflation [51–53] has been widely lauded as a 'proof' of the theory. However, in order to reconcile apparent discrepancies with measurements of the CMBR's anisotropy signature from the *Planck* satellite, BICEP2 researchers have suggested that ad hoc tweaks of the ΛCDM model may be necessary [54].

Details involving the emergence of dynamical matter in the SdS cosmology have not been worked out; however, there is no reason to suspect ab initio that the gravitational waves whose signature has potentially been preserved in the CMBR, would not also be produced in the scenario described here. More importantly, though, the SdS cosmology provides a description that precisely agrees with the observed large-scale expansion of our Universe, and does so without the need to invent any ad hoc hypotheses in order to 'save the appearances' that we have found to be so very different from our prior theoretical expectations. The theory simultaneously *solves* the expansion problem outlined in Sect. "On Cosmic Expansion" (since expansion *must* proceed at an absolute rate, regardless of the universe's dynamical matter-content) and *subverts* the major issues associated with the assumption that cosmic expansion is determined by the Universe's matter-content.

In the SdS cosmology, the vacuum *can* very well gravitate locally without affecting the cosmic expansion rate, and the universe *should* appear, from every point, to be isotropic out to the event horizon. Even the flatness problem—viz. the problem that the curvature parameter has been constrained very precisely around zero, when the expectation, again under the assumption that the Universe's structure should be determined by its matter-content, is that it should have evolved to be very far from that—which inflation is also meant to resolve, is subverted in this picture. For indeed, the universe described here, despite being closed, would be described by fundamental observers to expand exactly according to the *flat* ΛCDM scale-factor, Eq. (7.30).

Additionally, the SdS cosmology provides a fundamentally different perspective on the so-called 'coincidence problem'—viz. that the matter and dark energy density parameters of the standard FLRW model are nearly equal, when for all our knowledge they could easily be hundreds of orders of magnitude different. Indeed, if we write down the Friedman equation for a flat ΛCDM universe using Eq. (7.12) with dimensionality restored,

$$H^2 = \frac{\Lambda}{3}\coth^2\left(\frac{\sqrt{3\Lambda}}{2}\tau\right) = \frac{1}{3}(8\pi\rho + \Lambda), \qquad (7.36)$$

we find that the matter-to-dark energy density ratio, $\varepsilon \equiv 8\pi\rho/\Lambda$, while infinite at the big bang, should approach zero exponentially quickly. For example, given the measured value $\Lambda \sim 10^{-35}$ s^{-2}, we can write

$$\tau = \frac{2}{\sqrt{3\Lambda}}\text{arcoth}\sqrt{\varepsilon + 1} \approx 11 \text{ Gyr} \cdot \text{arcoth}\sqrt{\varepsilon + 1}. \qquad (7.37)$$

From here, we find that the dark energy density was 1 % of the matter density at $\tau(\varepsilon = 100) = 1$ billion years after the big bang, and that the matter density will be 1 % of the dark energy density at $\tau(\varepsilon = 0.01) = 33$ billion years after the big bang.

At first glance, these results may seem to indicate that it is not so remarkable that ε should have a value close to 1 at present. However, from an FLRW perspective, the value of Λ could really have been anything—and if it were only 10^4 larger than it is (which is indeed still far less than our theoretical predictions), ε would have dropped below 1 % already at $\tau = 0.3$ billion years, and the Universe should now, at 13.8 billion years, be nearly devoid of observable galaxies (so we would have trouble detecting Λ's presence). On the other hand, if Λ were smaller by 10^4, it would only become detectable after 100 billion years. Therefore, it is indeed a remarkable coincidence that Λ has the particular order of magnitude that it has, which has allowed us to conclusively detect its presence in our Universe.

In contrast to current views, within the FLRW paradigm, on the problem that we know of no good reason why Λ should have the particularly special, very small value that it has—which has often led to controversial discussions involving the anthropic principle and a multiverse setting—the SdS cosmology again does not so much 'solve' the problem by explaining why the particular value of Λ should be observed, but really offers a fundamentally different perspective in which the same problem simply does not exist. For indeed, while the mathematical form of the observed scale-factor in the SdS cosmology should equal that of a flat ΛCDM universe, the energy densities described in Friedmann's equations are only effective parameters within the former framework, and really of no essential importance. And in fact, as the analysis in the first part of this Appendix indicates (and again, cf. the sections by Eddington [4] and Dyson [43]), Λ fundamentally sets the scale in the SdS universe: from this perspective, $\sqrt{3/\Lambda}$, which has an empirical value on the order of 10 billion years, *is* the fundamental timescale. It therefore makes little sense to question what effect different values of Λ would have on the evolution of an SdS universe, since Λ sets the scale of everything a priori; thus, the observable universe should rescale with any change in the value of Λ. But for the same reason, it is interesting to note that the present order of things has arisen on roughly this characteristic timescale. From this different point-of-view, then, the more relevant question to ponder is: Why should the structure of the subatomic world be such that when particles did coalesce to form atoms and stars, those stars evolved to produce the atoms required to form life-sustaining systems, etc., all on roughly this characteristic timescale?

Despite the SdS cosmology's many attractive features, it may still be objected that the specific geometrical structure of the SdS model entails a significant assumption on the fundamental geometry of physical reality, for which there should be a reason; i.e., the question arises: if the geometry is not determined by the world-matter, then by what? While a detailed answer to this question has not been worked out (although, see Sect. 3.3 in [40]), it is relevant to note that the local form of the SdS solution— which is only parametrically different from the cosmological form upon which our analysis has been based; i.e. $\Lambda M^2 < 1/9$ rather than $\Lambda M^2 > 1/9$—is the space-time description outside a spherically symmetric, uncharged black hole, which is exactly the type that is expected to result from the eventual collapse of every massive cluster of galaxies in our Universe—even if it takes all of cosmic time for that collapse to finally occur. In fact, there seems to be particular promise in this direction, given that the singularity at $r = 0$ in the SdS cosmology is not a real physical singularity, but the artifact of a derivative metric that must be ill-defined there, since space must actually always have a finite radius according to the fundamental metric, Eq. (7.32). This is the potential advantage that was noted above, of the finite minimum radius of the foliation of de Sitter space defined in Eq. (7.3). And as far as that goes, it should be noted that the Penrose-Hawking singularity theorem 'cannot be directly applied when a *positive cosmological constant* λ is present' [55], which is indeed our case. For all these reasons, we might realistically expect a description in which gravitational collapse leads to universal birth, and thus an explanation of the Big Bang and the basic cosmic structure we've had to assume.

Along with such new possibilities as an updated description of collapse, and of gravitation in general, that may be explored in a relativistic context when the absolute background structure of cosmology is objectively assumed, the SdS cosmology, through its specific requirement that the observed rate of expansion *should be* described exactly by the flat ΛCDM scale-factor, has the distinct possibility to *explain* why our Universe should have expanded as it evidently has—and therein lies its greatest advantage.

References

1. A. Einstein, Ernst Mach. Phys. Z. **17**, 101–104 (1916)
2. A. Einstein, Maxwell's influence on the evolution of the idea of physical reality, in *James Clerk Maxwell: A Commemoration Volume*, ed. by J.J. Thomson (Cambridge University Press, 1931)
3. W. de Sitter, On Einstein's theory of gravitation, and its astronomical consequences. Third paper. Mon. Not. R. Astron. Soc. **78**, 3–28 (1917)
4. A. Eddington, *The Mathematical Theory of Relativity*, 2nd edn. (Cambridge University Press, 1923)
5. H. Weyl, Zur allgemeinen relativitätstheorie. Phys. Z. **24**, 230–232 (1923) (English translation: H. Weyl, Republication of: On the general relativity theory. Gen. Relativ. Gravitat. 35 1661–1666 (2009))
6. G. Lemaître, Note on de Sitter's universe. J. Math. Phys. **4**, 188–192 (1925)
7. H.P. Robertson, On relativistic cosmology. Philos. Mag. **5**, 835–848 (1928)
8. A. Einstein, Notiz zu der arbeit von A. Friedmann "Über die krümmung des raumes". Z. Phys. **16**, 228–228 (1923)
9. H. Weyl, Massenträgheit und kosmos. Naturwissenschaften. **12**, 197–204 (1924)
10. E. Hubble, A relation between distance and radial velocity among extra-galactic Nebulae. Proc. Nal. Acad. Sci. **15**, 168–173 (1929)
11. W. de Sitter, On the distances and radial velocities of extra-galactic nebulae, and the explanation of the latter by the relativity theory of inertia. Proc. Nal. Acad. Sci. **16**, 474–488 (1930)
12. A. Einstein, W. de Sitter, On the relation between the expansion and the mean density of the universe. Proc. Nal. Acad. Sci. **18**(3), 213–214 (1932)
13. A. Eddington, *The Expanding Universe* (Cambridge University Press, Cambridge, 1933)
14. A. Eddington, On the instability of Einstein's spherical world. Mon. Not. R. Astron. Soc. **90**, 668–678 (1930)
15. A.G. Riess et al., Observational evidence from supernovae for an accelerating universe and a cosmological constant. Astronom. J. **116**, 1009–1038 (1998)
16. S. Perlmutter et al., Measurements of Ω and Λ from 42 high-redshift supernovae. Astrophys. J. **517**, 565–586 (1999)
17. A.G. Riess et al., Type Ia supernovae discoveries at $z > 1$ from the Hubble Space Telescope: evidence for past deceleration and constraints on dark energy evolution. Astrophys. J. **607**, 665–687 (2004)
18. A.G. Riess et al., New Hubble Space Telescope discoveries of type Ia supernovae at $z \geq 1$: narrowing constraints on the early behavior of dark energy. Astrophys. J. **659**, 98–121 (2007)
19. W.M. Wood-Vasey et al., Observational constraints on the nature of dark energy: first cosmological results from the ESSENCE supernova survey. Astrophys. J. **666**, 694–715 (2007)
20. T.M. Davis et al., Scrutinizing exotic cosmological models using ESSENCE supernova data combined with other cosmological probes. Astrophys. J. **666**, 716–725 (2007)
21. M. Kowalski et al., Improved cosmological constraints from new, old, and combined supernova data sets. Astrophys. J. **686**, 749–778 (2008)
22. M. Hicken et al., Improved dark energy constraints from ~100 new CfA supernova type Ia light curves. Astrophys. J. **700**, 1097–1140 (2009)
23. A.G. Riess et al., A 3% solution: determination of the Hubble constant with the Hubble Space Telescope and Wide Field Camera 3. Astroph. J. **730**(119), 1–18 (2011)
24. N. Suzuki et al., The Hubble Space Telescope cluster supernova survey. V. Improving the dark-energy constraints above $z > 1$ and building an early-type-hosted supernova sample. Astrophys. J. **746**(85), 1–24 (2012)
25. D. Hinshaw et al., Nine-year Wilkinson microwave anisotropy probe (WMAP) observations: cosmological parameter results. Astrophys. J. Suppl. Ser. **208**(19), 1–25 (2013)
26. Planck Collaboration (P. A. R. Ade et al.), Planck 2013 results. XVI. Cosmological parameters (2013), ArXiv:1303.5076 [astro-ph.CO]
27. H. Bondi, *Cosmology* (Cambridge University Press, Cambridge, 1960)

28. A. Einstein, Kosmologische betrachtungen zur allgemeinen relativitätstheorie. Sitzungsber. K. Preuss. Akad. Wiss. pp. 142–152 (1917) (English translation in H.A. Lorentz et al., (eds.) The principle of relativity, pp. 175–188. Dover Publications (1952))

29. W. de Sitter, On the relativity of inertia. Remarks concerning Einstein's latest hypothesis. Proc. R. Acad. Amst. **19**, 1217–1225 (1917)

30. A.S. Eddington, *Space, Time, and Gravitation* (Cambridge University Press, Cambridge, 1920)

31. A. Einstein, Zum kosmologischen problem der allgemeinen relativitätstheorie. Sitzungsber. Preuss. Akad. Wiss. 235–237 (1931)

32. A. Einstein, *The Meaning of Relativity*, 2nd edn. (Princeton University Press, New Jersey, 1945)

33. A. Einstein, Zur Elektrodynamik bewegter Körper. Ann. der Phys. 17 (1905) (English translation in H.A. Lorentz et al., (eds.) The principle of relativity, pp. 35–65. Dover Publications (1952))

34. H. Putnam, Time and physical geometry. J. Philos. **64**, 240–247 (1967)

35. H. Minkowski, Space and time, in *The Principle of Relativity*, ed. by H.A. Lorentz, et al. (Dover Publications, New York, 1952), pp. 73–91

36. A. Fölsing, E. Osers, trans. *Albert Einstein: A Biography* (Viking, New York, 1997)

37. H. Weyl, *Philosophy of Mathematics and Natural Science* (Princeton University Press, New Jersey, 1949)

38. R. Geroch, *General Relativity From A to B* (University of Chicago Press, Chicago, 1978)

39. S.E. Rugh, H. Zinkernagel, Weyl's principle, cosmic time and quantum fundamentalism (2010), ArXiv:1006.5848 [gr-qc]

40. D. Janzen, A solution to the cosmological problem of relativity theory. Dissertation, University of Saskatchewan 2012, http://hdl.handle.net/10388/ETD-2012-03-384

41. G. Lemaître, Cosmological Application of Relativity. Rev. Mod. Phys. **21**, 357–366 (1949)

42. C.W. Misner, K.S. Thorne, J.A. Wheeler, *Gravitation*(W. H. Freeman and Company, 1973)

43. F.J. Dyson, Missed opportunities. Bull. Am. Math. Soc. **78**(5), 635–652 (1972)

44. T. Maudlin, *Philosophy of Physics: Space and Time* (Princeton University Press, New Jersey, 2012)

45. B. Greene, *The Fabric of the Cosmos* (Vintage Books, New York, 2004)

46. D. Janzen, Time is the denominator of existence, and bits come to be in it. Submitted to FQXi's 2013 Essay Contest, It From Bit or Bit From It? (2013)

47. S. Weinberg, The cosmological constant problem. Rev. Mod. Phys. **61**(1), 1–23 (1989)

48. A.H. Guth, Inflationary universe: A possible solution to the horizon and flatness problems. Phys. Rev. D **23**(2), 347–356 (1981)

49. C.P. Burgess, The cosmological constant problem: why it's hard to get dark energy from microphysics (2013), ArXiv: 1309.4133 [hep-th]

50. BICEP2 Collaboration (Ade, P. A. R., et al.): Planck 2013 results. XVI. Cosmological parameters (2013), ArXiv:1303.5076 [astro-ph.CO]

51. L.P. Grishchuck, Amplification of gravitational waves in an isotropic universe. Zh. Eksp. Theor. Fiz. **67**, 825–838 (1974)

52. A.A. Starobinsky, Spectrum of relict gravitational radiation and the early state of the universe. JETP Lett. **30**, 682–685 (1979)

53. V.A. Rubakov, M.V. Sazhin, A.V. Veryaskin, Graviton creation in the inflationary universe and the grand unification scale. Phys. Lett. **115B**(3), 189–192 (1982)

54. T. Commissariat, Neil Turok urges caution on BICEP2 results (2014), http://physicsworld.com/cws/article/news/2014/mar/18/neil-turok-urges-caution-on-bicep2-results. Accessed 18 March 2014

55. S.W. Hawking, R. Penrose, The singularities of gravitational collapse and cosmology. Proc. R. Soc. Londo. A, Math. Phys. Sci. **314** (1519) 529–548 (1970)

Chapter 8
Not on but of

Olaf Dreyer

Abstract In physics we encounter particles in one of two ways. Either as fundamental constituents of the theory or as emergent excitations. These two ways differ by how the particle relates to the background. It either sits *on* the background, or it is an excitation *of* the background. We argue that by choosing the former to construct our fundamental theories we have made a costly mistake. Instead we should think of particles as excitations of a background. We show that this point of view sheds new light on the cosmological constant problem and even leads to observable consequences by giving a natural explanation for the appearance of MOND-like behaviour. In this context it also becomes clear why there are numerical coincidences between the MOND acceleration parameter a_0, the cosmological constant Λ and the Hubble parameter H_0.

Which of Our Basic Physical Assumptions Are Wrong?

In theoretical physics we encounter particles in two different ways. We either encounter them as fundamental constituents of a theory or as emergent entities. A good example of the first kind is a scalar field. Its dynamics is given by the Lagrangian

$$\int d^4x \left(\frac{1}{2}(\partial\phi)^2 - m^2\phi^2 + \cdots \right).$$

(8.1)

An example of an emergent particle is a spin-wave in a spin-chain. It given by

$$|k\rangle = \sum_{n=1}^{N} \exp\left(2\pi i \frac{nk}{N} \right) |n\rangle$$

(8.2)

If $|0\rangle$ is the ground state of the spin-chain then $|n\rangle$ is the state where spin n is flipped.

O. Dreyer (✉)
Dipartimento di Fisica, Università di Roma "La Sapienza" and Sez. Roma1 INFN,
P.le A. Moro 2, 00185 Rome, Italy
e-mail: olaf.dreyer@gmail.com

© Springer International Publishing Switzerland 2015
A. Aguirre et al. (eds.), *Questioning the Foundations of Physics*,
The Frontiers Collection, DOI 10.1007/978-3-319-13045-3_8

Of all the differences between these two concepts of particles we want to stress one in particular: the relation these particles have to their respective backgrounds.

The scalar field is formulated as a field *on* spacetime. The above Lagrangian does not include the metric. If we include the metric we obtain

$$\int d^4x \sqrt{-g} \left(\frac{1}{2} g^{ab} \partial_a \phi \partial_b \phi - m^2 \phi^2 + \cdots \right). \tag{8.3}$$

This way the scalar field knows about a non-trivial spacetime. Einstein told us that this is not a one-way street. The presence of a scalar field in turn changes the background. The proper equations that describe this interaction between matter and background are obtained by adding the curvature tensor to the above Lagrangian. This interaction does not change the basic fact about the scalar field that it is sitting *on* spacetime. It is distinct from the spacetime that it sits on.

This is in contrast to the emergent particle. There is no clean separation possible between the particle and the background, i.e. the ground state in this case. The spin-wave above is an excitation *of* the background not an excitation *on* the background.

We can now state what basic assumption we think needs changing. Currently we built our fundamental theories assuming the matter-on-background model. We will argue here that this assumption is wrong and that instead all matter is of the emergent type and that the excitation-of-background model applies.

Our basic assumption that is wrong: Matter does not sit *on* a background but is an excitation *of* the background.

In the following we will do two things. First we will show that the assumption that matter sits on a background creates problems and how our new assumption avoids these problems. Then we will argue that the new assumption has observable consequences by showing how MOND like behavior arises naturally.

The Cosmological Constant

The basic assumption that matter sits on a background directly leads to one of the thorniest problems of theoretical physics: the cosmological constant problem. All matter is described by quantum fields which can be thought of as a collection of quantum harmonic oscillators. One feature of the spectrum of the harmonic oscillator is that it has a non-vanishing ground state energy of $\hbar\omega/2$. Because the quantum field sits on spacetime the ground state energy of all the harmonic oscillators making up the field should contribute to the curvature of space. The problem with this reasoning is that it leads to a prediction that is many orders of magnitude off. In fact if one assumes that there is some large frequency limit ω_∞ for the quantum field then the energy density coming from this ground state energy is proportional to the fourth power of this frequency:

$$\epsilon \simeq \omega_\infty^4 \tag{8.4}$$

If one chooses ω_∞ to be the Planck frequency then one obtains a value that is 123 orders of magnitude larger than the observed value of the cosmological constant. This has been called the worst prediction of theoretical physics and constitutes one part of the cosmological constant problem (see [1, 2] for more details).

Let us emphasize again how crucial the basic assumption of matter-on-background is for the formulation of the cosmological constant problem. Because we have separated matter from the background we have to consider the contributions coming from the ground state energy. We have to do this even when the matter is in its ground state, i.e. when no particles are present. This is to be contrasted with the excitation-of-background model. If there are no waves present there is also no ground state energy for the excitations to consider. The cosmological constant problem can not be formulated in this model.

There are two objections to this reasoning that we have to deal with. The first objection is that although the above reasoning is correct it is also not that interesting because there is no gravity in the picture yet. We will deal with this objection in the next section. The other objection concerns the ontological status of the vacuum fluctuations. Isn't it true that we have observed the Casimir effect? Since the Casimir effect is based on vacuum fluctuations the cosmological constant problem is not really a problem that is rooted in any fundamental assumption but in observational facts. For its formulation no theoretical foundation needs to be mentioned. We will deal with this objection now.

There are two complementary views of the Casimir effect and the reality of vacuum fluctuations. In the more well known explanation of the Casimir force between two conducting plates the presence of vacuum fluctuations is assumed. The Casimir force then arises because only a discrete set of frequencies is admissible between the plates whereas no such restriction exists on the outside. The Casimir effect between two parallel plates has been observed and this has led to claims that we have indeed seen vacuum fluctuations. This claim is not quite true because there is another way to think about the Casimir effect. In this view vacuum fluctuations play no role. Instead, the Casimir effect results from forces between the currents and charges in the plates. No vacuum diagrams contribute to the total force [3]. If the emergent matter is described by the correct theory, say quantum electrodynamics, we will find the Casimir effect even if there are no vacuum fluctuations.

We see that the cosmological constant problem can be seen as a consequence of us viewing matter as sitting on a background. If we drop this assumption we can not even formulate the cosmological constant problem.

Gravity

The above argument for viewing matter as an excitation of a background is only useful if we can include gravity in the picture. In [4] we have argued that this can be achieved by regarding the ground state itself as the variable quantity that is responsible for gravity. In the simplest case the vacuum is described by a scalar quantity θ. If we assume that the energy of the vacuum is given by

$$E = \frac{1}{8\pi} \int d^3x \, (\nabla\theta)^2, \tag{8.5}$$

then we can calculate the force between two objects m_i, $i = 1, 2$. If we introduce the gravitational mass of an object by

$$\mathsf{m} = \frac{1}{4\pi} \int\limits_{\partial\mathsf{m}} d\sigma \cdot \nabla\theta, \tag{8.6}$$

where $\partial\mathsf{m}$ is the boundary of the object m, then the force between them is given by

$$F = \frac{\mathsf{m}_1\mathsf{m}_2}{r^2}. \tag{8.7}$$

Usually we express Newton's law of gravitation not in terms of the gravitational masses m_i, $i = 1, 2$, but in terms of the inertial masses m_i, $i = 1, 2$. In [4] we have argued that the inertial mass of an object is given by

$$m = \frac{2\mathsf{m}}{3a} \, \mathsf{m}, \tag{8.8}$$

where a is the radius of the object. In terms of the inertial masses m_i Newton's law then takes the usual form

$$F = G\frac{m_1 m_2}{r^2}, \tag{8.9}$$

where G has to be calculated from the fundamental theory:

$$G = \left(\frac{3a}{2\mathsf{m}}\right)^2 \tag{8.10}$$

In this picture of gravity the metric does not play a fundamental role. Gravity appears because the ground state θ depends on the matter.

MOND as a Consequence

The picture of gravity that we have given in the last section is valid only for zero temperature. If the temperature is not zero we need to take the effects of entropy into account and instead of looking at the energy we have to look at the free energy

$$F = E - TS. \tag{8.11}$$

We thus have to determine the dependence of the entropy S on the temperature T and the ground state θ. The entropy should not depend on θ directly because every θ corresponds to the same ground state. The entropy should only be dependent on changes of θ. If we are only interested in small values of T we find

$$S = \sigma T (\nabla \theta)^2, \tag{8.12}$$

for some constant σ. The total free energy is thus

$$F = E - TS \tag{8.13}$$
$$= E(1 - 8\pi\sigma T^2). \tag{8.14}$$

We see that a non-zero temperature does not change the form of the force but just its strength. The new gravitational constant is given by

$$G_T = \frac{1}{1 - \sigma T^2} G_{T=0}. \tag{8.15}$$

The situation changes in an interesting way if there is a large maximal length scale $L_{\max}$ present in the problem. The contributions to the entropy of the form $(\nabla\phi)^2$ come from excitations with a wavelength of the order

$$L = |\nabla\phi|^{-1}. \tag{8.16}$$

If this wavelength L is larger than $L_{\max}$ than these excitations should not exist and thus not contribute to the entropy. Instead of a simple $(\nabla\phi)^2$ term we should thus have a term of the form

$$C(L_{\max} |\nabla\theta|) \cdot (\nabla\theta)^2, \tag{8.17}$$

where the function C is such that it suppresses contributions from excitations with wavelengths larger than $L_{\max}$. For wavelengths much smaller than the maximal wavelength we want to recover the usual contributions to the entropy. Thus, if $L_{\max} \cdot |\nabla\theta|$ is much bigger than unity we want C to be one:

$$C(x) = 1, \quad \text{for} \quad x \gg 1. \tag{8.18}$$

For $x \ll 1$ we assume that the function C possesses a series expansion of the form

$$C(x) = \alpha x + \beta x^2 + \ldots. \tag{8.19}$$

For small values of $L_{\max} \cdot |\nabla\theta|$ we thus find that the dependence of the entropy on ϕ is of the form

$$T^2 \sigma \int d^4 x \, \alpha L_{\max} |\nabla\phi|^3. \tag{8.20}$$

It is here that we make contact with the Lagrangian formulation of Milgrom's odd new dynamics (or MOND, see [5, 6] for more details). In [7] Bekenstein and Milgrom have shown that a Lagrangian of the form

$$\int d^3x \left(\rho\theta + \frac{1}{8\pi G} a_0^2 F\left[\frac{(\nabla\theta)^2}{a_0^2} \right] \right) \tag{8.21}$$

gives rise to MOND like dynamics if the function F is chosen such that

$$\mu(x) = F'(x^2). \tag{8.22}$$

Here $\mu(x)$ is the function that determines the transition from the classical Newtonian regime to the MOND regime. It satisfies

$$\mu(x) = \begin{cases} 1 & x \gg 1 \\ x & x \ll 1 \end{cases} \tag{8.23}$$

From this it follows that the function F satisfies

$$F(x^2) = \begin{cases} x^2 & x \gg 1 \\ 2/3\, x^3 & x \ll 1 \end{cases} \tag{8.24}$$

The behavior of the Lagrangian is then

$$a_0^2 F\left(\frac{|\nabla\theta|^2}{a_0^2} \right) = \begin{cases} |\nabla\theta|^2 & a_0^{-1}|\nabla\theta| \gg 1 \\ \frac{2}{3a_0}|\nabla\theta|^3 & a_0^{-1}|\nabla\theta| \ll 1 \end{cases} \tag{8.25}$$

This is exactly the behavior of the free energy that we have just derived if we make the identification

$$\frac{2}{3a_0} = \alpha\sigma T^2 L_{\max}. \tag{8.26}$$

There are currently two candidates for a maximal length scale $L_{\max}$. These are the Hubble scale

$$L_H = cH_0 \tag{8.27}$$

and the cosmic horizon scale

$$L_\Lambda = \sqrt{\frac{1}{\Lambda}}. \tag{8.28}$$

It is a remarkable fact of the universe that we live in that *both* of these length scales satisfy the relationship that we derived in (8.26) if we further assume that the constant $\alpha\sigma T^2$ is of order one. We have thus established a connection between the acceleration parameter a_0, the cosmological constant Λ, and the Hubble parameter H_0. In standard cosmology these coincidences remain complete mysteries.

Discussion

Particles are either fundamental or they are emergent. If they are fundamental they are sitting on a background; if they are emergent they are excitations of a background. Rather than being a purely philosophical issue we have argued that this distinction is important and that the assumption that particles are fundamental is wrong. Assuming instead that particles are emergent leads to the resolution of theoretical problems as well as having observational consequences. We have argued that the cosmological constant problem as it is usually formulated can not even be stated if we think of particles as excitations of a background. Also, we have shown that this picture gives a straight forward way of understanding the appearance of MOND like behavior in gravity. The argument also makes clear why there are numerical relations between the MOND parameter a_0, the cosmological constant Λ, and the Hubble parameter H_0.

Our derivation of MOND is inspired by recent work [8–17] that uses Verlinde's derivation of Newton's law of gravity [18] as a starting point. Our derivation differs from these in that it does not rely on holography in any way. Our formulae for the entropy are all completely three-dimensional.

Are there other possible consequences of this point of view that might be observable? In [19] we have argued that in fact there are. The backgrounds that the particles are excitations of appear in nature as the result of phase transitions. Was our cosmological background also a result of such a phase transition? If so, what would be the consequence? One feature that distinguishes phase transitions is the disappearance of scales. This manifests itself in the appearance of power laws that describe the system. Absence of scale and power laws have made an appearance in cosmology lately. The mechanism of inflation produces a scale-free spectrum of metric perturbations in the early universe that has been observed recently. In [19] we argue that this is no coincidence. In fact we argue that it is precisely such a spectrum that would be expected from a background-producing phase transition in the early universe. Furthermore the precise tilt of the spectrum is related to the anomalous scaling of the order parameter and thus to the fundamental discreteness of nature.

We thus see that the point of view advocated here is not just of philosophical interest. We shed light on the cosmological constant problem, find a connection with MOND phenomenology, and provide a simple alternative to inflation.

Acknowledgments I would like to thank Mamazim Kassankogno, and Stevi Roussel Tankio Djiokap for helpful discussion and the Foundational Questions Institute, FQXi, for financial support and for creating an environment where it is alright to play with unorthodox ideas.

References

1. S. Weinberg, The cosmological constant problem. Rev. Mod. Phys. **61**(1), 1–23 (1989)
2. S. Carroll, The cosmological constant. Living Rev. Relativ. **4**, 80 (2001)
3. R.L. Jaffe, Casimir effect and the quantum vacuum. Phys. Rev. D **72**(2), 5 (2005)
4. O. Dreyer, Internal relativity (2012). arXiv:1203.2641

5. M. Milgrom, A modification of the Newtonian dynamics as a possible alternative to the hidden mass hypothesis. Astrophys. J. **270**, 365–370 (1983)
6. B. Famaey, S. McGaugh, Modified Newtonian dynamics (MOND): observational phenomenology and relativistic extensions (2011). Arxiv preprint: arXiv:1112.3960v2
7. J. Bekenstein, M. Milgrom, Does the missing mass problem signal the breakdown of Newtonian gravity? Astrophys. J. **286**, 7–14 (1984)
8. C.M. Ho et al., Quantum gravity and dark matter. Gen. Relativ. Gravit. **43**(10), 2567–2573 (2011)
9. C.M. Ho et al., Cold dark matter with MOND scaling. Phys. Lett. B **693**(5), 567–570 (2010)
10. P.V. Pikhitsa, MOND reveals the thermodynamics of gravity (2010). Arxiv preprint: arXiv:1010.0318
11. F.R. Klinkhamer, Entropic-gravity derivation of MOND. Mod. Phys. Lett. A **27**, 6 (2012)
12. F.R. Klinkhamer, M. Kopp, Entropic gravity, minimum temperature, and modified Newtonian dynamics (2011). Arxiv preprint: arXiv:1104.2022v6
13. X. Li, Z. Chang, Debye entropic force and modified Newtonian dynamics. Commun. Theor. Phys. **55**, 733–736 (2011)
14. J.A. Neto, Nonhomogeneous cooling, entropic gravity and MOND theory. Int. J. Theor. Phys. **50**, 3552–3559 (2011)
15. V.V. Kiselev, S.A. Timofeev, The holographic screen at low temperatures (2010). Arxiv preprint: arXiv:1009.1301v2
16. E. Pazy, N. Argaman, Quantum particle statistics on the holographic screen leads to modified Newtonian dynamics. Phys. Rev. D **85**(10), 104021 (2012)
17. L. Modesto, A. Randono, Entropic corrections to Newton's law (2010). Arxiv preprint: arXiv:1003.1998v1
18. E. Verlinde, On the origin of gravity and the laws of Newton. JHEP **1104**, 029 (2011)
19. O. Dreyer, The world is discrete (2013). arXiv:1307.6169

Chapter 9
Patterns in the Fabric of Nature

Steven Weinstein

Abstract From classical mechanics to quantum field theory, the physical facts at one point in space are held to be independent of those at other points in space. I propose that we can usefully challenge this orthodoxy in order to explain otherwise puzzling correlations at both cosmological and microscopic scales.

> Nature uses only the longest threads to weave her patterns, so that each small piece of her fabric reveals the organization of the entire tapestry. (*The Character of Physical Law*, Richard Feynman)

Introduction

Despite radical differences in their conceptions of space, time, and the nature of matter, all of the physical theories we presently use—non-relativistic and relativistic, classical and quantum—share one assumption: the features of the world at distinct points in space are understood to be independent. Particles may exist anywhere, independent of the location or velocity of other particles. Classical fields may take on any value at a given point, constrained only by local constraints like Gauss's law. Quantum field theories incorporate the same independence in their demand that field operators at distinct points in space commute with one another.

The independence of physical properties at distinct points is a theoretical assumption, albeit one that is grounded in our everyday experience. We appear to be able to manipulate the contents of a given region of space unrestricted by the contents of other regions. We can arrange the desk in our office without concern for the location of the couch at home in our living room.

S. Weinstein (✉)
Perimeter Institute for Theoretical Physics, 31 Caroline St N,
Waterloo, ON N2L 2Y5, Canada
e-mail: sw@uwaterloo.ca; sweinstein@perimeterinstitute.ca

S. Weinstein
U. Waterloo Department of Philosophy, 200 University Ave West,
Waterloo, ON N2L 3G1, Canada

© Springer International Publishing Switzerland 2015
A. Aguirre et al. (eds.), *Questioning the Foundations of Physics*,
The Frontiers Collection, DOI 10.1007/978-3-319-13045-3_9

Yet there are realms of physical theory, more remote from everyday experience and physical manipulation yet accessible to observation, in which there appear to be striking correlations between the values of physical properties at different points in space. Quantum theory predicts (and experiment confirms) the existence of strongly correlated measurement outcomes apparently inexplicable by classical means. I refer, of course, to the measurements of the properties of pairs of particles originally envisioned by Einstein, Podolsky and Rosen (EPR) [7], measurements that suggested to EPR the incompleteness of the theory. Bell [2] showed that no theory satisfying two seemingly natural conditions could possibly account for these correlations. The condition known variously as *Bell locality, strong locality* or *factorizability* has been endlessly analyzed. It is violated by quantum mechanics, as well as alternative theories such as the deBroglie-Bohm theory. The other condition, *statistical independence* (a.k.a. *measurement independence* or *setting Independence*), has only rarely been called into question. A theory that violates statistical independence is one in which contains non-local degrees of freedom, which is to say that the properties of a physical system at one point may be constrained in a lawlike manner by the properties at other points which are not in causal contact with the first point.

On a completely different scale, the electromagnetic radiation that pervaded the early universe—the remnants of which form the cosmic microwave background—appears to have been extraordinarily homogeneous. It is strikingly uniform, yet the theories that describe the early universe—classical electrodynamics (for the radiation) and general relativity (for the expanding spacetime the radiation fills)—do not stipulate any sort of restrictions or correlations that would go anywhere near explaining this. To the extent that they have been explained at all, it has been through the postulation of an as-yet unobserved field known as the inflaton field.

What I want to do here is raise the possibility that there is a more fundamental theory possessing nonlocal constraints that underlies our current theories. Such a theory might account for the mysterious nonlocal effects currently described, but not explained, by quantum mechanics, and might additionally reduce the extent to which cosmological models depend on finely tuned initial data to explain the large scale correlations we observe. The assumption that spatially separated physical systems are entirely uncorrelated is a parochial assumption borne of our experience with the everyday objects described by classical mechanics. Why not suppose that at certain scales or certain epochs, this independence emerges from what is otherwise a highly structured, nonlocally correlated microphysics?

Nonlocal Constraints

All physical theories in current use assume that the properties of physical systems at different points in space are independent. Correlations may emerge dynamically—many liquids crystallize and develop a preferred orientation when cooled, for example—but the fundamental theories permit any configuration as an initial condition.

For example, consider the straightforward and simple theory of the free massless scalar field $\phi(\vec{x})$. A scalar field is simply an assignment of a single number (a "scalar" rather than a vector) to every point in space and time. The evolution of the field is given by the well-known wave equation

$$\frac{\partial^2 \phi(\vec{x}, t)}{\partial t^2} = c^2 \nabla^2 \phi(\vec{x}, t) ,$$

in conjunction with initial data $\phi(\vec{x})$ and $\partial \phi(\vec{x})/\partial t$ giving the value of the field and its rate of change at some initial time. This initial data can be specified arbitrarily—it is unconstrained.

A more realistic field theory is the classical electrodynamics of Maxwell, which *does* feature constraints. In Maxwell's theory, we have a pair of coevolving fields, the electric field $\vec{E}$ and the magnetic field $\vec{B}$. The fields are described by vectors at each point rather than scalars. The significant difference between the electromagnetic field and the free scalar field is that the electric and magnetic fields may not be specified arbitrarily. They are subject to constraints $\nabla \cdot \vec{E}(\vec{x}) = 4\pi\rho(\vec{x})$ and $\nabla \cdot \vec{B}(\vec{x}) = 0$ which hold at every point $\vec{x}$ in space. The divergence of the electric field at any given point must be equal to a multiple of the charge density at that point, and the divergence of the magnetic field must be zero. The divergence is a measure of the outflow of the field in the neighborhood of a point, and the two constraints tell us respectively that any such outflow of the electric field is due to the presence of a charge at that point acting as a source, while the magnetic field can have no sources (there are no magnetic charges). These constraints are *local* in that they provide a constraint on values of the field at each point that does not depend on values of the field or the charge distribution at other points.

What would a nonlocal constraint look like? Here's a candidate: $\nabla \cdot \vec{E}(\vec{x}) = 4\pi\rho(\vec{x} - (1, 1, 1))$. This says that the divergence of the electric field at one point is equal to a constant times the charge density at a point which is one unit away in the x, y and z directions. But this constraint is hardly worthy of the name, since it only holds at a single time: unlike the constraint $\nabla \cdot E(\vec{x}) = 4\pi\rho(\vec{x})$, it is not preserved by the equations of motion (Maxwell's equations for the field and the Lorentz force law for the charge distribution). I.e., it will not continue to hold as the field evolves. Since it is not preserved, it does not hold at arbitrary moments of time, hence it is not a true regularity or law.

Let's return to simple classical particle mechanics for an example of a true nonlocal constraint, one that is conserved in time. The particles are characterized by their positions and their momenta. The constraint we will impose is that the total momentum (the sum of the momenta of each of the particles) is zero. This is a constraint because we cannot specify the momentum freely for each particle; if we know the momentum of all but one of the particles, the momentum of the other particle is fixed. It is nonlocal, because the momentum of that particle is a function of the momenta of particles some distance away. Unlike the first nonlocal constraint we considered, it is conserved, since total momentum is a conserved quantity in particle mechanics.

Fig. 9.1 Timelike
compactification

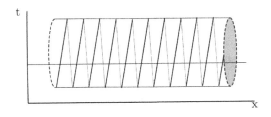

But it is not a particularly interesting constraint, because all but one of the momenta may be freely specified. Whereas the two constraints in electromagnetism reduce the number of degrees of freedom from 6 to 4 at each point in space (so that there are only two-thirds the number of degrees of freedom), this constraint only reduces the total number of degrees of freedom by one.

A more interesting nonlocal constraint may be obtained by considering once more the wave equation, this time in one space dimension (for simplicity). Suppose that the spacetime on which the field takes values is compactified in the time direction, so that the entirety forms a cylinder (see Fig. 9.1).

The solutions must clearly be periodic, and this amounts to imposing a nonlocal constraint. More specifically, whereas in the ordinary initial value problem, initial data may be any smooth functions $\phi(x, 0)$ and $\phi_t(x, 0)$ (where ϕ_t stands for $\partial\phi/\partial t$), we now require that $\phi(x, 0) = \phi(x, T)$ and $\phi_t(x, 0) = \phi_t(x, T)$, where T is the circumference of the cylinder. This is just to say that the time evolution from 0 to T must return us to the same starting point. What are the constraints, then, on this initial data? They are essentially those data that can be written as sums of sine or cosine waves with wavelength $\frac{T}{2\pi n}$ (for any integer value of n).[1]

The restriction to a discrete (though infinite) set of plane waves means that initial data do not have compact support; they are periodic. However, for sufficiently small

---

[1] Solutions to the wave equation can be written as sums of plane waves, with Fourier space representation $\hat{\phi}(k, t) = \hat{F}(k)e^{-ikt} + \hat{G}(k)e^{ikt}$. Since these plane waves must have period T (in the preferred frame dictated by the cylinder), we have a constraint $k = \frac{2\pi n}{T}$ (where n is a positive or negative integer), so that initial data are no longer arbitrary smooth functions of k

$$\hat{\phi}(k, 0) = \hat{F}(k) + \hat{G}(k)$$
$$\hat{\phi}_t(k, 0) = -ik(\hat{F}(k) - \hat{G}(k))$$

but are rather constrained by the requirement $k = \frac{2\pi n}{T}$. Thus the initial data are the functions

$$\phi(x, 0) = \frac{1}{\sqrt{T}} \sum_{n=-\infty}^{\infty} \hat{\phi}(\frac{2\pi n}{T}, 0)e^{i\frac{2\pi n}{T}x} dk$$

$$\phi_t(x, 0) = \frac{1}{\sqrt{T}} \sum_{n=-\infty}^{\infty} \hat{\phi}_t(\frac{2\pi n}{T}, 0)e^{i\frac{2\pi n}{T}x} dk$$

i.e., they consist of arbitrary sums of plane waves with wave number $k = \frac{2\pi n}{T}$, for any integer value of n.

Δx, the local physics is indistinguishable from the local physics in ordinary space-time. Only at distance scales on the order of T does the compact nature of the time direction become evident in the repetition of the spatial structure. Thus we have here an example of a nonlocal constraint which can give the appearance of unconstrained local degrees of freedom.

Now, this spacetime obviously has closed timelike curves, and it is interesting to note that under such conditions, classical computers are as powerful as quantum computers [1]. Thus there is some reason to think that a nonlocal constraint might allow one to mimic other quantum phenomena using classical physics. In any event, we will now proceed to a discussion of the way in which the presence of nonlocal constraints opens the door to a little-explored loophole in Bell's theorem, in that their presence undermines the *statistical independence* assumption required for the proof of the theorem.

Bell's Theorem

Einstein believed quantum theory to be an incomplete description of the world, and he and his collaborators Podolsky and Rosen attempted to show this in their 1935 paper [7]. The argument involves a pair of particles specially prepared in an entangled state of position and momentum.[2] Quantum mechanics makes no definite predictions for the position and momentum of each particle, but does make unequivocal predictions for the position or momentum of one, given (respectively) the position or momentum of the other. EPR argued on this basis that quantum mechanics must be incomplete, since the measured properties of one particle can be predicted with certainty following measurement of the properties of the other, even when these measurements are spatially separated and thus causally independent.[3]

In 1964, John Bell proved a result based on David Bohm's streamlined version of the EPR experiment [2, 4]. Instead of positions and momenta, Bohm focuses on the spins of a pair of particles (in this case fermions). Prepared in what has come to be known as a Bell state,

$$\psi = \frac{1}{\sqrt{2}}(|+x\rangle_A |-x\rangle_B - |-x\rangle_A |+x\rangle_B), \tag{9.1}$$

quantum mechanics predicts that a measurement of the component of spin of particle A in any direction (e.g., the $\hat{z}$ direction) is as likely to yield $+1$ as -1 (in units of $\hbar/2$), and so the average value $\bar{A}$ is 0. However, quantum mechanics also indicates that an outcome of $+1$ for a measurement of the spin of A in the $\hat{z}$ direction is guaranteed to

---

[2] The state used by EPR is an eigenstate of the operators representing the sum of the momenta and the difference of the positions of the two particles.

[3] The argument of the EPR paper is notoriously convoluted, but I follow [9] in regarding this as capturing Einstein's understanding of the core argument.

yield an outcome of -1 for B for a measurement of the spin of B in the $\hat{z}$ direction, etc. This is directly analogous to the correlations between position and momentum measurements in the original EPR experiment.

In and of themselves, these phenomena offer no barrier to a hidden-variable theory, since it is straightforward to explain such correlations by appealing to a common cause—the source—and postulating that the particles emanate from this source in (anti)correlated pairs. However, one must also account for the way that the anticorrelation drops off as the angle between the components of spin for the two particles increases (e.g., as A rotates from $\hat{x}$ toward $\hat{z}$ while B remains oriented along the $\hat{x}$ direction). It was Bell's great insight to note that the quantum theory implies that the anticorrelation is held onto more tightly than could be accounted for by any "local" theory—that is, any theory satisfying the seemingly natural condition known in the literature variously as *strong locality, Bell locality*, or *factorizability*. Bell showed that the predictions of a local theory must satisfy a certain inequality, and that this inequality is violated by quantum theory for appropriate choices of the components of spin to be measured. Bell's result was widely understood to provide a barrier to the sort of "completion" of quantum mechanics considered by Einstein. That is, Einstein's hope for a more fundamental theory underlying quantum theory would have to violate *strong locality*, of which more below.

However, there is a further assumption known as the *statistical independence* assumption (also known as *measurement independence*) that is necessary for Bell's result. This assumption is quite closely related to the assumption of local degrees of freedom, or the absence of nonlocal constraints. Without the assumption, Bell's result does not go through, and the possibility re-emerges of a local completion of quantum theory after all.[4]

Rather than repeat the derivation of Bell's result, let me just focus on the meaning of the two crucial assumptions of *strong locality* and *statistical independence*. The physical situation we are attempting to describe has the following form:

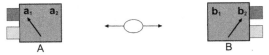

A source (represented by the ellipse) emits a pair of particles, or in some other way causes detectors A and B to simultaneously (in some reference frame) register one of two outcomes. The detectors can be set in one of two different ways, corresponding, in Bohm's version of the EPR experiment, to a measurement of one of two different components of spin.

Let us now suppose that we have a theory that describes possible states of the particles and which gives rise to either probabilistic or deterministic predictions as to the results of various measurements one might make on the particles. The state of the particle will be represented by a parameter λ, describing either a discrete set of states $\lambda_1, \lambda_2 \ldots$ or a continuous set. The expressions $\bar{A}(a, \lambda)$ and $\bar{B}(b, \lambda)$ correspond

---

to the expected (average) values of measurements of properties a and b at detectors A and B (respectively) in a given state λ. (The appeal to *average* values allows for stochastic theories, in which a given λ might give rise to any number of different outcomes, with various probabilities.)

In general, one might suppose that $\bar{A}$ also depends on either the detector setting b or the particular outcome B (i.e., $\bar{A} = \bar{A}(a, \lambda, b, B)$), and one might suppose the same for $\bar{B}$. That it does not, that the expectation value $\bar{A}$ in a given state λ does *not* depend on what one chooses to measure at B, or on the value of the distant outcome B (and vice-versa) is Bell's *strong locality* assumption. Given this assumption, one can write the expression $E(a, b, \lambda)$ for the expected product of the outcomes of measurements of properties a and b in a given state λ as

$$E(a, b, \lambda) = \bar{A}(a, \lambda)\bar{B}(b, \lambda). \tag{9.2}$$

This strong locality is also known as 'factorizability', deriving as it does from the fact that the joint probability of a pair of outcomes can be factorized into the product of the marginal probabilities of each outcome. We can thus represent the analysis of the experimental arrangement in this way, where the expression for $E(a, b, \lambda)$ in the center encodes the assumption of strong locality:

$$E(a_1, b_2, \lambda) = \bar{A}(a_1, \lambda)\bar{B}(b_2, \lambda)$$

Now the further assumption required for Bell's result is that the probability of a given state λ is independent of the detector settings. In other words, Bell assumes that the theory will be one in which

$$P(\lambda|a, b) = P(\lambda). \tag{9.3}$$

This is *statistical independence*. For example, we might suppose that the theory tells us that one of three states λ_1, λ_2, λ_3 will be generated by our particle preparation procedure. The statistical independence condition tells us that the likelihood of obtaining any one of these states is independent of how the detectors will be set at the time of detection. In other words, knowledge of the future settings of the detectors (their settings at the time the particles arrive) does not provide any further information as to which of the three states was emitted.[5]

The assumption of *statistical independence* has been called into question only infrequently, but when it has, the critique has often been motivated by an appeal to the plausibility of Lorentz-invariant "backward causation" (also known as "retro-causation") whereby the change of detector settings gives rise to effects that propagate along or within the backward lightcone and thereby give rise to nontrivial initial

---

[5] Actually, a slightly weaker condition than *SI* is sufficient to derive the CHSH inequality. See [8] and the discussion thereof in Sect. 3.3.1 of [14].

Fig. 9.2 EPR: Spacetime
diagram

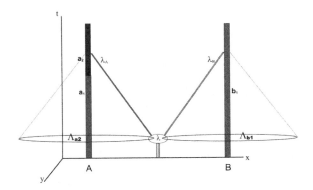

correlations in the particle properties encoded in λ (e.g., [6, 13, 16]). In my [18] I offer a critique of this way of thinking. Here instead I would like to offer a rather different sort of motivation for thinking that statistical independence might be violated, coming as promised from the possibility of nonlocal constraints.

Depicted in Fig. 9.2 is a run of the EPR-Bohm experiment in which the setting of A is changed from a_1 to a_2 while the particles (or whatever it is that emanates from the source) are in flight. What we have here is a pair of particles traveling toward detectors A and B, with detector A switching from setting a_1 to a_2 while the particles are in flight, and detector B simply set to b_1.

Let's again suppose that the particles are in one of three states $\lambda_1, \lambda_2, \lambda_3$. According to classical, relativistic physics, the detector settings a_2 and b_1 are determined by the goings-on in their past lightcones, which include the particle preparation event but also far more. Suppose that setting a_2 is compatible with a variety of initial data at the time of preparation, and the same for b_1. Let Λ_{a2} be the presumably large subset of microscopic states (in the past lightcone of the detection event) consistent with a final detector setting of a_2, and let Λ_{b1} be those states compatible with b_1. Though the particle preparation is contained in the past lightcones of the detection events, let us suppose that the state of the particles, λ_1, λ_2, or λ_3, does not play a *dynamical* role in determining the setting of either detector. The question at hand is whether there is any reason to think that, nevertheless, the state of the particles is *correlated* with the detector setting, which is to say whether the theory constrains the state of the particles on the basis of Λ_{a2} and Λ_{b1}.

Now, if the underlying theory is one in which local degrees of freedom are independent, there is no reason to think that knowledge of Λ_{a2} and Λ_{b1} should tell us anything about which of $\lambda_1, \lambda_2, \lambda_3$ are realized. On the other hand, if there are nonlocal constraints, then it may well be otherwise. Suppose that Λ_{a2} is compatible with λ_1 and λ_2 but incompatible with λ_3. In other words, suppose that there are no initial microstates that generate a_2 which are consistent with the particle pair starting in state λ_3. Then we already have a violation of the statistical independence condition, without even bothering yet to consider correlations with the other detector B.

Of course, there may be, and typically are, many things going on in the past lightcone of a detection event at the time the particle pair is produced. Most of these will at least have the appearance of being irrelevant to the final setting of the detector. There is certainly no guarantee that a nonlocal constraint will generate the kind of correlations between detector settings and specially prepared particles that we are talking about. The precise nature of the nonlocal constraint or constraints that could explain quantum correlations is a decidedly open question.

Superdeterminism, Conspiracy and Free Will

The idea that the rejection of *statistical independence* involves preexisting and persisting correlations between subsystems has been broached before, under the headings 'conspiracy theory', 'hyperdeterminism', and 'superdeterminism'. From here on, let us adopt 'superdeterministic' as a generic term for this way of thinking about theories that violate this condition. Bell [3], Shimony [15], Lewis [11] and others have suggested that superdeterministic theories imply some sort of conspiracy on the part of nature. This is frequently accompanied by the charge that the existence of such correlations is a threat to "free will". Let me briefly address these worries before returning to the big picture.

The idea that postulating a correlation between detector settings and particle properties involves a "conspiracy" on the part of nature appears to derive from the idea that it amounts to postulating that the initial conditions of nature have been set up by some cosmic conspirator in anticipation of our measurements. It seems that the conspiracy theorist is supposing that violations of *statistical independence* are not lawlike, but rather are ad hoc. But nonlocal constraints are lawlike, since (by definition), we require them to be consistent with the dynamical evolution given by the laws of motion. If they exist, they exist at every moment of time. This is no more a conspiracy than Gauss's law is a conspiracy.

Another worry about giving up *statistical independence* and postulating generic nonlocal, spacelike correlations has to do with a purported threat to our "free will". This particular concern has been the subject of renewed debate in the last couple of years, prompted in part by an argument of Conway and Kochen [5]. The core of the worry is that if detector settings are correlated with particle properties, this must mean that we cannot "freely choose" the detector settings. However, as 't Hooft [17] points out, this worry appears to be based on a conception of free will which is incompatible with ordinary determinism, never mind superdeterminism. Hume [10] long ago argued that such a conception of free will is highly problematic, in that it is essential to the idea that we freely exercise our will that our thoughts are instrumental in bringing about, which is to say determining, our actions.

The Cosmos

So much for the possible role of nonlocal constraints in underpinning quantum phenomena. The other point of interest is early universe cosmology. Our universe appears to have emanated from a big bang event around 14 billion years ago, and to have been highly homogeneous for quite some time thereafter. The cosmic microwave background radiation is a fossil remnant of the time, around 400,000 years into the universe's existence, when radiation effectively decoupled from matter, and this radiation appears to be quite evenly distributed across the sky, with slight inhomogeneities which presumably seeded later star and galaxy formation.

The task of explaining the homogeneity of the early matter distribution is known as the horizon problem. This, along with the flatness problem and monopole problem, were for some time only explained by fine-tuning, which is to say that they were not really explained at all. Later, inflationary models entered the picture, and these provide a mechanism for generating inhomogeneity in a more generic fashion. However, these models are still speculative—there is no direct evidence for an 'inflaton' field—and moreover inflation itself requires rather special initial conditions [12].

The existence of a nonlocal constraint on the matter distribution and on the state of the gravitational field might address one or more of these problems without recourse to inflation. Certainly, a detailed description of the very early universe requires few variables, since the universe looks essentially the same from place to place with respect to both matter distribution (high temperature, homogeneous) and spatial structure (flat). A reduction in the number of variables is what we would expect from a constrained system, and any constraint demanding that the matter distribution is identical from place to place is indeed nonlocal. On the face of it, this constraint is not preserved under dynamical evolution because of the action of gravity. One might speculate, though, that the constraint holds between matter and gravitational degrees of freedom, and that the early universe is simply a demonstration of one way to satisfy it. The interplay of gravity and matter mix up the degrees of freedom as time goes on, and the current remnants of these macroscopic correlations are the quantum correlations discussed above.

Conclusion

The idea of using nonlocal constraints to account for the large-scale matter distribution in the universe and the large-scale spacetime structure of the universe is interesting but highly speculative, and the idea that these same constraints might account for quantum correlations as well is even more speculative. The most conservative strategy of exploration would be to ignore cosmological scenarios and instead focus on the persistent and experimentally repeatable correlations in the quantum realm. But I think it is worth considering a connection between the two, if for no other reason than the fact that it has proven difficult to construct a testable and sensible quantum theory of gravity, suggesting that the relation between gravitation and quantum phenomena might be different from anything heretofore explored.

A more conservative approach focusing just on quantum phenomena might ponder the way in which the ordinarily superfluous gauge degrees of freedom of modern gauge theories might serve as nonlocal hidden variables. The vector potential in electrodynamics, for example, ordinarily plays no direct physical role: only derivatives of the vector potential, which give rise to the electric and magnetic fields, correspond to physical "degrees of freedom" in classical and quantum electrodynamics. The Aharonov-Bohm effect shows that the vector potential does play an essential role in the quantum theory, but the effect is still gauge-invariant. One might nevertheless conjecture that there is an underlying theory in which the potential *does* play a physical role, one in which the physics is *not* invariant under gauge transformations. It may be impossible for us to directly observe the vector potential, and the uncertainties associated with quantum theory may arise from our ignorance as to its actual (and nonlocally constrained) value. From this perspective, quantum theory would be an effective theory which arises from "modding out" over the gauge transformations, with the so-called gauge degrees of freedom being subject to a nonlocal constraint and accounting for the correlations we observe in EPR-type experiments.

I would conclude by reminding the reader that the sort of nonlocality under discussion in no way violates either the letter or the spirit of relativity. No influences travel faster than light. The idea is simply that there are correlations between spatially separate degrees of freedom, and thus that the fabric of nature is a more richly structured tapestry than we have heretofore believed.

References

1. S. Aaronson, J. Watrous, Closed timelike curves make quantum and classical computing equivalent. arXiv:0808.2669v1
2. J.S. Bell, On the Einstein-Podolsky-Rosen paradox. Physics **1**, 195–200 (1964)
3. J.S. Bell, La nouvelle cuisine, in *John S. Bell on the Foundations of Quantum Mechanics*, ed. by M. Bell, K. Gottfried, M. Veltman (World Scientific, Singapore, 2001), pp. 216–234
4. D. Bohm, *Quantum Theory* (Prentice-Hall, Englewood Cliffs, 1951)
5. J. Conway, S. Kochen, The free will theorem. Found. Phys. **36**, 1441–1473 (2006)
6. O. De Costa Beauregard, S-matrix, Feynman zigzag and Einstein correlation. Phys. Lett. **A67**, 171–174 (1978)
7. A. Einstein, B. Podolsky, N. Rosen, Can quantum-mechanical description of reality be considered complete? Phys. Rev. **47**, 777–780 (1935)
8. A. Fahmi, Non-locality and classical communication of the hidden variable theories (2005). arxiv:quant-ph/0511009
9. A. Fine, *The Shaky Game: Einstein Realism and the Quantum Theory* (University of Chicago Press, Chicago, 1986)
10. D. Hume, *A Treatise of Human Nature* (Oxford University Press, Oxford, 2000). (original publication)
11. P. Lewis, Conspiracy theories of quantum mechanics. Br. J. Philos. Sci. **57**, 359–381 (2006)
12. R. Penrose, Difficulties with inflationary cosmology, in *Proceedings of the 14th Texas Symposium on Relativistic Astrophysics* (New York Academy of Sciences 1989), pp. 249–264

13. H. Price, *Time's Arrow and Archimedes' Point: New Directions for the Physics of Time* (Oxford University Press, Oxford, 1996)
14. M.P. Seevinck, *Parts and Wholes: An Inquiry into Quantum and Classical Correlations*. PhD thesis, Utrecht University (2008). arxiv:0811.1027
15. A. Shimony, An exchange on local beables, in *Search for a Naturalistic World View*, vol. II (Cambridge University Press, Cambridge, 1993), pp. 163–170
16. R.I. Sutherland, Bell's theorem and backwards in time causality. Int. J. Theor. Phys. **22**, 377–384 (1983)
17. G. 't Hooft. On the free will postulate in quantum mechanics (2007). arXiv:quant-ph/0701097
18. S. Weinstein, Nonlocality without nonlocality. Found. Phys. **39**, 921–936, (2009). arXiv.org:0812.0349

Chapter 10
Is Quantum Linear Superposition an Exact Principle of Nature?

Angelo Bassi, Tejinder Singh and Hendrik Ulbricht

Abstract The principle of linear superposition is a hallmark of quantum theory. It has been confirmed experimentally for photons, electrons, neutrons, atoms, for molecules having masses up to ten thousand amu, and also in collective states such as SQUIDs and Bose-Einstein condensates. However, the principle does not seem to hold for positions of large objects! Why for instance, a table is never found to be in two places at the same time? One possible explanation for the absence of macroscopic superpositions is that quantum theory is an approximation to a stochastic nonlinear theory. This hypothesis may have its fundamental origins in gravitational physics, and is being put to test by modern ongoing experiments on matter wave interferometry.

The Absence of Macroscopic Superpositions

In the year 1927, American physicists Clinton Davisson and Lester Germer performed an experiment at Bell Labs, in which they scattered a beam of electrons off the surface of a nickel plate. In doing so, they accidentally discovered that the scattered electrons exhibited a diffraction pattern analogous to what is seen in the Bragg diffraction of X-rays from a crystal. This experiment established the wave nature of electrons, and confirmed de Broglie's hypothesis of wave-particle duality. An electron can be in

This essay received the Fourth Prize in the FQXi Essay Contest, 2012.

A. Bassi (✉)
Department of Physics, University of Trieste, Strada Costiera 11,
34151 Trieste, Italy
e-mail: bassi@ts.infn.it

T. Singh
Tata Institute of Fundamental Research, Homi Bhabha Road,
Mumbai 400005, India
e-mail: tpsingh@tifr.res.in

H. Ulbricht
School of Physics and Astronomy, University of Southampton,
Southampton SO17 1BJ, UK
e-mail: h.ulbricht@soton.ac.uk

© Springer International Publishing Switzerland 2015
A. Aguirre et al. (eds.), *Questioning the Foundations of Physics*,
The Frontiers Collection, DOI 10.1007/978-3-319-13045-3_10

more than one position at the same time, and these different position states obey the principle of quantum linear superposition: the actual state of the electron is a linear sum of all the different position states.

The principle of linear superposition is the central tenet of quantum theory, an extremely successful theory for all observed microscopic phenomena. Along with the uncertainty principle, it provides the basis for a mathematical formulation of quantum theory, in which the dynamical evolution of a quantum system is described by the Schrödinger equation for the wave function of the system.

The experimental verification of linear superposition for electrons heralded a quest for a direct test of this principle for larger, composite particles and objects. Conceptually, the idea of the necessary experimental set up is encapsulated in the celebrated double slit interference experiment. A beam of identical particles is fired at a screen having two tiny slits separated by a distance of the order of the de Broglie wavelength of the particles, and an interference pattern is observed on a photographic plate at the far end of the screen. Such an experiment was successfully carried out for Helium ions, neutrons, atoms, and small molecules, establishing their wave nature and the validity of linear superposition.

When one considers doing such an interference experiment for bigger objects such as a beam of large molecules the technological challenges become enormous. The opening of a slit should be larger than the physical size of the molecule (say a few hundred nanometres) and the separation between slits should be smaller than the coherence length (few micrometres) [1]. The first experiment in this class was performed in 1999 in Vienna with C_{60} molecules (fullerene, having nearly 700 nucleons per molecule) and the observed interference, as predicted by quantum theory, dramatically vindicated the validity of linear superposition for such large systems [2]. Today, a little over a decade later, experimentalists have succeeded in pushing this limit to molecules with about 10,000 nucleons, and are aiming to push it higher to particles with a million nucleons in the next few years. This is an extraordinary feat, considering the great difficulties involved in maneuvering such large particles as they travel from the source, through the slits, and on to the detector where the interference pattern forms.

But will the principle of linear superposition continue to hold for larger and larger objects? The answer is indeed yes, according to the modern outlook towards quantum theory. The theory does not say that superposition should hold only in the microscopic world—in fact, a molecule with ten thousand nucleons, for which the principle has been confirmed, isn't exactly microscopic!

However when we look at the day to day world around us linear superposition does not seem to hold! A table for instance is never found to be 'here' and 'there' at the same time. In other words, superposition of position states does not seem to hold for macroscopic objects. In fact already at the level of a dust grain, which we can easily see with the bare eye, and which has some 10^{18} nucleons, the principle breaks down. What could be going on in the experimentally untested desert between 10^4 nucleons, where linear superposition is valid, and 10^{18} nucleons, where it is not valid?

The observed absence of macroscopic superpositions is the source of the quantum measurement problem, as illustrated by the double slit experiment for a beam of electrons. The interference pattern which forms when an electron passes through both the slits is destroyed when a detector (measuring apparatus) is placed close to and behind one of the slits. The detector clicks (does not click) with a probability proportional to the square of the complex amplitude for the electron to pass through the upper slit (lower slit). This is the so-called Born probability rule. The presence of the detector breaks the superposition of the two states 'electron going through upper slit' and 'electron going through lower slit' and the interference pattern is lost. One says that the quantum state of the electron has collapsed, either to 'electron going through upper slit only', or to 'electron going through lower slit only'. The collapse of the superposition of the two electron states (up and down), and the emergence of probabilities contradict the Schrödinger equation, since this equation is deterministic and linear. It predicts that a superposition should be preserved during the measurement process, that there should be no probabilities, and that position superpositions should be observed for macroscopic objects as well, such as a measuring apparatus.

Why are superpositions of different position states for macroscopic objects not observed, in contrast with what quantum theory predicts? Possible explanations include reinterpretations of quantum theory such as the many worlds interpretation or consistent histories, and mathematical reformulations of the theory such as Bohmian mechanics. By introducing one or more additional assumptions, these reinterpretations and reformulations ensure that macroscopic superpositions will not be observed, while making the same experimental predictions as quantum theory.

Quantum Theory Is Approximate

Here we explore another possibility, which has been investigated extensively during the last three decades or so [3–5]. What if quantum linear superposition is an approximate principle of nature? This means the following: consider an object (say an atom, a molecule, or a table) which consists of N nucleons and is in a superposition of two quantum states of an observable one is trying to measure (for example spin or position). According to quantum theory the life time of such a superposition is infinite! However, no experiment to date rules out the possibility that the lifetime τ of such a superposition is finite, and is a monotonically decreasing function of N. The dependence of τ on N should be such that for microscopic systems such as atoms the superposition life time is astronomically large (longer than the age of the Universe). Furthermore, for large objects such as a table or a macroscopic apparatus the superposition life time should be so small that linear superposition is simply not observed on the scale of current experiments.

Somewhere in between the microworld and the macroworld, i.e. between objects of 10^4 and 10^{18} nucleons, the superposition lifetime would be neither too large, nor too small, but just right enough to be detectable in the laboratory. In principle, this could be achieved as follows. Suppose one has prepared in a controlled manner a beam of very large identical molecules which are such that the superposition life time between two different position states is τ. Let such a beam be passed through two slits in a double slit experiment, thereby creating a superposition of two different position states. Let the distance of the photographic plate from the slits be large enough that the time of travel of a molecule from the slits to the plate far exceeds τ. Then the superposition will exponentially decay before the molecule reaches the plate, and no interference pattern will be seen. And this will happen even though no detector has been placed behind either of the slits!

The requirements on a mathematically consistent generalization of quantum theory in which superposition is an approximate principle are extremely stringent. To begin with the theory should be nonlinear: superposition of two allowed quantum states of the system should not be a stable allowed state. And yet for microsystems this theory should be in excellent agreement with standard quantum theory so that results of all known experiments are reproduced. The nonlinear process responsible for the breakdown of superposition should be stochastic (i.e. random) in nature because the outcome of a particular quantum measurement cannot be predicted. All the same, the nonlinear process should be such that a collection of outcomes of repeated measurements obey the Born probability rule. Another reason the nonlinear effect should be stochastic is that this helps avoid faster-than-light signalling. A deterministic (i.e. non-stochastic) nonlinear quantum mechanics is known to allow for superluminal communication. The nonlinear mechanism should be amplifying in nature, being negligible for microscopic systems, but becoming more and more important for large many particle systems, so that breakdown of superposition becomes more and more effective.

It is a remarkable achievement that such a theory could be developed, and was given shape in the eighties by physicists Ghirardi, Pearle, Rimini and Weber [6, 7]. This theory, which has come to be known as Continuous Spontaneous Localization (CSL) has been extensively investigated in the following two decades, with regard to its properties and solutions, constraints from laboratory and astrophysical observations, and direct tests of its predictions in the laboratory [5]. It is perhaps the only well studied generalization of quantum theory whose experimental predictions differ markedly from those of quantum theory in the macroworld and which hence provides an excellent benchmark against which the accuracy of quantum theory can be tested.

In its original form (known as the GRW model), the CSL model is based on the following two principles:

1. Given a system of n distinguishable particles, each particle experiences a sudden spontaneous localization (i.e. collapse of position state) with a mean rate λ, to a spatial region of extent r_C.
2. In the time interval between two successive spontaneous localizations, the system evolves according to the Schrödinger equation.

Thus two new fundamental constants of nature have been introduced, and these are assumed to take definite numerical values, in order to successfully reproduce the observed features of the microscopic and the macroscopic world. The constant λ, assumed to be of the order $10^{-16}\,\mathrm{s}^{-1}$, determines the rate of spontaneous localization for a single particle. Amplification is achieved by showing that for a composite object of n particles, the collapse rate is $(\lambda n)^{-1}$ s. The second constant r_C is a length scale assumed to be about 10^{-5} cm, and indicates that a widely spread quantum state collapses to a size of about r_C during localization.

In its modern version, the CSL model consists of coupling a randomly fluctuating classical field with the particle number density operator of a quantum system, so as to produce collapse towards spatially localized states. The collapse process is continuous and dynamics is described by a single nonlinear stochastic differential equation which contains both aspects: standard Schrödinger evolution as well the nonlinear stochastic terms which result in collapse and breakdown of superposition. The fundamental constants λ and r_C continue to be retained. Today, experiments are being devised to test CSL and are approaching the ballpark range where the preferred values of λ and r_C will be confirmed or ruled out.

The CSL dynamical equation can be used to show that macroscopic superpositions last for extremely short time intervals, thus explaining why such position superpositions are not observed. Similarly the superposition lifetime for microstates is extremely large. The same CSL equation elegantly solves the measurement problem. Suppose a microscopic quantum system is prepared in a superposition of two quantum states, say A and B, of an observable which is to be measured. The CSL equation for this system is essentially the same as the Schrödinger equation, the nonlinear effect being negligible, and evolution proceeds in the standard deterministic linear fashion, preserving superposition of the states A and B. When the quantum system comes into interaction with the measuring apparatus, which let us say is characterized by a pointer, one must now write down the CSL equation for the combined system consisting of the quantum system and the measuring apparatus, and now the nonlinear stochastic terms come into play. The state A would cause the pointer to be at a position P_1 and the state B would cause the pointer to be at a different position P_2. The superposition of A and B would cause the pointer to be in superposition of the macroscopically different states P_1 and P_2. This however is not permitted by CSL, since the pointer is macroscopic: the superposition very quickly decays to the pointer state P_1 or P_2. This is precisely what is interpreted as the collapse of the superposed quantum state to either A or B. Furthermore, the CSL equation shows that the outcome of an individual measurement is random, but repeated measurements are shown to result in outcomes P_1 or P_2 with relative frequency as given by the Born probability rule [8].

Testing the Idea with Experiments

The nonlinear stochastic modification of the Schrödinger equation and the introduction of the constants λ and r_C modify the predictions of quantum theory for various standard results from known experiments and astrophysics. This allows bounds to be put on these constants. These include bounds coming from decay of supercurrents in SQUIDs, spontaneous X-ray emission from Germanium, absence of proton decay, heating of the intergalactic medium, dissociation of cosmic hydrogen, heating of interstellar dust grains, and temperature distortions in the cosmic microwave background. A novelty of the CSL mechanism is that the process of spontaneous localization produces a very tiny increase in the energy of the localized particle, thus violating energy conservation. Some of the bounds on the fundamental constants come from the lack of observation of any such violation. Using other arguments based on latent image formation in photography [9] and formation of image in the human eye [10], it has been suggested that the theoretical value of λ should be as high as $10^{-8}\,s^{-1}$. We thus already see a fascinating debate taking place between theory and experiment: the principle of quantum linear superposition is being confronted by experiment in the true sense of the word.

However the most direct tests of CSL and linear superposition will come from interference experiments for large objects. Unlike quantum theory, CSL predicts that under certain suitable conditions an interference pattern will not be seen. Thus an accurate experiment carried out under these conditions will definitely establish whether quantum linear superposition is an exact or approximate principle. If interference is seen with a large molecule, it sets an upper bound on the value of λ. This experimental field, known as matter wave interferometry, has made great strides in recent years, and is one of the most important sources for testing proposed alternatives to quantum theory, such as CSL [1, 11–14].

The experiments involve overcoming technical challenges to prepare intense beams of molecules in gas phase, to preserve spatial and temporal coherence of the beam, and to detect the particles efficiently. Beam splitters for molecules are typically highly ordered periodic diffraction gratings of nanowires made from metal or semiconductors, or they are standing light fields realized using the so-called Kapitza-Dirac effect. The 1999 Vienna experiment with C_{60} fullerene used far-field Fraunhofer molecular diffraction with a grating constant of 100 nm. For larger molecules, where it becomes imperative to effectively increase the beam intensity, more promising results have been achieved through near field interference using the Talbot-Lau interferometer (TLI).

A TLI operating in the near-field (i.e. the spatial period of the gratings and the interference pattern are on the same scale) was specifically invented to deal with beams of low intensity and low collimation in interference experiments [15]. A three grating TLI operates with weakly collimated beams: the first grating prepares beam coherence and imprints a spatial structure on the beam. The second grating is thus simultaneously illuminated by some 10^4 individual coherent molecular beams and creates a self-image on a third grating, on which the interference pattern forms. Effectively, the number of molecules contributing to the final interference pattern is multiplied by the number of illuminating slits of the first grating, and all the coherent beams from the 10^4 source slits are incoherently summed to contribute to the same interference pattern. More recently, a modified version, known as the Kapitza-Dirac TLI [16] has been employed, in which the second—the diffraction grating is replaced by an optical phase grating (Fig. 10.1). Molecules are diffracted at periodic optical potentials due to the Kapitza-Dirac effect. The KDTLI has been used to demonstrate interference with a molecule known as perfluoro-alkylated C_{60},

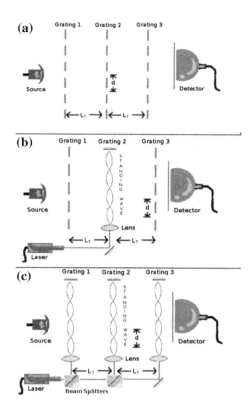

having a mass 7,000 amu [17]. This is the largest molecule on which a successful
matter wave experiment has been carried out so far, and sets an upper bound of
$10^{-5}\,\mathrm{s}^{-1}$ on the CSL parameter λ. This is only three orders of magnitude away from
the theoretical value 10^{-8} for λ predicted by Adler, and this latter value could be
confronted by an experiment with molecules having a mass of about 5,00,000 amu.

However, for molecules above 10^5 amu, their electromagnetic interactions with
the material gratings disable the interference pattern, and new technologies must
be sought to efficiently control and manipulate the center of mass motion of heavy
particles. Experiments are performed in ultra-high vacuum conditions to prevent
decoherence by collisions. Beams should be slow, so that the de Broglie wavelength
is not too low, they should be highly collimated and should have a high phase space
density. These features can be achieved through promising cooling techniques currently under development. Another important aspect is whether to use neutral or
charged particles. All large particle interference experiments thus far have been performed with neutrals—they have the advantage that they suffer lesser decohering
effects from interaction with the environment. On the other hand charged particles
are easier to manipulate and control, especially while preparing coherent particle
beams of heavy molecules [5].

A clever new proposal combines the best of both worlds: manipulation of charged particles during the preparation of the beam, and interference after neutralization. This novel three light grating TLI aims towards the interference of particles up to 10^9 amu and is known as the optical time-domain matter-wave (OTIMA) interferometer [18]. Charged particles will be provided by a mass filtered cluster aggregation source. The charged clusters are neutralized at the first grating using light-matter effects, diffracted at the second grating, and ionized again for detection at the third grating.

An alternative approach to testing quantum superposition and CSL is optomechanics, which involves coupling micromechanical devices to light fields with which they interact [19, 20]. A tiny mechanical object such as a micromirror is cooled to extremely low temperatures and prepared to be, say, in a quantum superposition of the vibrational ground state and first excited state. This superposed state is coupled to a sensitive optical interferometer: if the superposed state decays in a certain time, as predicted by CSL, the optical response of the interferometer will be different, as compared to when the superposition does not decay (Fig. 10.2). Optomechanics aims to test superposition for very massive particles in the range 10^6–10^{15} amu, but the vibrational amplitude is very small compared to the desired amplitude, which is the fundamental length scale r_C of CSL. This makes it a challenge for optomechanics to reach the expected regime where new physics is expected [21–24].

However, very promising progress can be expected by combining techniques from optomechanics and matter wave interferometry. Massive particles trapped in optical

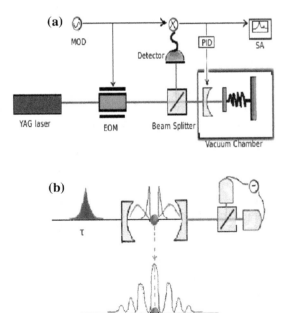

Fig. 10.2 Optomechanics: **a** Prototype of optomechanically cooled cantilever. Quantum optical detection techniques enable the sensitive read out of vibrations as they couple to light fields. **b** Mechanical resonator interference in a double slit (MERID). The centre of mass motion of a single optically trapped nanoparticle is first cooled and then superimposed by an optical double potential. The interference pattern evolves in free fall after switching off the trapping field. Figure courtesy: Kinjalk Lochan. Figure source [5]

traps are analogues of optomechanical systems, with the significant difference being that by suitable choice of the trapping potentials, superposition of position states can be created with a separation comparable to r_C. After the superposed state has been created the trapping potential is swiched off; the particles are allowed to fall freely, and their spatial density distribution can be studied to verify the existence of an interference pattern. Such an experiment can be carried out for very massive objects (such as polystyrene beads of about 30 nm diameter and mass 10^6 amu). Known as MERID (mechanical resonator interference in a double slit), it is a promising future technique for testing the CSL model [25]. It can be expected that within the next two decades matter wave interferometry and optomechanics experiments will reach the ballpark of 10^6–10^9 amu where predictions of CSL will differ from quantum theory sufficiently enough for them to be discriminated in the laboratory.

Why is Quantum Theory Approximate?

Continuous Spontaneous Localization has been proposed as a phenomenological modification of quantum theory which is successful in explaining the observed absence of macroscopic superpositions, and the Born rule. It is beyond doubt though that underlying CSL there ought to be deep physical principles which compel us to accept such a radical modification of quantum theory. Fascinating progress has been taking place towards unravelling these underlying principles, mainly along three different directions, all of which point to an involvement of gravity, and a revision of our understanding of spacetime structure at the deepest level.

It has been suggested independently by Karolyhazy et al. [26], Diosi [27] and Penrose [28] that gravity is responsible for the absence of macroscopic superpositions. While their arguments differ somewhat at the starting point they all come to approximately the same conclusion that the absence of superpositions will become apparent around 10^6–10^9 amu, a range that agrees well with the prediction of CSL! The key idea is that every object obeys the uncertainty principle and hence there is an intrinsic minimal fluctuation in the spacetime geometry produced by it. When the quantum state describing this object propagates in this fluctuating spacetime, it loses spatial coherence beyond a certain length scale after a critical time. Such a loss of spatial coherence is an indicator of breakdown of superposition of different position states. This length and time scale is shown to be astronomically large for microscopic particles, and very small for macroscopic objects, thus demonstrating macroscopic localization. There is tantalizing evidence in the literature [29] that the stochastic mechanism of CSL is provided by spacetime fluctuations, and that the fundamental constants λ and r_C derive from gravity. An ongoing optomechanics experiment is specifically devoted to testing the role of gravity in causing breakdown of superposition [20].

A second remarkable line of development, known as Trace Dynamics, has come from the work of Stephen Adler and collaborators [30] who suggest that it is unsatisfactory to arrive at a quantum theory by quantizing its very own limit, namely classical dynamics. Instead, quantum theory is derived as an equilibrium statistical thermodynamics of an underlying unitarily invariant classical theory of matrix dynamics. It is fascinating that the consideration of Brownian motion fluctuations around the equilibrium theory provides a nonlinear stochastic modification of the Schrödinger equation of the kind proposed in CSL. This is another clue that the absence of macroscopic superpositions may have to do with quantum theory being an approximation to a deeper theory.

Thirdly, the absence of macroscopic superpositions is possibly related to another deep rooted but little appreciated incompleteness of quantum theory. In order to describe dynamical evolution the theory depends on an external classical time, which is part of a classical spacetime geometry. However, such a geometry is produced by classical bodies, which are again a limiting case of quantum objects! This is yet another sense in which the theory depends on its classical limit. There hence ought to exist a reformulation of quantum theory which does not depend on a classical time [31]. Such a reformulation has been developed by borrowing ideas from Trace Dynamics and there is evidence that there exist stochastic fluctuations around such a reformulated theory, which have a CSL type structure, and are responsible for the emergence of a classical spacetime and breakdown of superposition in the macroscopic world [32–34].

The CSL model, as it is known today, is nonrelativistic in character, which means that it is a stochastic generalization of the nonrelativisitic Schrödinger equation. It is intriguing that the model has thus far resisted attempts at a relativisitic generalization. On the other hand it is known that the collapse of the wave function during a quantum measurement is instantaneous and non-local. This has been confirmed by experimental verification of Bell's inequalities. Thus it is certainly true that there is a need for reconciliation between CSL induced localization, and the causal structure of spacetime dictated by special relativity. In a remarkable recent paper [35] it has been shown that if one does not assume a predefined global causal order, there are multipartite quantum correlations which cannot be understood in terms of definite causal order and spacetime may emerge from a more fundamental structure in a quantum to classical transition.

For nearly a century the absence of macroscopic superpositions, in stark contradiction with what quantum theory predicts, has confounded physicists, and led Schrödinger to formulate his famous cat paradox. The quantum measurement problem, a direct consequence of this contradiction, has been debated endlessly, and very many solutions proposed, by physicists as well as philosophers. However up until recent times the debate has remained largely theoretical, for no experiment has ever challenged the phenomenal successes of quantum theory. Times have changed now. There is a phenomenological model which proposes that quantum linear superposition is an approximate principle; there are serious underlying theoretical reasons which suggest why this should be so, and most importantly, experiments and technology have now reached a stage where this new idea is being directly tested in the laboratory. Perhaps after all it will be shown that the assumption that linear superposition is exact is a wrong assumption. We will then have nothing short of a revolution, which will have been thrust on us by experiments which disagree with quantum mechanics, thus forcing a radical rethink of how we comprehend quantum theory, and the structure of spacetime.

Acknowledgments This work is supported by a grant from the John Templeton Foundation.

Technical Endnotes (*For Details See* [5])

The Physics of Continuous Spontaneous Localization

The essential physics of the CSL model can be described by a simpler model, known as QMUPL (Quantum Mechanics with Universal Position Localization), whose dynamics is given by the following stochastic nonlinear Schrödinger equation

$$d\psi_t = \left[-\frac{i}{\hbar} H dt + \sqrt{\lambda}(q - \langle q \rangle_t)dW_t - \frac{\lambda}{2}(q - \langle q \rangle_t)^2 dt \right] \psi_t, \qquad (10.1)$$

where q is the position operator of the particle, $\langle q \rangle_t \equiv \langle \psi_t | q | \psi_t \rangle$ is the quantum expectation, and W_t is a standard Wiener process which encodes the stochastic effect. Evidently, the stochastic term is nonlinear and also nonunitary. The collapse constant λ sets the strength of the collapse mechanics, and it is chosen proportional to the mass m of the particle according to the formula $\lambda = \frac{m}{m_0} \lambda_0$, where m_0 is the nucleon's mass and λ_0 measures the collapse strength. If we take $\lambda_0 \simeq 10^{-2}$ m^{-2} s^{-1} the strength of the collapse model corresponds to the CSL model in the appropriate limit.

The above dynamical equation can be used to prove position localization; consider for simplicity a free particle ($H = p^2/2m$) in the gaussian state (analysis can be generalized to other cases):

$$\psi_t(x) = \exp\left[-a_t(x - \overline{x}_t)^2 + i\overline{k}_t x + \gamma_t \right]. \qquad (10.2)$$

By substituting this in the stochastic equation it can be proved that the spreads in position and momentum

$$\sigma_q(t) \equiv \frac{1}{2}\sqrt{\frac{1}{a_t^R}}; \qquad \sigma_p(t) \equiv \hbar\sqrt{\frac{(a_t^R)^2 + (a_t^I)^2}{a_t^R}}, \qquad (10.3)$$

do not increase indefinitely but reach asymptotic values given by

$$\sigma_q(\infty) = \sqrt{\frac{\hbar}{m\omega}} \simeq \left(10^{-15}\sqrt{\frac{\text{Kg}}{\text{m}}} \right) \text{m}, \qquad \sigma_p(\infty) = \sqrt{\frac{\hbar m\omega}{2}} \simeq \left(10^{-19}\sqrt{\frac{\text{m}}{\text{Kg}}} \right) \frac{\text{Kg m}}{\text{s}}, \qquad (10.4)$$

such that: $\sigma_q(\infty)\,\sigma_p(\infty) = \hbar/\sqrt{2}$ which corresponds to almost the minimum allowed by Heisenberg's uncertainty relations. Here, $\omega = 2\sqrt{\hbar\lambda_0/m_0} \simeq 10^{-5}\,\mathrm{s}^{-1}$.

Evidently, the spread in position does not increase indefinitely, but stabilizes to a finite value, which is a compromise between the Schrödinger's dynamics, which spreads the wave function out in space, and the collapse dynamics, which shrinks it in space. For microscopic systems, this value is still relatively large ($\sigma_q(\infty)\sim 1$ m for an electron, and ~ 1 mm for a C_{60} molecule containing some 1,000 nucleons), such as to guarantee that in all standard experiments—in particular, diffraction experiments—one observes interference effects. For macroscopic objects instead, the spread is very small ($\sigma_q(\infty)\sim 3 \times 10^{-14}$ m, for a 1 g object), so small that for all practical purposes the wave function behaves like a point-like system. This is how localization models are able to accommodate both the "wavy" nature of quantum systems and the "particle" nature of classical objects, within one single dynamical framework.

The same stochastic differential equation solves the quantum measurement problem and explains the Born probability rule without any additional assumptions. For illustration, consider a two state microscopic quantum system $\mathcal{S}$ described by the initial state

$$c_+|+\rangle + c_-|-\rangle \tag{10.5}$$

interacting with a measuring apparatus $\mathcal{A}$ described by the position of a pointer which is initially in a 'ready' state ϕ_0 and which measures some observable O, say spin, associated with the initial quantum state of $\mathcal{S}$. As we have seen above, the pointer being macroscopic (for definiteness assume its mass to be 1 g), is localized in a gaussian state ϕ^G, so that the initial composite state of the system and apparatus is given by

$$\Psi_0 = \left[c_+|+\rangle + c_-|+\rangle\right] \otimes \phi^G. \tag{10.6}$$

According to standard quantum theory, the interaction leads to the following type of evolution:

$$\left[c_+|+\rangle + c_-|-\rangle\right] \otimes \phi^G \quad \mapsto \quad c_+|+\rangle \otimes \phi_+ + c_-|-\rangle \otimes \phi_-, \tag{10.7}$$

where ϕ_+ and ϕ_- are the final pointer states corresponding to the system being in the collapsed state $|+\rangle$ or $|-\rangle$ respectively. While quantum theory explains the transition from the entangled state (10.7) to one of the collapsed alternatives by invoking a new interpretation or reformulation, the same is achieved dynamically by the stochastic nonlinear theory given by (10.1).

It can be proved from (10.1) that the initial state (10.6) evolves, at late times, to

$$\psi_t = \frac{|+\rangle \otimes \phi_+ + \epsilon_t|-\rangle \otimes \phi_-}{\sqrt{1 + \epsilon_t^2}}. \tag{10.8}$$

The evolution of the stochastic quantity ϵ_t is determined dynamically by the stochastic equation: it either goes to $\epsilon_t \ll 1$, with a probability $|c_+|^2$, or to $\epsilon_t \gg 1$, with a probability $|c_-|^2$. In the former case, one can say with great accuracy that the state vector has 'collapsed' to the definite outcome $|+\rangle \otimes \phi_+$ with a probability $|c_+|^2$. Similarly, in the latter case one concludes that the state vector has collapsed to $|-\rangle \otimes \phi_-$ with a probability $|c_-|^2$. This is how collapse during a quantum measurement is explained dynamically, and random outcomes over repeated measurements are shown to occur in accordance with the Born probability rule. The time-scale over which ϵ_t reaches its asymptotic value and the collapse occurs can also be computed dynamically. In the present example, for a pointer mass of 1 g, the collapse time turns out to be about 10^{-4} s.

Lastly, we can understand how the modified stochastic dynamics causes the outcome of a diffraction experiment in matter wave-interferometry to be different from that in quantum theory. Starting from the fundamental Eq. (10.1) it can be shown that the statistical operator $\rho_t = \mathbb{E}[|\psi_t\rangle\langle\psi_t|]$ for a system of N identical particles evolves as

$$\rho_t(x, y) = \rho_0(x, y)e^{-\lambda N(x-y)^2 t/2}. \qquad (10.9)$$

Experiments look for a decay in the density matrix by increasing the number of the particles N in an object, by increasing the slit separation $|x - y|$, and by increasing the time of travel t from the grating to the collecting surface. The detection of an interference pattern sets an upper bound on λ. The absence of an interference pattern would confirm the theory and determine a specific value for λ (provided all sources of noise such as decoherence are ruled out.)

A detailed review of the CSL model and its experimental tests and possible underlying theories can be found in [5].

References

1. K. Hornberger, S. Gerlich, P. Haslinger, S. Nimmrichter, M. Arndt, Rev. Mod. Phys. **84**, 157 (2011)
2. M. Arndt, O. Nairz, J. Vos-Andreae, C. Keller, G. Van der Zouw, A. Zeilinger, Nature **401**, 680 (1999)
3. S.L. Adler, A. Bassi, Science **325**, 275 (2009)
4. A. Bassi, G.C. Ghirardi, Phys. Rep. **379**, 257 (2003)
5. A. Bassi, K. Lochan, S. Satin, T.P. Singh, H. Ulbricht, Rev. Mod. Phys. **85**, 471 (2013)
6. G.C. Ghirardi, A. Rimini, T. Weber, Phys. Rev. D **34**, 470 (1986)
7. G.C. Ghirardi, P. Pearle, A. Rimini, Phys. Rev. A **42**, 78 (1990)
8. A. Bassi, D.G.M. Salvetti, J. Phys. A **40**, 9859 (2007)
9. S.L. Adler, J. Phys. A **40**, 2935 (2007)
10. A. Bassi, D.-A. Deckert, L. Ferialdi, Europhys. Lett. **92**, 50006 (2010)
11. W. Feldmann, R. Tumulka, J. Phys. A: Math. Theor. **45**, 065304 (2012)
12. S. Nimmrichter, K. Hornberger, P. Haslinger, M. Arndt, Phys. Rev. A **83**, 043621 (2011)
13. O. Romero-Isart, Phys. Rev. A **84**, 052121 (2011)
14. M. Arndt, A. Ekers, W. von Klitzing, H. Ulbricht, New J. Phys. 14 (2011)
15. J. Clauser, M. Reinsch, Appl. Phys. B: Lasers Opt. **54**, 380 (1992)

16. K. Hornberger, S. Gerlich, H. Ulbricht, L. Hackermüller, S. Nimmrichter, I. Goldt, O. Boltalina, M. Arndt, New J. Phys. **11**, 043032 (2009)

17. S. Gerlich, S. Eibenberger, M. Tomandl, S. Nimmrichter, K. Hornberger, P.J. Fagan, J. Tüxen, M. Mayor, M. Arndt, Nat. Commun. **2**, 263 (2011)

18. S. Nimmrichter, P. Haslinger, K. Hornberger, M. Arndt, New J. Phys. **13**, 075002 (2011)

19. S. Bose, K. Jacobs, P. Knight, Phys. Rev. A **56**, 4175 (1997)

20. W. Marshall, C. Simon, R. Penrose, D. Bouwmeester, Phys. Rev. Lett. **91**, 130401 (2003)

21. T. Kippenberg, K. Vahala, Science **321**, 1172 (2008)

22. M. Aspelmeyer, S. Groeblacher, K. Hammerer, N. Kiesel, J. Opt. Soc. Am. B **27**, A189 (2010)

23. J. Chan, T. Alegre, A. Safavi-Naeini, J. Hill, A. Krause, S. Groeblacher, M. Aspelmeyer, O. Painter, Nature **478**, 89 (2011)

24. A. O'Connell, M. Hofheinz, M. Ansmann, R. Bialczak, M. Lenander, E. Lucero, M. Neeley, D. Sank, H. Wang, M. Weides et al., Nature **464**, 697 (2010)

25. O. Romero-Isart, A. Panzer, F. Blaser, R. Kaltenbaek, N. Kiesel, M. Aspelmeyer, J. Cirac, Phys. Rev. Lett. **107**, 020405 (2011)

26. F. Karolyhazy, A. Frenkel, B. Lukács, in *Quantum Concepts in Space and Time*, ed. by R. Penrose, C.J. Isham (Clarendon, Oxford, 1986)

27. L. Diósi, Phys. Lett. A **120**, 377 (1987)

28. R. Penrose, Gen. Relativ. Gravit. **28**, 581 (1996)

29. L. Diósi, Phys. Rev. A **40**, 1165 (1989)

30. S.L. Adler, *Quantum Theory as an Emergent Phenomenon* (Cambridge University Press, Cambridge, 2004) pp. xii + 225

31. T.P. Singh, J. Phys. Conf. Ser. **174**, 012024 (2009)

32. T. P. Singh, (2011). arXiv:1106.0911

33. K. Lochan, T.P. Singh, Phys. Lett. A **375**, 3747 (2011)

34. K. Lochan, S. Satin, T.P. Singh, Found. Phys. **42**, 1556 (2012)

35. O. Oreshkov, F. Costa, C. Brukner, Nat. Commun. **3**, 1092 (2012)

Chapter 11
Quantum-Informational Principles for Physics

Giacomo Mauro D'Ariano

Abstract It is time to to take a pause of reflection on the general foundations of physics, for re-examining the logical solidity of the most basic principles, as the relativity and the gravity-acceleration equivalence. The validity at the Planck scale of such principles is under dispute. A constructive criticism engages us in seeking new general principles, which reduce to the old ones only in the already explored domain of energies. At the very basis of physics there are epistemological and operational rules for the same formulability of the physical law and for the computability of its theoretical predictions. Such rules give rise to new solid principles, leading us to a quantum-information theoretic formulation, that hinges on the logical identification of the experimental protocol with the quantum algorithm.

The information-theoretic program for physics foundations has already been advocated in the past by several authors [1]. Recently the program succeeded in deriving the full structure of quantum theory from informational principles [2–5], and we will very briefly examine them here, as exemplars of good principles. The problem is now to extend the informational program to relativistic quantum field theory, the most fundamental theoretical structure of physics. The plan here proposed is to ground quantum field theory on two new principles pertaining only the formulability and computability of the physical law: (1) the Deutsch-Church-Turing principle, and (2) the topological homogeneity of interactions. As we will see, in conjunction with the principles of quantum theory, these two new principles entail a quantum cellular automata extension of quantum field theory.

The quantum automaton extends field theory in the sense that it includes localized states and measurements, for whose description quantum field theory is largely inadequate. The quantum automaton doesn't suffer any formal violation of causality, e.g.

The following dissertation is a minimally updated version of the original essay presented at the FQXi Essay Contest 2012 "Questioning the foundations". A short summary of the follow-ups and recent research results is given in the Postscriptum.

G.M. D'Ariano (✉)
Dipartimento di Fisica dell'Università degli Studi di Pavia, via Bassi 6, 27100 Pavia, Italy
e-mail: dariano@unipv.it

G.M. D'Ariano
Istituto nazionale di Fisica Nucleare, Gruppo IV, Sezione di Pavia, Italy

© Giacomo Mauro D'Ariano 2015
A. Aguirre et al. (eds.), *Questioning the Foundations of Physics*,
The Frontiers Collection, DOI 10.1007/978-3-319-13045-3_11

superluminal tails of the probability distributions. It is not afflicted by any kind of divergence, being exactly computable by principle. Relativistic covariance and other symmetries are violated, but are recovered at the usual scale of energy.

The generality of the new principles does not deplete them of physical content. On the contrary, the Dirac automaton—the most elementary theory of this kind—despite its simplicity leads us to unexpected interesting predictions, e.g. it anticipates a maximum mass for the Dirac particle just as a consequence of the unitarity of quantum evolution, without invoking black-hole general-relativity arguments. It also opens totally unexpected routes for redefining mechanical notions. As regards gravity, the theory seems to suggest the route of the emergent thermodynamic force of Jacobson-Verlinde [6, 7], here, specifically, as a purely quantum-digital effect of the Dirac automaton.

Good and Bad Principles

Which kinds of principles are good and which are bad? We can limit ourself to four main different types of principles: (1) dogmatic, (2) empirical, (3) simplifying (or conventional), and (4) epistemological.

The dogma. This is definitely the worst case. Do we have dogmas in physics? We have few subtle ones. It is not a blasphemy to regard the non existence of an absolute reference frame as a dogma. What about the reference frame of the background radiation? We indeed always invoke the frame of "fixed stars" for establishing if a frame is inertial. The denial of the existence of an absolute frame is a relic of the anthropocentrism repudiation that followed the Keplerian revolution. We will come back to this dogma later.

The empirical principle. A principle is empirical if it has no logical motivation other than its empirical evidence. A typical example is the Einstein's equivalence principle: the identity between inertial and gravitational mass is an observed fact. But do we have a good reason for it? The principle implies that the trajectory of a mass in a gravitational field is independent on the mass, and this leads us to reinterpret gravity as a property of space—the starting point of general relativity theory, which is then a re-interpretation of the principle, not a logical motivation. Another relevant example of empirical principle is the invariance of the speed of light with the reference frame—quite an odd one, isn't it? This lead Einstein to his first formulation of special relativity. The principle was later recognized by Einstein himself to be only an instance of the more general Galilei principle (the invariance of the physical law with the reference frame) upon including the laws of electromagnetism: this was definitely a great logical improvement. The empirical ones are good temporary practical principles when we relinquish further explanation.

The simplifying principle. A simplifying principle is an unfalsifiable conventional assumption that abridges the formulation of the physical law. An example of such kind of principle is the assumption of homogeneity of time (it is impossible to compare two different time-intervals in temporal sequence). But a purported non-homogeneity of time would introduce only an unnecessary functional time-parametrization in the physical law. Another example is the assumption that the speed of light be isotropic in space. Reichenbach [8] correctly argued that in order to determine simultaneity of distant events we need to know the light speed, but in order to measure the light speed we need to establish simultaneity of two different events for synchronizing clocks, and this leaded to a logical loop. What we can do? Using a single clock we can only determine the two-way average speed of light on a closed path. Reichenbach wrote indeed unconventional Lorentz transformations for non-isotropic light speed, with the only result of introducing an additional anisotropy parameter that is utterly irrelevant in practice. In conclusion: the simplifying principles are good ones, but we must keep in mind their conventional nature.

The epistemological principle. This is the most solid kind of principle: a principle that cannot be violated, even "in-principle", because its violation would involve contradicting the scientific method itself. Somebody would argue that claiming principles only of this kind would be equivalent to claiming an "ultimate theory". This is true. But should this be a good reason for not seeking principles of this kind, and for evaluating their ultimate logical consequences? Clearly, to be a principle for physics it cannot involve only pure logic: it must also incorporate the basic axiomatic of the physical experiment. G. Ludwig has been a great advocate of such kind of principles [9]. Einstein himself formulated special relativity in terms of precise protocols for synchronizing clocks in order to establish coordinate systems. In the recent literature operational axiomatic frameworks of this kind have emerged for quantum theory, later converging to a unified framework [2, 10–12]. The basic notions—tests, events and systems—make the framework equivalent to a category theory for physics [13]. At the same time, it is also the skeleton axiomatization of a general information theory, and, as mentioned, it ultimately leads to the informational axiomatization of quantum theory [2–5]. A remarkable fact about the operational approach is that it logically identifies the experimental protocol with the computer algorithm, providing a stronger logical connection between theory and experiment.

The Relativity Principle

The relativity principle of Galilei and Einstein seems to possess a definite epistemological character, since it establishes the independence of the physical law from the reference system, apparently a necessary requirement for the law formulation and experimentation. The principle instead is based on the "no-absolute" dogma, and nothing forbids defining the law within an absolute frame as long as we are able to translate it to any other frame (which is what we actually do when we invoke the "fixed stars" frame). This viewpoint may look as a sacrilege, it is the only logical possibility for violations of Lorentz covariance.

The Causality Principle

Causality has been always a taboo in physics. It is a principle underlying all modern physics, and has been central to debates on the foundations of relativity and quantum mechanics for over a century. Despite this, there is still a philosophical train of thought arguing that the causality notion should be removed from physics. B. Russell was one of the major advocates of this opinion [14].

On the other hand, causality is such a natural assumption that is often overlooked as an axiom (see e.g. the first quantum axiomatization work of Hardy [10]). Instead, it is the first of the informational axioms of quantum theory [2], also referred to as "no signaling from the future". In simple words it says: in a cascade of measurements on the same system, the outcome probability of a measurement does not depend on the choice of the measurement performed at the output. The principle also implies no-signaling without interaction—shortly "no-signaling", and also commonly known as "Einstein causality". I should make now clear that, being causality an axiom of quantum theory, any information purportedly originated in the future, as a time travel, would logically constitute a falsification of the theory. For example, it would mean to require nonlinearities in state evolution, or other variations of the theory.

As we will see later, in the present informational context special relativity emerges as an approximate principle due to the joint implication of three principles: (1) the causality principle, (2) the Deutsch-Church-Turing principle, and (3) the principle of topological homogeneity of interactions.

The problem of physical causation is a huge topic in philosophy, and a thorough discussion would take a thick volume. For the philosopher disbeliever I just want to add that the reconciliation with the Humean position (that causality is just a human way of looking at phenomena) passes through the probabilistic nature of the causal link stated in the axiom, which involve the comparison between two probabilities: the Humean viewpoint corresponds to the Bayesian interpretation of probability [15].

If causality cannot be proved, it can be falsified, as for any other scientific theory. How? By considering any binary test that is granted to be deterministic, namely to have zero probability for one outcome: if operating at the output of the test we can make this same outcome to happen, then we can logically claim a signaling from the future, given for granted the apparatus and its preparation.

Causal reasoning has always been a basic methodology in physics and in science generally, but the romantic dream of a time travel keeps a sentiment against it alive.

Informational Principles for Quantum Theory

In addition to causality, there are five other informational principles that are needed for deriving quantum theory [2]: (ii) local tomography, (iii) perfect distinguishability, (iv) atomicity of composition, (v) ideal compressibility, and (vi) purification. All six principles apart from (vi) hold for both classical and quantum information: only the purification one singles out quantum theory.

The information-theoretical framework hinges around the notion of event, which can occur probabilistically and has inputs and outputs systems. A complete collection of such events with overall unit probability is what is called *test*—physically a measurement instrument. The systems are just the usual physical systems. Informationally, tests and events represent subroutines, whereas the systems are registers on which information is read and written. Axiom (ii) (stating that joint states of multiple systems can be discriminated by measurements on single systems) has become popular [16], since it reconciles the holism of quantum theory with the reductionism of the experimental approach [17]. Axiom (iii) is crucial for hypothesis falsification, and reconciles probabilism with logic. Axiom (iv) establishes that maximal knowledge of two transformations implies maximal knowledge of their composition, a requirement that seems obvious indeed. The compression axiom (v) is the one that leads to the notion of sub-systems (e.g. the *qubit* is a subsystem of the *qutrit*). It entails the possibility of addressing separately the unknown from the perfectly known. Finally, the purification postulate (vi) informally speaking is the principle of "conservation of information". In simple words it says that irreversibility and mixing can be always regarded as the result of discarding an environment, otherwise everything is describable in terms of pure states and reversible transformations. Another informal way of stating the principle is that ignorance about a part is always compatible with the maximal knowledge about the whole.

The six principles for quantum theory have nothing of "mechanical" nature: what I call "quantum theory" is just the "theory of systems", i.e. the mathematical framework of Hilbert spaces, algebra of observables, unitary transformations. It has no bearing on the "mechanics", namely particles, dynamics, quantization rules: for these the name "quantum mechanics" would be more appropriate. Quantum mechanics, however, is just a small portion of the more general quantum field theory, which itself is a theory of systems: the quantum fields. The only mechanical elements remaining in quantum field theory are the so-called "quantization rules" (or the path-integral) that one may want to avoid in order to make the theory completely autonomous from the classical theory, whereas, reversely, it should be classical mechanics to be derived as an approximation of quantum field theory via a "classicalization" rule. But, how can we formulate a field theory that is quantum *ab initio*? We need more informational principles, in addition to the six ones of quantum theory. Those principles, which will substitute the relativity principles, are: the Deutsch-Church-Turing principle, and the principle of topological homogeneity.

Substitutes for the Relativity Principle

The Deutsch-Church-Turing principle. Rephrasing Deutsch [18]: "Every physical process describable in finite terms must be perfectly simulated by a quantum computer made with a finite number of qubits and a finite number of gates". In the logic of specularity between experimental protocols and algorithms (both include also outcomes), I would say: Every finite experimental protocol is perfectly simulated by a finite quantum algorithm. It is immediate to see that the principle implies two sub-principles: (a) the density of information is finite, and (b) interactions are local. The kind of information that we are considering here is quantum, whence the assertion that the density of information is finite means that the dimension of the Hilbert space for any bounded portion of reality is finite. This means that e.g. there are no Bosons, and the bosonic particle is only an asymptotic approximate notion. Richard Feynman himself is reported to like the idea of finite information density, because he felt that "There might be something wrong with the old concept of continuous functions. How could there possibly be an infinite amount of information in any finite volume?"[1]. The finite dimension of the Hilbert space also implies locality of interactions, namely that the number of quantum systems connected to each gate is finite.

Topological homogeneity of interactions. The principle states that the quantum algorithm describing a physical law is a periodic quantum network. In the informational paradigm the physical law is represented by a finite set of connected quantum gates, corresponding to a finite protocol, theoretically specular of a finite quantum algorithm. Thus locality of interactions is required in order to define a physical law in terms of a finite protocol under the local control of the experimenter, whereas homogeneity represents the universality of the law, which is assumed to hold everywhere and ever. It follows that algorithmically the physical law is represented by a quantum unitary cellular automaton[19]. The "space"-period and the "time"-period of the automaton correspond to the minimum space and time units l_P and t_P—the Planck distance and the Planck time, respectively. At some very small scale—the Planck scale—the world is discrete.

The Quantum Cellular Automaton

Causality together with the Deutsch-Church-Turing principle imply that information propagates at finite speed, the maximum speed being the "speed of light" $c = l_P/t_P$—the causal speed of the automaton. The two principles together thus imply that the state of any finite set of systems can be evaluated exactly as the evolution for of finite number of time-steps of a larger but still finite number of systems in the past causal cone, regardless the quantum network being unbounded. We take as vacuum state any state that is locally invariant under the automaton evolution. The localized states are then those that differ from the vacuum, only for a finite number of systems. The future causal cone of these state-supporting systems is then the place where only we need to evaluate the evolution, again with no need of boundary conditions. We do not have any divergence, nor ultraviolet (no continuum), nor infrared (no calculation for infinite extension): the Deutsch-Church-Turing principle excludes tout court the continuum and the infinite dimension.

Recovering the old quantum field theory. The old field theory is re-covered as an approximation via an analytical asymptotic evaluation of the automaton evolution in the relativistic limit of small wave vectors and for delocalized states, which correspond to the customary quantum particles. In this way one can both derive the Dirac equation in the relativistic regime, but also describe the physics of very large Planckian masses and in the ultra-relativistic regime of huge momenta [20].

Emerging physics. It must be stressed that the homogeneity of interactions is a purely topological property, not a metrical one: "to be near" for systems means just "to be interacting", and the length of the graph links has no physical meaning. Space-time metric emerges from the pure topology by event counting, and the Planck length l_P and time t_P conceptually are only digital-analog conversion factors. Also the particle mass m of the Dirac automaton is a pure number $0 \leq m \leq 1$, and the Planck mass m_P is the conversion factor to customary kilograms.

Universal automata constants. The three quantities l_P, t_P, m_P, are the irreducible universal constants of the automata theory, and the adimensional mass is the only free parameter of the Dirac automaton. The Planck constant can be rewritten in terms of the automata universal constants as $\hbar = m_P l_P^2 t_P^{-1}$.

Inertial mass. As I already explained in my previous FQXi essay [21, 22], the inertial mass is reinterpreted as the slowing down of the information flow via the coupling between the modes flowing along the directions in the network at maximal speed c (for $d > 1$ space-dimensions is a coupling between different chiralities [23]).

Particle speed and Planck mass as bound on mass. The speed of a zero-mass particle depends on the wave-length, and approaches zero at Planckian wavelengths anisotropically in space (see Fig. 11.1). For massive particles the speed of light in the Dirac equation decreases also versus the mass for very large Planckian masses, the automaton evolution becoming stationary at the Planck mass [24], since for larger masses the evolution would be non unitary. It follows that the particle mass is mounded by the Planck mass, at which it behaves essentially as a mini black hole. It is remarkable how these conclusions are reached without using general relativity, just a result of quantum theory.

Energy and momenta are finite in the digital world. The maximum momentum is the De Broglie relation $\hbar \pi / l_P$. We can have only one particle and one antiparticle per Planck cell, and the bound on how much energy can be crammed into a unit of

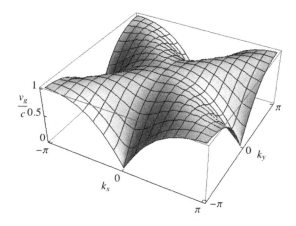

Fig. 11.1 Group velocity v_g (normalized to c) for a zero-mass particle automaton versus the adimensional momentum (k_x, k_y) (from Ref. [23]). The speed is approximately isotropic for low momentum (relativistic regime), and becomes anisotropic for very large momenta (ultra-relativistic regime)

space is determined by the maximum energy per particle, which cannot be more that $\hbar\pi t p^{-1} = 6.14663*10^9$ J, a huge energy! This is the energy for achieving 2 ops [25] of the automaton during the Planck time, as given by the Margulus-Levitin theorem [26] (each step of the automaton is obtained with two rows of quantum gates).

A Quantum-Digital Space-Time

The quantum nature of the automaton is crucial for the emergence of space-time. There are two main points against using a classical automaton.

First point against a classical automaton. With a classical automaton one cannot have isotropic space emerging from an homogeneous classical causal network, due to the Weyl Tile argument [27]: we count the same number of tiles in a square lattice both along the diagonal and in the direction of the square sides: where the $\sqrt{2}$ comes from? Indeed, the maximal speed of information in bcc-lattice automaton, as in the Dirac case, would be faster by a factor $\sqrt{2}$ or $\sqrt{3}$ along diagonals than along lattice axes, ending up with an anisotropic space for any homogeneous lattice [28], (the problem is not cured by the continuum limit). Instead, in a quantum network isotropy is recovered through quantum superpositions of different paths (see e.g. Fig. 11.2c), and we have again isotropy of max-speed in the relativistic regime of small momenta (Fig. 11.1), whereas anisotropy would be in principle visible only in the ultra-relativistic regime of huge momenta (Figs. 11.1,11.2b) or for ultra-localized states (Fig. 11.2d). In a similar manner the quantum nature of the network provides the mechanism for restoration of all continuum symmetries in the relativistic regime. The digital version of Lorentz transformations for a classical homogeneous causal network can be found in Ref. [29]: the usual Lorentz covariance cannot be restored from them. Recovering Lorentz covariance from a classical causal network (i.e.

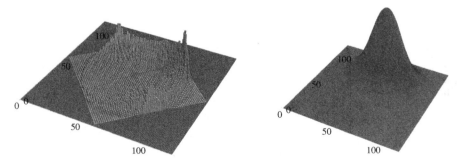

Fig. 11.2 How particles would look in a digital world made by a quantum automaton: the Dirac automaton for $d = 2$ space dimensions. The height of the plot is proportional to the absolute amplitude of finding a particle with up-spin. Colors represents the spin state. The two figures depict the evolved state after 60 steps of an initial state centered in the center of the plane. *Left* spin-up localized state. *Right* Gaussian spin-up particle state, with $\Delta_{x^2} = 2\Delta_{y^2} = 8lp$. (Theory in Ref. [23])

describing a causal ordering partial relation) conflicts with the homogeneity principle, and needs a random topology, as in the causal set program of Sorkin.

Second point against a classical automaton. The second reason against classical automata is that quantum superposition of localized states provides a mechanism for directing information in space, in a continuum of directions, by superimposing localized states at neighboring locations with constant relative phase between them, thus giving momentum to the information flow. Such mechanism is not possible in a classical causal network with finite information density. It is the interplay between quantum coherence and nonlocality that plays the crucial role of keeping information going along a desired direction with minimal spreading, a task that cannot be accomplished by a classical automaton.

Emergence of classical mechanics. The Hamiltonian for the classical field theory corresponding to the quantum automaton can be reversely derived from the unitary operator of the automaton [21, 22]. Customary quantum particles are Gaussian coherent superposition of single-system states with constant relative phase between neighboring systems, corresponding to the particle momentum: the classical trajectory is the "typical path" along the quantum network, namely the path with maximum probability of the Gaussian packet.

Where Is Gravity?

The big question is now where gravity comes from. I still don't have a definite answer, but I believe that the equivalence principle must be rooted in the automaton mechanism: the gravitational force must emerge at the level of the Dirac free theory, which itself defines the inertial mass. This does not occur in customary quantum field theory, but may happen in the quantum automaton theory, in terms of a tiny "thermodynamic" effect that can occur even for few particles: a purely quantum-digital effect. Indeed, the digital nature of the quantum automaton seems to make it the natural scenario for the generalized holographic principle at the basis of the Jacobson-Verlinde idea of gravity as entropic force [6, 7]. The hypothesis of gravity as a quantum-digital effect is very fascinating: it would mean we are indeed experiencing the quantum-digital nature of the world, in everyday experience: through gravity!

Postscriptum

All predictions contained in this Essay has been later derived, and are now available in technical papers. The reader should look at Ref. [23]. Other results can be found in Ref. [20, 30, 31].

The main result is contained in manuscript [23], entitled "Derivation of the Dirac equation from informational principles". There it is proved the remarkable result that from the only general assumptions of locality, homogeneity, isotropy, linearity and unitarity of the interaction network, only two quantum cellular automata follow that have minimum dimension two, corresponding to a Fermi field. The two automata are connected by CPT, manifesting the breaking of Lorentz covariance. Both automata converge to the Weyl equation in the relativistic limit of small wave-vectors, where Lorentz covariance is restored. Instead, in the ultra-relativistic limit of large wave-vectors (i.e. at the Planck scale), in addition to the speed of light one has extra invariants in terms of energy, momentum, and length scales. The resulting distorted Lorentz covariance belongs to the class of the *Doubly Special Relativity* of Amelino-Camelia/Smolin/Magueijo. Such theory predicts the phenomenon of *relative locality*, namely that also coincidence in space, not only in time, depends on the reference frame. In terms of energy and momentum covariance is given by the group of transformations that leave the automaton dispersion relations unchanged. Via Fourier transform one recovers a space-time of quantum nature, with points in superposition. All the above results about distorted Lorentz covariance are derived in the new Ref. [32].

The Weyl QCA is the elementary building block for both the Dirac and the Maxwell field. The latter is recovered in the form of the de Broglie neutrino theory of the photon. The Fermionic fundamental nature of light follows from the minimality of the field dimension, which leads to the Boson as an emergent notion [33].

The discrete framework of the theory allows to avoid all problems that plague quantum field theory arising from the continuum, including the outstanding problem of localization. Most relevant, the theory is quantum *ab initio*, with no need of quantization rules.

References

1. *Feynman and Computation* (Westview Press, Boulder, 2002)
2. G. Chiribella, G.M. D'Ariano, P. Perinotti, Phys. Rev. A **84**, 012311 (2011)
3. C. Brukner, Physics **4**, 55 (2011)
4. L. Hardy, (2011). arXiv:1104.2066
5. L. Masanes, M.P. Müller, New J. Phys. **13**, 063001 (2011)
6. T. Jacobson, Phys. Rev. Lett. **75**, 1260 (1995)
7. E. Verlinde, J. High Energy Phys. **4**, 029 (2011)
8. H. Reichenbach, *The Philosophy of Space and Time* (Dover, New York, 1958)
9. G. Ludwig, G. Thurler, *A New Foundation of Physical Theories* (Springer, New York, 2006)
10. L. Hardy, (2001). arXiv:quant-ph/0101012
11. H. Barnum, J. Barrett, L. Orloff Clark, M. Leifer, R. Spekkens, N. Stepanik, A. Wilce, R. Wilke, New J. Phys. **12**, 3024 (2010)
12. B. Dakic, C. Brukner, in *Deep Beauty: Understanding the Quantum World Through Mathematical Innovation*, ed. by H. Halvorson (Cambridge University Press, 2011)
13. B. Coecke, Contemp. Phys. **51**, 59 (2010)
14. B. Russell, Proc. Aristot. Soc. **13**, 1 (1912)
15. C.A. Fuchs, (2010). arXiv:1003.5209
16. L. Hardy, W.K. Wootters, Found. Phys. **42**, 454 (2012)
17. G.M. D'Ariano, in *Philosophy of Quantum Information and Entanglement*, ed. by A. Bokulich, G. Jaeger (Cambridge University Press, Cambridge, 2010)

18. D. Deutsch, R. Soc. Lond. **400**, 97 (1985)
19. B. Schumacher, R.F. Werner, (2004). arXiv:quant-ph/0405174
20. A. Bisio, G.M. D'Ariano, A. Tosini, (2012). arXiv:1212.2839
21. G.M. D'Ariano, FQXi Essay Contest Is Reality Digital or Analog? (2011)
22. G.M. D'Ariano, Il Nuovo Saggiatore **28**, 13 (2012)
23. G.M. D'Ariano, P. Perinotti, Phys. Rev. A **90**, 062106 (2014)
24. G.M. D'Ariano, Phys. Lett. A **376**, 678 (2012). arXiv:1012.0756
25. S. Lloyd, (2012). arXiv:1206.6559
26. N. Margolus, L.B. Levitin, Phys. D **120**, 188 (1998)
27. H. Weyl, *Philosophy of Mathematics and Natural Sciences* (Princeton University Press, Princeton, 1949)
28. T. Fritz, Discret. Math. **313**, 1289 (2013)
29. G.M. D'Ariano, A. Tosini, Stud. Hist. Philos. Mod. Phys. **44**
30. G.M. D'Ariano, F. Manessi, P. Perinotti, A. Tosini, (2014). arXiv:1403.2674
31. A. Bisio, G.M. D'Ariano, A. Tosini, Phys. Rev. A **88**, 032301 (2013)
32. A. Bibeau-Delisle, A. Bisio, G.M. D'Ariano, P. Perinotti, A. Tosini, (2013). arXiv:1310.6760
33. A. Bisio, G.M. D'Ariano, P. Perinotti, arXiv:1407.6928

Chapter 12
The Universe Is Not a Computer

Ken Wharton

Abstract When we want to predict the future, we compute it from what we know about the present. Specifically, we take a mathematical representation of observed reality, plug it into some dynamical equations, and then map the time-evolved result back to real-world predictions. But while this computational process can tell us what we want to know, we have taken this procedure too literally, implicitly assuming that the universe must compute itself in the same manner. Physical theories that do not follow this computational framework are deemed illogical, right from the start. But this anthropocentric assumption has steered our physical models into an impossible corner, primarily because of quantum phenomena. Meanwhile, we have not been exploring other models in which the universe is not so limited. In fact, some of these alternate models already have a well-established importance, but are thought to be mathematical tricks without physical significance. This essay argues that only by dropping our assumption that the universe is a computer can we fully develop such models, explain quantum phenomena, and understand the workings of our universe. (This essay was awarded third prize in the 2012 FQXi essay contest; a new afterword compares and contrasts this essay with Robert Spekkens' first prize entry.)

Introduction: The Newtonian Schema

Isaac Newton taught us some powerful and useful mathematics, dubbed it the "System of the World", and ever since we've assumed that the universe actually runs according to Newton's overall scheme. Even though the details have changed, we still basically hold that the universe is a computational mechanism that takes some initial state as an input and generates future states as an output. Or as Seth Lloyd says, "It's a scientific fact that the universe is a big computer" [1].

Such a view is so pervasive that only recently has anyone bothered to give it a name: Lee Smolin now calls this style of mathematics the "Newtonian Schema" [2]. Despite

K. Wharton (✉)
Department of Physics and Astronomy, San José State University,
San José, CA 95192-0106, USA
e-mail: kenneth.wharton@sjsu.edu

© Springer International Publishing Switzerland 2015
A. Aguirre et al. (eds.), *Questioning the Foundations of Physics*,
The Frontiers Collection, DOI 10.1007/978-3-319-13045-3_12

the classical-sounding title, this viewpoint is thought to encompass all of modern physics, including quantum theory. This assumption that we live in a Newtonian Schema Universe (NSU) is so strong that many physicists can't even articulate what other type of universe might be conceptually possible.

When examined critically, the NSU assumption is exactly the sort of anthropocentric argument that physicists usually shy away from. It's basically the assumption that the way we humans solve physics problems must be the way the universe actually operates. In the Newtonian Schema, we first map our knowledge of the physical world onto some mathematical state, then use dynamical laws to transform that state into a new state, and finally map the resulting (computed) state back onto the physical world. This is useful mathematics, because it allows us humans to predict what we don't know (the future), from what we do know (the past). But is it a good template for guiding our most fundamental physical theories? Is the universe effectively a quantum computer? This essay argues "no" on both counts; we have erred by assuming the universe must operate as some corporeal image of our calculations.

This is not to say there aren't good arguments for the NSU. But it is the least-questioned (and most fundamental) assumptions that have the greatest potential to lead us astray. When quantum experiments have thrown us non-classical curveballs, we have instinctively tried to find a different NSU to make sense of them. Thanks to this deep bias, it's possible that we have missed the bigger picture: the mounting evidence that the fundamental rules that govern our universe cannot be expressed in terms of the Newtonian Schema. It's evidence that we've so far found a way to fold back into an NSU, but at a terrible cost—and without debate or recognition that we've already developed the core framework of a promising alternative.

The next section will detail the problems that arise when one tries to fit quantum phenomena into an NSU. The following sections will then outline the alternative to the NSU and show how it naturally resolves these same problems. The conclusion is that the best framework for our most fundamental theories is not the Newtonian Schema, but a different approach that has been developed over hundreds of years, with ever-growing importance to all branches of physics. It seems astounding that we have not recognized this alternate mathematics as a valid Schema in its own right, but *no* alternative makes sense if we've already accepted Lloyd's "fact" that the universe is a (quantum) computer. Only by recognizing that the NSU is indeed an *assumption* can we undertake an objective search for the best description of our universe.

Challenges from the Quantum

Until the 20th century, the evidence against the NSU was circumstantial at best. One minor issue was that (fundamental) classical laws can equally well be run forward and backward—say, to retrodict the historical locations of planets. So there's nothing in the *laws* to imply that the universe is a forward-running computer program, calculating the future from some special initial input. Instead, every moment is just as special as every other moment.

Of course, the same is true for a deterministic and reversible computer algorithm—from the data at any time-step, one can deduce the data at all other time-steps. Combined with a special feature of the Big Bang (its status as an ordered, low-entropy boundary condition), this concern mostly vanishes.[1]

But *quantum* phenomena raise three major challenges to the NSU. Standard quantum theory deals with each of them in basically the same way—by assuming the NSU must be correct, and using suspiciously anthropocentric reasoning to recast the universe in an image of our quantum calculations.

First, we have Heisenberg's uncertainty principle (HUP). In the classical context of Heisenberg's original paper [3], this means we can never know the initial state of the universe with enough precision to compute the future. This would not alone have challenged the NSU—a universal computer could potentially use the full initial state, even if we did not know it. But it weakens the above argument about how the Big Bang is special, because not even the Big Bang can beat the HUP—as confirmed by telltale structure in the cosmological microwave background. The special low-entropy order in the universe's initial state is accompanied by random, non-special, disorder.

But conventional quantum theory rejects the above reading of the HUP. In spirit with the NSU, the unknown quantities are no longer even thought to exist. Note the implication: if we humans can't possibly know something, then the universe shouldn't know it either. The Big Bang is restored as the universe's special "input", and the NSU is saved. But this step leads to new problems—namely, we can't use classical laws anymore, because we don't have enough initial data to solve them. To maintain an NSU, we're forced to drop down from classical second-order differential equations to a simpler first-order differential equation (the Schrödinger equation).

This leads to the second major challenge—the Schrödinger equation yields the wrong output. Or more accurately, the future that it computes is not what we actually observe. Instead, it merely allows us to (further) compute the probabilities of different possible outcomes. This is a huge blow to the NSU. Recall the three steps for the Newtonian Schema: (1) Map the physical world onto a mathematical state, (2) Mathematically evolve that state into a new state, and (3) Map the new state back onto the physical world. If one insists on a universe that computes itself via the Schrödinger equation, the only way to salvage the NSU is to have step 3 be a probabilistic map. (Even though the inverse of that map, step 1, somehow remains deterministic.)

Once again, since we are restricted from knowing the exact outcome, conventional quantum theory puts the same restrictions on the NSU itself. In step 3, the story goes, not even the *universe* knows which particular outcome will occur. And yet one particular outcome *does* occur, at least when one looks. Even worse, the measurement process blurs together steps 2 and 3, affecting the state of the universe itself in a manner manifestly inconsistent with the Schrödinger equation. The question of

---

[1] Although it does raise questions, such as *why* the laws happen to be time-symmetric, if the boundary conditions are so time-asymmetric.

exactly where (and how) the universe stops using the Schrödinger equation is the infamous "measurement problem" of quantum theory. It becomes harder to think of the universe as computing itself if the dynamical laws are not objectively defined.

So it's perhaps unsurprising that many physicists imagine an NSU that ignores step 3 altogether; the universe is simply the computation of the ever-evolving Schrödinger equation, the mismatch with reality notwithstanding. The only consistent way to deal with this mismatch is to take the Everettian view that our entire experience is just some small, subjective sliver of an ultimate objective reality—a reality that we do *not* experience [4].

Which brings us to the third challenge to the NSU: the dimensionality of the quantum state itself. The phenomenon of quantum entanglement—where the behaviors of distant particles are correlated in strikingly non-classical ways—seems to require a quantum state that does not fit into the spacetime we experience. The quantum state of a N-particle system formally lives in a "configuration space" of $3N$ dimensions. If the universe is the self-computation of such a state, then we live in a universe of enormous dimensionality. Any consistent, NSU view of quantum theory (not merely the Everettians) must maintain that Einstein's carefully-constructed spacetime is fundamentally incorrect. Instead, one must hold that Schrödinger accidentally stumbled onto the correct mathematical structure of the entire universe.

Of course, configuration space was not an invention of Schrödinger's; it continues to be used in statistical mechanics and other fields where one does not know the exact state of the system in question. Poker probabilities, for example, are computed in such a space. Only after the cards are turned face up does this configuration space of possibilities collapse into one actual reality.

In the case of cards, though, it's clear that the underlying reality was there all along—configuration space is used because the players lack information. In the case of a theory that underlies everything, that's not an option. Either the quantum state neglects some important "hidden variables", or else reality is actually a huge-dimensional space. Conventional thinking denies any hidden variables, and therefore gives up on ordinary spacetime. Again, note the anthropocentrism: we use configuration spaces to calculate entangled correlations, so the universe must *be* a configuration space.[2]

The NSU becomes almost impossible to maintain in the face of all these challenges. Treating the universe as a computer requires us to dramatically alter our dynamical equations, expand reality to an uncountable number of invisible dimensions, and finesse a profound mismatch between the "output" of the equations and what we actually observe.

Of course, no one is particularly happy with this state of affairs, and there are many research programs that attempt to solve each of these problems. But almost none of these programs are willing to throw out the deep NSU assumption that may be at ultimate fault. This is all the more surprising given that there *is* a well-established alternative to the Newtonian Schema; a highly regarded mathematical

---

[2] Like a poker player that denies any reality deeper than her own knowledge, imagining the face-down cards literally shifting identities as she gains more information.

framework that is in many ways superior. The barrier is that practically no one takes this mathematics literally, as an option for how the universe might "work". The next sections will outline this alternative and reconsider the above challenges.

The Lagrangian Schema

While a first-year college physics course is almost entirely dominated by the Newtonian Schema, some professors will include a brief mention of Fermat's Principle of least time. It's a breathtakingly simple and powerful idea (and even pre-dates Newton's *Principia*)—it just doesn't happen to fit in with a typical engineering-physics curriculum.

Fermat's Principle is easy to state: Between any two points, light rays take the quickest path. So, when a beam of light passes through different materials from point X to point Y, the path taken will be the fastest possible path, as compared to all other paths that go from X to Y. In this view, the reason light bends at an air/water interface is not because of any algorithm-like chain of cause-and-effect, but rather because it's globally more efficient.

However elegant this story, it's not aligned with the Newtonian Schema. Instead of initial inputs (say, position and angle), Fermat's principle requires logical inputs that are both initial and final (the positions of X and Y). The initial angle is no longer an input, it's a logical *output*. And instead of states that evolve in time, Fermat's principle is a comparison of entire paths—paths that cannot evolve in time, as they already cover the entire timespan in question.

This method of solving physics problems is not limited to light rays. In the 18th century, Maupertuis, Euler, and Lagrange found ways to cast the rest of classical physics in terms of a more general minimization[3] principle. In general, the global quantity to be minimized is not the time, but the "action". Like Fermat's Principle, this so-called Lagrangian Mechanics lies firmly outside the Newtonian Schema. And as such, it comprises an alternate way to do physics—fully deserving of the title "Lagrangian Schema".

Like the Newtonian Schema, the Lagrangian Schema is a mathematical technique for solving physics problems. In both schemas, one first makes a mathematical representation of physical reality, mapping events onto parameters. On this count, the Lagrangian Schema is much more forgiving; one can generally choose any convenient parameterization without changing the subsequent rules. And instead of a "state", the key mathematical object is a scalar called the Lagrangian (or in the case of continuous classical fields, the Lagrangian density, $\mathcal{L}$), a function of those parameters and their local derivatives.

There are two steps needed to extract physics from $\mathcal{L}$. First, one partially constrains $\mathcal{L}$ on the boundary of some spacetime region (e.g., fixing X and Y in Fermat's Principle). For continuous fields, one fixes continuous field parameters. But only

---

[3] Actually, extremization.

the boundary parameters are fixed; the intermediate parameters and the boundary derivatives all have many possible values at this stage.

The second step is to choose one of these possibilities (or assign them probabilistic weights). This is done by summing the Lagrangian (densities) everywhere inside the boundary to yield a single number, the action S. The classical solution is then found by minimizing the action; the quantum story is different, but it's still a rule that involves S.

To summarize the Lagrangian Schema, one sets up a (reversible) two-way map between physical events and mathematical parameters, partially constrains those parameters on some spacetime boundary *at both the beginning and the end*, and then uses a global rule to find the values of the unconstrained parameters. These calculated parameters can then be mapped back to physical reality.

Newton Versus Lagrange

There are two fairly-widespread attitudes when it comes to the Lagrangian Schema. The first is that the above mathematics is just that—mathematics—with no physical significance. Yes, it may be beautiful, it may be powerful, but it's not how our universe *really* works. It's just a useful trick we've discovered. The second attitude, often held along with the first, is that action minimization is provably equivalent to the usual Newtonian Schema, so there's no point in trying to physically interpret the Lagrangian Schema in the first place.

To some extent, these two attitudes are at odds with each other. If the two schemas are equivalent, then a physical interpretation of one should map to the other. Still, the arguments for "schema-equivalence" need to be more carefully dismantled. This is easiest in the quantum domain, but it's instructive to first consider a classical case, such as Fermat's Principle.

A typical argument for schema-equivalence is to use Fermat's principle to derive Snell's law of refraction, the corresponding Newtonian-style law. In general, one can show that action minimization always implies such dynamic laws. (In this context, the laws are generally known as the Euler-Lagrange equations.) But a dynamical law is not the whole Newtonian Schema—it's merely step 2 of a three-step process. And the input and output steps differ: Snell's law takes different inputs than Fermat's Principle and yields an output (the final ray position) that was *already constrained* in the action minimization. Deriving Newtonian results from a Lagrangian premise therefore requires a bit of circular logic.

Another way to frame the issue is to take a known embodiment of the Newtonian Schema—a computer algorithm—and set it to work solving Lagrangian-style problems with initial and final constraints. The only robust algorithms for solving such problems are iterative,[4] with the computer testing multiple histories, running back and forth in time. And this sort of algorithm doesn't sound like a universe that com-

---

[4] As in the Gerchberg-Saxton algorithm [5].

putes itself—the most obvious problem being the disconnect between algorithmic time and actual time, not to mention the infinite iterations needed to get an exact answer.

Still, conflating these two schemas in the classical domain where they have some modest connection is missing the point: These are still two different ways to solve problems. And when *new* problems come around, different schemas suggest different approaches. Tackling every new problem in an NSU will therefore miss promising alternatives. This is of particular concern in quantum theory, where the connection between the two schemas gets even weaker. Notably, in the Feynman path integral (FPI), the classical action is no longer minimized when calculating probabilities, so it's no longer valid to "derive" the Euler-Lagrange equations using classical arguments.[5]

So what should we make of the Lagrangian Schema formulations of quantum theory? (Namely, the FPI and its relativistic extension, Lagrangian quantum field theory, LQFT.) Feynman never found a physical interpretation of the FPI that didn't involve negative probabilities, and LQFT is basically ignored when it comes to interpretational questions. Instead, most physicists just show these approaches yield the same results as the more-typical Newtonian Schema formulations, and turn to the latter for interpretational questions. But this is making the same mistake, ignoring the differences in the inputs and outputs of these two schemas. It's time to consider another approach: looking to the Lagrangian Schema not as equivalent mathematics, but as a different framework that can provide new insights.

Quantum Challenges in a Lagrangian Light

Previously we outlined outlined three challenges from quantum theory, and the high cost of answering them in the NSU framework. But what do these challenges imply for an *LSU*? How would the founders of quantum theory have met these challenges if they thought the universe ran according to the mathematics of the Lagrangian Schema—not as a computer, but rather as a global four-dimensional problem that was solved "all at once"? Surprisingly, the quantum evidence hardly perturbs the LSU view at all.

The first challenge was the uncertainty principle, but the classical LSU had this built in from the start, because it never relied on complete initial data in the first place. Indeed, for classical particles and fields, there's a perfect match between the initial data one uses to constrain the Lagrangian and the amount of classical data one is permitted under the HUP. In Fermat's principle, if you know the initial light ray position, the HUP says you *can't* know the initial angle.

---

[5] It's only when one combines the quantum wave equations with the probabilistic Born rule that FPI probabilities are recovered; see the discussion of Eq. (1) in [6].

Curiously, this "perfect match" is only one-way. The HUP allows more ways to specify the initial data than is seemingly permitted by the Lagrangian Schema. For example, the HUP says that one can know the initial position *or* the angle of the light ray, but Fermat's principle only works with constrained initial positions.

But this is not a problem so much as a suggested research direction, evident only to a Lagrangian mindset. Perhaps the HUP is telling us that we've been too restricted in the way we've fixed the initial and final boundaries on classical Lagrangians. The natural question becomes: What would happen if we required action-minimization for *any* HUP-compatible set of initial and final constraints? For classical fields, the answer turns out to be that such constraints must be roughly quantized, matching equations that look like quantum theory [7].

Because the LSU doesn't need complete initial data to solve problems, there's nothing wrong with the second-order differential equations of classical physics (including general relativity, or GR). With this change, one can revive Heisenberg's original interpretation of the HUP, yielding a natural set of initially-unknown "hidden variables" (such as the ray angles in Fermat's Principle). In simple cases [8], at least, these hidden variables can not only explain the probabilistic nature of the outcomes, but can actually be computed (in hindsight, after the final boundary becomes known). Furthermore, there's no longer a compelling reason to drop to the first-order Hamiltonian equations, the standard Newtonian Schema version of quantum theory. And since it's this leap from Lagrangian to Hamiltonian that introduces many of the deepest problems for quantum gravity (the "problem of time", etc.), there are good reasons to avoid it if at all possible.

The Lagrangian Schema also provides a nice perspective on the second challenge: the failure of Newtonian-style equations to yield specific, real-world outcomes (without further probabilistic manipulations). Recall this was the most brutal challenge to the NSU itself, raising the still-unresolved measurement problem and breaking the symmetry between the past and future. But the LSU doesn't utilize dynamical equations, so it dodges this problem as well. The temporal outcome is not determined by an equation, it's imposed as an *input* constraint on $\mathcal{L}$. And because of the time-symmetric way in which the constraints are imposed, there's no longer any mathematical difference between the past and future; both constraints directly map to the real world, without further manipulation. In fact, the Lagrangian procedure of "fixing" the future boundary looks remarkably like quantum measurements, providing a new perspective on the measurement problem [9].

A common complaint at this point is that the above feature is a bug, in that it somehow makes the Lagrangian Schema unable to make predictions. After all, what we usually want to know is the outcome B given the input A, or at least the conditional probability $P(B_i|A)$ (the probability of some possible outcome B_i given A). But if one particular outcome (say, B_1) is itself an external constraint imposed on $\mathcal{L}$, a logical input rather than an output, then we can't solve the problem without knowing the temporal outcome. Furthermore, since in this case B_1 is 100 % certain, the other possibilities (B_2, B_3, etc.) can never happen, contrary to quantum theory.

But like the NSU, this complaint conflates our useful calculations with objective reality. In truth, any particular observed event does indeed have a single outcome, with after-the-fact 100 % certainty. If we don't yet know that outcome, we can still *imagine* fixing different outcome constraints B_i, and using $\mathcal{L}$ to compute an expected joint probability $P(A, B_i)$ for each possibility. It's then a simple matter to normalize subject to some particular initial condition A and generate the conditional probabilities $P(B_i|A)$. These probabilities live in our heads until the actual outcome appears and show us what has been the case all along, at which point we update our incomplete knowledge. This is basic Bayesian probability (see the above poker example), and many have noted that it is a more natural interpretation of the standard quantum "collapse" [10, 11].

Finally, consider the challenge of quantum entanglement. The problem with the NSU mindset is that it demands an input state that can compute all possible outputs, even if we don't know what type of measurement will eventually be made. In N-particle systems, the number of possible future measurements goes up exponentially with N. Keeping track of *all* possible future measurements requires a state that lives in an enormous configuration space. It simply doesn't "fit" in the universe we observe, or in Einstein's GR.

But as we've seen, the NSU conflates the information we humans need to solve a problem and the data that must actually correspond to reality. In any particular case, a vast portion of this traditional quantum state turns out to be needless—it never gets mapped to reality and is erased by the so-called "collapse". That's because all possible measurements *don't* occur; only the actual measurement occurs. Once the future measurement choice is known, the joint probabilities take on familiar forms, with descriptions that have exact mathematical analogies to cases that *do* fit in spacetime [6, 12].

Which brings us to the key point: If one wants to "fit" quantum theory into the spacetime of GR, one *must* use the Lagrangian Schema, solving the problem "all at once". Only then can the solution take into account the actual future measurement— which, recall, is imposed as a boundary constraint on $\mathcal{L}$. So an LSU-minded physicist, when encountering entanglement, would have no reason to add new dimensions. The "spooky" link between entangled particles would merely be joint correlations enforced by virtue of both particles contributing to the same global action [12].

When viewed from a Lagrangian Schema mindset, the transition from classical to quantum phenomena is not only less jarring, but arguably a natural extension. Sure, some things have to change—perhaps extending the principle of action minimization [7]—but they're changes that only make sense in an LSU, with no NSU translation. Classical physics provided a few cases where the two Schemas seemed to almost overlap, perhaps lulling us into a feeling that these two approaches must *always* overlap. But the fact that quantum phenomena are so incomprehensible in an NSU, and more natural in an LSU, should make us consider whether we've been using a deeply flawed assumption all along.

Conclusions: Our Lagrangian Universe

The best reasons for taking the Lagrangian Schema seriously lie in quantum theory, but there are other reasons as well. It's the cleanest formulation of general relativity, with the automatic parameter-independence that GR requires, and bypasses problematic questions such as how much initial data one needs to solve the Newtonian-style version. The LSU blends time and space together just like GR, while the NSU has to grapple with a dynamic evolution that seems to single out time as "special". The standard model of particle physics is not a set of dynamic equations, but is instead a Lagrangian density, with deep and important symmetries that are only evident in such a framework. Even NSU-based cosmological mysteries, such as why causally-disconnected regions of the universe are so similar, no longer seem as problematic when viewed in an LSU light.

But from the computational perspective of the NSU, any description of an LSU seems baffling and unphysical. When trying to make sense of the LSU, a NSU-minded physicist might ask a number of seemingly-tough questions. *Which past events cause the future boundary constraint? How do objects in the universe "know" what future boundary they're supposed to meet? Doesn't Bell's Theorem* [13] *prove that quantum correlations can't be caused by past hidden variables?* A close look reveals these questions are already biased—they all implicitly assume that we live in an NSU. But without the mentality that the past "causes" the future by some algorithmic process, the above questions are no longer well-posed.

Constructing a complete theory built upon the Lagrangian Schema is a vast project, one that has barely even begun. The necessary first step, though, is to recognize that the NSU is an assumption, not a statement of fact. Even then, it will be difficult to put such a deep bias behind us completely, to distinguish our most successful calculations from our most fundamental physical models. But it also wasn't easy to fight other anthropocentric tendencies, and yet the Earth isn't the center of the universe, our sun is just one of many, there is no preferred frame of reference. Now there's one last anthropocentric attitude that needs to go, the idea that the computations we perform are the same computations performed by the universe, the idea that the universe is as 'in the dark' about the future as we are ourselves.

Laying this attitude to one side, at least temporarily, opens up a beautiful theoretical vista. We can examine models that have no Newtonian Schema representation, and yet nicely incorporate quantum phenomena into our best understanding of spacetime. We can treat the universe as a global, four-dimensional boundary-value problem, where each subset of the universe can be solved in exactly the same manner, with exactly the same rules. Stories can be told about what happens between quantum measurements, and those very measurements can be enfolded in a bigger region, to simultaneously tell a bigger story. And most importantly, such models will suggest further models, with alterations that only make sense in a Lagrangian framework—perhaps a local constraint like $\mathcal{L} = 0$, or treating the Euler-Lagrange equations as just an approximation to a fundamentally underdetermined problem.

It is *these* models, the balance of the evidence suggests, that have a chance of representing how our universe *really* works. Not as we humans solve problems, not as a computer, but as something far grander.

Afterword: On Spekkens' Winning Essay

"The Universe is not a Computer" was awarded third prize in the 2012 FQXi Essay Contest, ranked behind other excellent essays—most notably the first prize winner[14], written by the Perimeter Institute physicist Robert Spekkens. Like the above essay, Spekkens zeroed in on how most physical theories are framed using a states-plus-dynamics Newtonian Schema, although with a different focus and conclusion. With these essays now being presented in the same volume, this afterword is an opportunity to compare and constrast these two viewpoints.

Spekkens' essay begins by noting that physical theories generally are divided into "dynamics" (laws that implement time evolution) and "kinematics" (the space of physical states permitted by a theory). These, of course, are the key components of any theory that falls under the Newtonian Schema, described above. After noting how very different theories are framed in precisely this manner, Spekkens' essay states that physicists "typically agree that any proposal must be described in these terms".

Both of our essays are in general agreement that it is *this framing of physical theories*, in terms of dynamics + kinematics, that contains a widespread mistaken assumption—but we are in disagreement as to the precise nature of the mistake. Spekkens' essay makes the excellent point that seemingly-different theories (which postulate different kinematics and dynamics) can in fact be empirically indistinguishable when these two components are taken together. I agree with Spekkens that two such theories should not be viewed as competitive explanations but rather as essentially identical. (One possible lesson for theorists might be that they are proposing too many different theories, and can focus on a bare few.)

On the other hand, the above essay argues that theorists have been far too conservative in postulating different theories, in that they almost exclusively are couched in the Newtonian Schema. One can make a case that framing theories in terms of kinematics + dynamics is more an instinctive habit than a well-thought-out "agreement", and that new Lagrangian-schema approaches are needed. The mistake, in this view, is that the kinematics + dynamics framework is too restrictive, not too permissive.

These different conclusions are not mutually exclusive. Take Spekkens' example of how classical physics can be couched in terms of forces and Newton's laws (on one hand) and Hamiltonian dynamics (on the other). These two theories are indeed empirically indistinguishable, and should not be thought of as essentially different. But they also both fall under the Newtonian Schema. It is notable that classical Lagrangian mechanics does not specify *any* dynamics, and therefore lies in a different category of theory altogether (a category unaddresed in Spekkens' essay). In this sense, our essays are both making the case that Newtonian Schema theories are more similar than they might appear, and the above essay is making the additional case that Lagrangian Schema theories are different and under-explored.

One counterpoint to this claim might be to note that Lagrangian Schema theories can *also* be expressed in terms of dynamics + kinematics; namely, there are *no* dynamical laws, and the allowed kinematical "states" are merely four-dimensional *histories* that obey certain restrictions. In other words, Lagrangian Schema theories are all kinematics, no dynamics.

Might Spekkens' claim for empirical indistinguishability perhaps be extended to apply to Lagrangian Schema theories as well, showing them to all be essentially equivalent to a class of Newtonian Schema counterparts? After all, classical Lagrangian mechanics is empirically equivalent to Newtonian mechanics (if leaving aside the input/output differences discussed above), and the probabilities generated by the Feynman path integral are empirically equivalent to the probabilities generated by the combination of the Schrödinger equation and the Born Rule [6]. Combined with the many inter-Newtonian-Schema examples in Spekkens' essay, this may make it seem like such an argument might be successfully developed.

But the essential differences between three-dimensional states governed by dynamics and four-dimensional "histories" with no dynamics is far more dramatic than these examples imply. Indeed, counter-examples have recently been published [15, 16] demonstrating simple Lagrangian Schema toy models with no dynamical counterpart whatsoever. And far from being some unimportant curiosity, it is this precise style of model that most naturally maps to the very quantum phenomena that defy Newtonian Schema explanations.

For example, consider the discussion concerning kinematical- and dynamical-locality in Spekkens' essay. There, the point was that since fully-local Newtonian Schema accounts run afoul of the Bell inequalities, trying to rescue kinematical locality was essentially impossible: Any such theory would necessarily have dynamical nonlocality, and would therefore always be empirically indistinguishable from a theory with kinematical nonlocality. But in the case of the Lagrangian Schema, *there is no dynamics*, local, nonlocal, or otherwise. The promise of rescuing kinematical locality (as discussed in Sect. "Quantum Challenges in a Lagrangian Light") is now far more than just an empty redefinition of terms—indeed, it is one of the primary motivations for pursuing Lagrangian Schema explanations in the first place.

So despite my general agreement with almost everything in Spekkens' winning essay, that essay is still framed in the Newtonian Schema mindset that is arguably a deep and mistaken assumption in its own right. The claim in Spekkens' abstract that "A change to the kinematics of a theory... can be compensated by a change to its dynamics without empirical consequence" is not always true when there are no dynamics in the original theory (as per the counter-examples in [15, 16]). Still, since it does appear that this claim is true for Newtonian Schema theories, Spekkens' essay will hopefully help to focus the debate where it is needed: not between empirically indistinguishable Newtonian Schema explanations of quantum phenomena, but rather between dynamical and "all at once" explanatory accounts of our universe.

References

1. S. Lloyd, http://www.edge.org/3rd_culture/lloyd06/lloyd06_index.html. Accessed 16 Aug 2012
2. L. Smolin, Phys. World, p. 21 (2009)
3. W. Heisenberg, Zeitschrift fur Physik **43**, 172 (1927)
4. H. Everett, Rev. Mod. Phys. **29**, 454 (1957)
5. R.W. Gerchberg, W.O. Saxton, Optik **35**, 237 (1972)
6. K.B. Wharton, D.J. Miller, H. Price, Symmetry **3**, 524 (2011)
7. K.B. Wharton, arXiv:0906.5409
8. K.B. Wharton, Found. Phys. **40**, 313 (2010)
9. K. Wharton, arXiv:1106.1254
10. R.W. Spekkens, Phys. Rev. A **75**, 32110 (2007)
11. N. Harrigan, R.W. Spekkens, Found. Phys. **40**, 125 (2010)
12. P.W. Evans, H. Price, K.B. Wharton, Brit. J. Found. Sci. (2012). doi:10.1093/bjps/axr052
13. J.S. Bell, Rev. Mod. Phys. **38**, 447 (1966)
14. R.W. Spekkens, arXiv:1209.0023
15. H. Price, Stud. Hist. Philos. Mod. Phys. **39**, 752 (2008)
16. K. Wharton, Information **5**, 190 (2014)

Chapter 13
Against Spacetime

Giovanni Amelino-Camelia

Abstract The notion of "location" physics really needs is exclusively the one of "detection at a given detector" and the time for each such detection is most primitively assessed as the readout of some specific material clock. The redundant abstraction of a macroscopic spacetime organizing all our particle detections is unproblematic and extremely useful in the classical-mechanics regime. But I here observe that in some of the contexts where quantum mechanics is most significant, such as quantum tunneling through a barrier, the spacetime abstraction proves to be cumbersome. And I argue that in quantum-gravity research we might limit our opportunities for discovery if we insist on the availability of a spacetime picture.

A 19th-century Alice could have asked "how do we know we all share the same time?" To my knowledge nobody asked the question then. And if some Alice did ask, Bob's reaction surely would have been going something like "what a stupid question! of course we all share the same time!"
We now know that this question is meaningful and actually the answer is no: we established experimentally that observers in relative motion do not share the same time.

A similar example of apparently stupid 19th-century question can be structured around double-slit experiments. And sometimes the question we are not asking is about the meaningfulness of a notion we are taking for granted: particularly in the second half of the 19th century we were very busy attempting to establish the properties of the ether, but we then figured out that there is no place in physics for any property of the ether.

To me the most defining mission of science is to establish which of the questions we are not even asking (implicitly assuming their answer is "evidently yes") is actually a meaningful question whose answer is no. And these notes are about my latest speculation of this sort: in modern physics, and particularly in quantum-gravity research, we are very busy trying to establish the fundamental properties of spacetime, and I start to wonder whether, after all, there might be no properties of spacetime to worry about.

G. Amelino-Camelia (✉)
Dipartimento di Fisica, Sapienza Università di Roma and INFN Sez. Roma1,
P.le A. Moro 2, 00185 Rome, Italy
e-mail: giovanni.amelino-camelia@roma1.infn.it

© Springer International Publishing Switzerland 2015   191
A. Aguirre et al. (eds.), *Questioning the Foundations of Physics*,
The Frontiers Collection, DOI 10.1007/978-3-319-13045-3_13

I am aware of the fact that this speculation is very audacious. Spacetime appears toplay a robust reliable role in our conceptualization of physics. This robustness is something we appreciate from very early on in life, since our "resident devices" (eyes, ears, nose..., of course particle detectors) inform us of the reliability of our spacetime inferences. And this reliability so far extends all the way to the most powerful devices we invented, including our best particle colliders (see Fig. 13.1). But in some sense we never actually "see" spacetime, we never "detect spacetime", and over the last century we gradually came to appreciate that there would be no spacetime without particles. And if we contemplate the possibility of reducing our description of Nature to its most primitive building blocks, the most "minimalistic" description of physics that is afforded to us, then it becomes evident that the notion of "location" physics really needs is exclusively the one of "detection at a given detector" and that the time for each such detection is most primitively assessed as the readout of some specific material clock at the detector. At least in this sense, the abstraction/inference of a macroscopic spacetime organizing all such timed detections is redundant. We could (and perhaps should) build all of our description of physics, including the so-called "spacetime observables", using as primitive/most-fundamental notions the ones of material detectors and clocks.

Of course, the spacetime abstraction is unrenounceably convenient for organizing and streamlining our description of observations done in the classical-mechanics regime. But I here highlight some aspects of quantum mechanics, such as tunneling through a barrier, for which the spacetime abstraction proves to be cumbersome. And

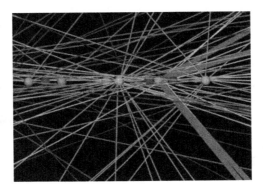

Fig. 13.1 Image courtesy of the ATLAS collaboration; an aspect of Event 71902630, Run 204769 at ATLAS-LHC, a 4-muon event relevant for the Higgs search. The most primitive ingredients for such analyses are of course timed particle detections, each one at a specific physical detector among the very many that compose ATLAS. But key for the success of such experiments is the fact that it happens to be the case that these *observed* localizations at detectors all combine to allow us to *infer* a location for the interaction vertex. It is indeed crucial for the analysis of this event to establish whether the four muon tracks (*thick red lines*) do originate from the same vertex, and this is not a trivial task, especially because of "pile up". [Each time two bunches of protons cross at ATLAS several proton-proton collisions occur (shown in figure as *dots* where particle tracks meet) within a region whose size is of only a few centimeters.]

I argue that, as we try to get ready for going even beyond quantum mechanics, in the context of quantum-gravity research, we must contemplate even more virulent departures from the "spacetime paradigm". So my thesis is that as we get to know Nature better our insistence on the consistency of the abstraction of a macroscopic spacetime picture may be gradually transforming from a source of valuable inspiration into an obstruction to knowledge.

Reliability of the Spacetime Abstraction in Classical Mechanics

Emission-Detection Setups and Wasteful Redundancies

It is useful for my purposes to start with some comments on emission-detection-type measurements in a background Minkowski spacetime. I consider "observers" Alice and Bob, which are (emitters/) detectors, equipped with clocks. Alice and Bob establish that they are in relative rest at a distance L with synchronized clocks, by using standard Einsteinian procedures relying on the time of flight between them of some reference particles (in practice they would use low-energy photons). Once this is done, Alice might write the following equation to describe the "trajectory in spacetime" of a certain specific particle:

$$x = \mathsf{v}t \tag{13.1}$$

Of the infinitely many "potential truths" codified in this equation only two facts are established experimentally: the particle is emitted at Alice at time $t = 0$ of Alice's clock and the particle is detected at Bob at time $t = L/\mathsf{v}$ of Bob's clock.

The redundancy of the spacetime abstraction for emission-detection-type measurements is particularly clear in relativistic theories, where it plays a role in the relationship between active and passive relativistic transformations. There the redundancy is in characterizations such as "the particle was at a distance of 10 m from Alice, where Bob is", which evidently can be fully replaced by "the particle was at Bob".

Bubble Chambers and Convenient Redundancies

If all we had were emission-detection-type measurements the abstraction of a spacetime would have probably never been made, since in those measurements it is wastefully redundant. But our "spacetime measurements" are not all of emission-detection type. Many of them are such that several particles are detected all attributable to a single spacetime event, and in such cases the spacetime abstraction is very valuable. I already showed a striking example of this sort in Fig. 13.1. And I should stress how this is important in astrophysics: when a star bursts we detect bunches of particles (mainly photons), and the discussion of the physics content of these measurements could in principle be limited to those timed particle detections. But it is evidently advantageous to recognize that in these instances the sequence of particle detections can be organized to infer a localized explosion "far away in spacetime".

And the spacetime abstraction acquires an added element of tangibility when we perform sequences of measurements of localization of the same particle (or body). As representative of this vast class of measurements let me consider the case of bubble-chamber measurements. Of course also in a bubble-chamber setup the primitive measurements are timed particle detections, typically photographic-camera detections of photons. That collection of photons however proves to be reliably describable in terms of inferences for the positions of certain bubbles, and in turn the collection of positions of bubbles allows us to build an inference of "trajectory" for a charged microscopic particle. Evidently here too the spacetime abstraction is redundant, but the advantages of adopting the spacetime-trajectory inference in such cases are undeniable (see Fig. 13.2).

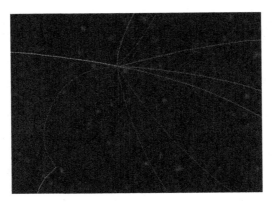

Fig. 13.2 A cartoonist impression of the type of information we gather with bubble-chamber experiments (the choice of specific example is based on an image shown in *The Particle Odyssey: A Journey to the Heart of the Matter* by F. Close, M. Marten and C. Sutton). The color-coded particle trajectories show an antiproton (*grey*) colliding with a proton at rest (not visible), and thereby producing four π^+ (positive pions; *red*) and four π^- (negative pions; *green*). One of the π^+ decays into a muon (*yellow*) and a neutrino (not visible). A magnetic field in the chamber causes the trajectories of positively-charged particles to bend in direction opposite to the one for negatively-charged particles. All this information is of course also coded in the "primitive measurements" here relevant, which are photographic-camera detections of photons from the bubbles. But the description in terms of that huge number of photon detections at the photographic camera is far less advantageous than the streamlined description in terms of "spacetime trajectories" of a few charged particles

Spacetime and the Ether

I have so far only highlighted some aspects of the redundancy of our spacetime inferences.

What is then spacetime?

Does spacetime "exist"?

I shall leave these questions to the appetites of philosophers. Physics can confine their discussion to two simple observations:

(i) the fact that we can add reference to a "spacetime" without adding any new item to our list of elementary/primitive measurement procedures (still only timed particle detections) is a complete proof of the redundancy of spacetime in science;
(ii) but, while awareness of its redundancy may at some point be valuable, the abstracted notion of spacetime is tangibly useful, and as long as this is the situation there would be no reason for us to change the way we use this notion.

It will be clear from these notes that I am speculating about the possibility that the status of spacetime in our current theories might resemble the status of the ether at the beginning of the 20th century. This was nicely summarized by Poincaré [1]: *Whether the ether exists or not matters little—let us leave that to the metaphysicians; what is essential for us is that everything happens as if it existed, and that this hypothesis is found to be suitable for the explanation of phenomena. After all have we any other reason for believing in the existence of material objects? That too is only a convenient hypothesis; only it will never cease to be so, while some day, no doubt, the ether will be thrown aside as useless.*

Spacetime in Quantum Mechanics

Is our insistence on the availability of consistent spacetime inferences always useful, or at least harmless?

Interesting challenges for the spacetime abstraction are already found within quantum mechanics, even though it may appear that one can list very many quantum-mechanics applications where the spacetime abstraction proves to be reliable. In most cases quantum mechanics is tested in experimental contexts where (because of the physical scales involved) most of the analysis is still conducted within classical mechanics, with only a few agents in the measurement procedure needing quantum-mechanical treatment. It is therefore legitimate to suspect that much of the success of our spacetime inferences in standard applications of quantum mechanics might be simply inherited from classical mechanics. And there are hints that this might be the case. I have a long-term project attempting to substantiate rather broadly the limitations that the spacetime abstraction faces within quantum mechanics [2]. For the purposes of these notes it is sufficient for me to comment, in this section, on two of these examples of open challenges for our understanding of quantum mechanics which may suggest that the spacetime abstraction is already turning from useful to cumbersome.

Troubles with Special-Relativistic Position Observable

Let me start by looking back at Sect. "Emission-Detection Setups and Wasteful Redundancies", where I commented on the redundancy of the spacetime abstraction for certain emission-detection-type measurements. Now let us imagine that the particle sent out by Alice reaches Bob in a case where quantum mechanics is very significant. For example we could consider a suitable "double slit" between Alice and Bob. Famously the spacetime description must be creatively adapted to such contexts. At least within the Galilean-relativistic version of quantum mechanics there still is a legitimate place for the spacetime abstraction: we cannot infer anymore a definite spacetime trajectory for a particle but we have a position observable $\hat{X}(t)$ which encodes in probabilistic manner the information on the spacetime trajectory of the particle. However, in spite of a stubborn effort by the community lasting for some 80 years, we have not found a suitable position observable generalizing to the special-relativistic theory this luxury of the Galilean limit. Several attempts have been made primarily inspired by early work by Newton and Wigner [3], but the outcome remains largely unsatisfactory [4, 5].

Our understanding of the origin of this challenge has improved significantly over the last decade through work [6–9] establishing a covariant formulation of quantum mechanics. There one sees (as here summarized in Appendix A) that both the spatial localization X and the time localization T are not good ("Dirac") observables of the theory, because they do not commute with the Hamiltonian constraint. It is increasingly clear that the observable aspects of special-relativistic quantum mechanics are all codified in properties of the asymptotic states, with the properties of the "prepared" (incoming) asymptotic state being linked to the properties of the "measured" (outgoing) asymptotic state by an "S-matrix". And evidently this is a good match for the conceptual perspective I already introduced in Sect. "Emission-Detection Setups and Wasteful Redundancies", where I observed that even within classical mechanics all that physics really needs can be coded in properties of the "prepared" system (at Alice) and properties of the "measured" system (at Bob). Matters being as this we do need material detectors and material clocks but we do not need spacetime.

Troubles with Quantum Tunneling

The observations I reported so far may appear to suggest that the spacetime abstraction is robust in the Galilean limit of quantum mechanics, but looses some of its reliabiilty in the special-relativistic case. But actually already in the Galilean limit of quantum mechanics the spacetime abstraction runs occasionally into troubles, especially when quantum-mechanical effects are dominant. For the purposes of these notes (I shall discuss more examples in Ref. [2]) let me clarify what I mean by this characterization by focusing on the specific example of quantum tunneling through a barrier.

In the classical limit a particle encountering a potential barrier higher than its kinetic energy simply cannot manage its way to the other side of the barrier. Quantum-mechanical effects provide such a particle with a small probability of ending up on the other side of the barrier. This is well known and well understood. But how should one describe the position of the particle when it is formally "inside" the barrier? And especially how much time does it take a particle to quantum-tunnel through a barrier? These are tough questions, whose importance has been clear since the early days of quantum mechanics [10, 11], and remain the main focus of a very active area of both theory and experimental research [12–15].

For speeds much smaller than the speed of light we express the speed v of a particle in terms of its kinetic energy K and its mass m, and the kinetic energy is in turn obtained subtracting to the total "nonrelativistic energy" $\mathcal{E}$ the potential energy U:

$$v = \sqrt{\frac{2K}{m}} = \sqrt{\frac{2(\mathcal{E} - U)}{m}} .$$

Since in quantum tunneling $\mathcal{E} - U < 0$ this recipe for the speed (and therefore the corresponding derivation of the travel time) becomes meaningless. We are dealing with a pure quantum-mechanical effect, the best of cases for exploring the role of the spacetime abstraction within quantum mechanics.

And (also see Appendix B) what we find experimentally in trying to determine the tunneling time does challenge the spacetime abstraction. We have at this point growingly robust evidence of the fact that the results for the tunneling time depend on the type of clock used. About a dozen different ways (clocks) for determining the travel time through the barrier are being used, all of which would agree if used as timing devices in classical-limit contexts, but their determinations of tunneling times differ [12–15].

A useful organizing notion for at least some of these tunneling-time measurements is the "Feynamn-path time" (see Ref. [12] and references therein) obtained by averaging the durations of all relevant Feynman paths with the weighting factor $exp(iS/\hbar)$ (where S here of course denotes the action). But some actual timing procedures (clocks) turn out to agree with the real part of the Feynamn-path time, others agree with its imaginary part, and others agree with the modulus of the Feynamn-path time [12]. Consistently with the thesis of these notes there appears to be no "time of spacetime" but only "time of a specific clock" [12].

Relevance for Quantum Gravity?

We often learn physics "going upstream". A good example is the program of "quantization of theories": from Nature's perspective theories start off being quantum and happen to be amenable to description in terms of classical mechanics only in some peculiar limiting cases, but our condition is such that we experience more easily those limiting cases rather than the full quantum manifestation of the laws.

I have here argued that also the spacetime abstraction might be a result of our "going upstream". There is no way to introduce spacetime operatively without clocks and detectors. And yet it is standard to develop theories by introducing the spacetime picture as first ingredient, then introducing a long list of formal properties of fields (or particles) in that spacetime, and only in the end we worry about actually having detectors and clocks in our theory. This worked so far. But there is no guarantee it will continue to work.

In quantum-gravity research there is a long-standing effort of understanding how spacetime should be described when both Planck's constant $\hbar$ and Newton's constant G_N are nonnegligible. We cannot claim much success addressing this issue.

We could perhaps try attacking the problem from the completely different perspective I am here advocating: we could look for candidate theories of the exchange of signals among (physical, material) emitters/detectors, now allowing for such theories an interplay between $\hbar$ and G_N, and without insisting on the availability of a spacetime abstraction suitable for organizing exactly all such exchanges of signals.

"Detectors First" and Black-Hole Holography

Perhaps the most natural opportunity for finding first applications of this new perspective could be the context of studies concerning the holographic description of black holes. Such holographic descriptions appear to be puzzling if one conceptualizes the physics of black holes as "contained" in the spacetime region determined by the black hole. And it should be noticed that the puzzle associated with the entropy-area law [16] becomes more severe for black holes of larger size (the mismatch between volume scaling and area scaling is increasingly severe as the size of the region increases).

But according to the perspective I am here advocating the presence of an horizon should not be described in terms of the structure of the abstracted spacetime but rather in terms of the network of emitter/detectors that can be setup. Ultimately horizons would be described as limitations to the network of detectors that can exchange information. One cannot include in the analysis the notion of "detector inside the black-hole horizon" because of the limitations on signal exchange produced by the horizon. And holography could reflect the fact that the reduction of meaningful detectors produced by the event horizon also reduces the amount of possible channels for information exchange among detectors.

The Possibility of Relative Locality

There are also some results recently obtained in the quantum-gravity literature which, while not getting rid of spacetime altogether, do already provide frameworks for describing a spacetime abstraction which is weaker than presently assumed, less capable of organizing comprehensively all phenomena. These are results challenging the absoluteness of locality.

In our current theories, when observer Alice detects particles from an event, and uses those particle detections as identification of a "distant point in spacetime", then, and this is absolute locality, all other observers can also analogously determine the position of the event. The point of spacetime inferred through such procedures carries different coordinates in the different reference frames of different observers, but the laws of transformation among reference frames ensure that the different observers "see the same spacetime point".

It is emerging [17, 18] that in theories with certain types of Lie-algebra spacetime noncommutativity [19, 20] and in the quantum-gravity approaches based on "group field theory" [21] this absoluteness of locality might be lost. The feature of these theories that is primarily responsible [17, 18] for the "relativity of spacetime locality" is the fact that the translation-symmetry generators, the total-momentum operators, are not obtained as a linear sum of single-particle momenta (also see Appendix C). For example, for processes $1 + 2 \rightarrow 3 + 4$ (two incoming, two outgoing particles) one can have conservation laws of the form

$$p_\mu^{[1]} \oplus p_\mu^{[2]} = p_\mu^{[3]} \oplus p_\mu^{[4]}$$

with $k_\mu \oplus q_\mu \neq k_\mu + q_\mu$, though of course one does recover $k_\mu \oplus q_\mu \simeq k_\mu + q_\mu$ when all momenta involved are small with respect to the Planck scale [17–21].

Then let us consider the determination of the interaction point obtainable by finding the intersection of the worldlines of the outgoing particles. And let me denote by $x_{[*,A]}^\mu$ the intersection point thereby determined by observer Alice, i.e. there is a value s_* of the worldlines affine parameter such that $x_{[3,A]}^\mu(s_*) = x_{[4,A]}^\mu(s_*) = x_{[*,A]}^\mu$. If one now acts on Alice with a translation of parameters b^μ, as a way to test how this point $x_{[*,A]}^\mu$ is viewed by observers distant from Alice, the mentioned nonlinearities of the sum rule that gives the total-momentum generators affect the analysis nontrivially [17, 18]. For example one finds that

$$x_{[3,B]}^\mu(s) = x_{[3,A]}^\mu(s) + b^\nu \{p_\nu^{[3]} \oplus p_\nu^{[4]}, x_{[3,A]}^\mu(s)\} \neq x_{[3,A]}^\mu(s) + b^\mu$$

and in particular $x_{[4,B]}^\mu(s_*) - x_{[4,A]}^\mu(s_*) \neq x_{[3,B]}^\mu(s_*) - x_{[3,A]}^\mu(s_*)$ (unless the momenta are much smaller than the Planck scale). In our current theories a spacetime point is absolutely marked by an intersection of worldlines. In relative-locality theories processes are still objective [17, 18] but their association to points of a spacetime is observer dependent, with the familiar spacetime abstraction emerging only for processes involving particles of energies much smaller than the Planck scale.

A Challenge for Experimentalists

To me it is irresistibly intriguing to speculate that our insistence on the availability of the spacetime abstraction might at this point be limiting our opportunities for discovery. And I am contemplating a meaningful question: it is for experiments to decide whether or not the reliability of our spacetime inferences is truly universal. I here stressed that opportunities for such experimental scrutiny are found already within ordinary quantum mechanics. For example, we should continue to investigate whether indeed the tunneling time can only be determined as "time of some specific clock", with different outcomes for different clocks (and therefore no emerging notion of a clock-independent "time of spacetime").

And if my "detectors-first perspective" does turn out to be applicable to quantum-gravity research, there will be (though perhaps in a distant future [22]) other opportunities for experiments looking for evidence against spacetime.

Appendix A: More on Covariant Quantum Mechanics

Within the manifestly-covariant formulation of special-relativistic quantum mechanics, which matured significantly over the last decade [6–8], the spatial coordinates and the time coordinate play the same type of role. And there is no "evolution", since dynamics is codified within Dirac's quantization as a constraint, just in the same sense familiar for the covariant formulation of classical mechanics (see, e.g., Chap. 4 of Ref. [23]).

Spatial, $\hat{X}$, and time, $\hat{T}$, coordinates are well-defined operators on a "kinematical Hilbert space", which is just an ordinary Hilbert space of normalizable wave functions [8, 9], where they act multiplicatively: $\hat{X}\Psi(x, t) = x\Psi(x, t)$, $\hat{T}\Psi(x, t) = t\Psi(x, t)$. And one has a standard description on this kinematical Hilbert space of their conjugate momenta [8, 9]:

$$[\hat{P}_0, \hat{T}] = i \,, \quad [\hat{P}, \hat{X}] = -i \,, \quad [\hat{P}, \hat{T}] = [\hat{P}_0, \hat{X}] = 0$$

Observable properties of the theory are however formulated on the "physical Hilbert space", obtained from the kinematical Hilbert space by enforcing the constraint of vanishing covariant-Hamiltonian, which in the case of a free special-relativistic particle takes the form

$$(\hat{P}_0^2 - \hat{P}^2)\Psi_{physical} = 0$$

The observables of the theory, the "Dirac observables", must commute with the constraint, and this is where one sees the root of the localization problem for special-relativistic quantum mechanics, which in different fashion had already been noticed by Newton and Wigner [3]: both $\hat{T}$ and $\hat{X}$ are not good observables, even within the free-particle theory, since they evidently do not commute with $\hat{P}_0^2 - \hat{P}^2$.

Appendix B: More on Quantum Tunneling

The study of quantum tunneling has a very long history. However, the quality of related experimental results has improved significantly over the last two decades [12, 13], particularly starting with the measurements reported in Ref. [14], where a two-photon interferometer was used to measure the time delay for a single photon (i.e. one photon at a time) to tunnel across a well-measured barrier.

And also the understanding of quantum tunneling, and particularly of the tunneling time, has improved significantly in recent times. Previously there was much controversy particulary revolving around relativistic issues. Under appropriate conditions [12–14] a particle prepared at time t_i in a quantum state with peak of the probability distribution located at a certain position to one side of a barrier is then found on the other side of the barrier at time t_f with distribution peaked at a distance L from the initial position, with L bigger than $c(t_f - t_i)$. It is by now well established that this apparently "superluminal" behavior is not in conflict with Einstein's relativity. Key for this emerging understanding is appreciating that in such measurement setups at first there is a large peaked distribution approaching the barrier from one side, and then a different (and much smaller, transmitted) peaked distribution is measured on the other side of the barrier. Contrary to our classical-limit-based intuition, as a result of quantum-mechanical effects (such as interference) the peak observed after the barrier is not some simple fraction of the peak that was approaching the barrier. This sort of travel times of distribution peaks, are not travel times of any signal, and indeed it is well known that for smooth, frequency-band limited, distributions the precursor tail of the distribution allows one to infer by analytic continuation [12, 14, 15] the structure of the peak. Even for free propagation, by the time the peak reaches a detector it carries no "new information" [12] with respect to the information already contained in the precursor tail. An example of "new information" is present in modified distributions containing "abrupt" signals [12, 15], and indeed it is found that when these new-information features are sufficiently sharp they never propagate superluminally [15].

It is important for the thesis advocated in these notes that, as these theoretical issues get clarified, and theoretical results get in better agreement with experiments, we are also getting more an more robust evidence of the fact that in quantifying the analysis of quantum tunneling we do not have the luxury of referring to some objective spacetime picture. In particular, there is no single "time of spacetime" but rather several possibilities for a "time of a specific clock" [12].

Appendix C: Relative Locality and Curved Momentum Space

As mentioned in the main text of these notes the main sources of interest in relative locality originate from results obtained in studies of certain types of Lie-algebra spacetime noncommutativity [19, 20] and of "group field theory" [21], where the generators of translation-symmetry transformations are not described as linear sums of single-particle momenta. In this appendix I want to highlight a particularly powerful formulation of relative locality that emerges in these theories by formally taking the limit [17, 18] of both $\hbar \to 0$ and $G_N \to 0$ but keeping their ratio $\hbar/G_N$ fixed. In this regime both quantum mechanics and gravity are switched off, but the mentioned nonlinearities for the composition of momenta are found to survive [17, 18] and to take the form of a manifestation of a nontrivial geometry for momentum space.

The implications for the phase space associated with each particle in this regime are found to be rather striking [17, 18]: this phase space is the cotangent bundle over momentum space, which one may denote by $\Gamma^{RL} = \mathcal{T}^*(\mathcal{P})$. So this regime is, at least in this respect, dual to the standard classical-gravity regime, where the single-particle phase space is the cotangent bundle of the spacetime $\mathcal{M}$, which one may denote by $\Gamma^{GR} = \mathcal{T}^*(\mathcal{M})$. And just like in the general-relativistic formulation of classical gravity momenta of particles at different points of spacetime, x and y, can only be compared by parallel-transporting along some path from x to y, using the spacetime connection, one finds on Γ^{RL} an analogous problem for the comparison of spacetime coordinates on the worldlines of two particles, A and B, with different momenta. These coordinates x_A^μ and x_B^μ live in different spaces and they can be compared only in terms of a parallel transport on momentum space. All this is formalized in Refs. [17, 18] where the relative-locality features mentioned in the main text of these notes are shown to admit a fully geometric description (in terms of the geometry of momentum space).

References

1. H. Poincaré, *Science and Hypothesis*, Chapter 12 (Walter Scott Publishing, London, 1905)
2. G. Amelino-Camelia, Spacetime in special-relativistic quantum theory, in preparation
3. T.D. Newton, E.P. Wigner, Rev. Mod. Phys. **21**, 400 (1949)
4. B. Schroer, arXiv:0711.4600
5. R.-A. Alemañ-Berenguer, arXiv:philsci-archive.pitt.edu/4313
6. J.J. Halliwell, Phys. Rev. **D64**, 04408 (2001)
7. R. Gambini, R.A. Porto, Phys. Rev. **D63**, 105014 (2001)
8. M. Reisenberger, C. Rovelli, Phys. Rev. **D65**, 125016 (2002)
9. L. Freidel, F. Girelli, E.R. Livine, Phys. Rev. **D75**, 105016 (2007)
10. L.A. MacColl, Phys. Rev. **40**, 621 (1932)
11. E.P. Wigner, Phys. Rev. **98**, 145 (1955)
12. A.M. Steinberg, Lect. Notes Phys. **734**, 333 (2008)
13. H.G. Winful, Phys. Rep. **436**, 1 (2006)
14. A.M. Steinberg, P. Kwiat, R. Chiao, Phys. Rev. Lett. **71**, 708 (1993)
15. M.D. Stenner, D.J. Gauthier, M.A. Neifeld, Nature **425**, 695 (2003)
16. S. Carlip, Lect. Notes Phys. **769**, 89 (2009)
17. G. Amelino-Camelia, L. Freidel, J. Kowalski-Glikman, L. Smolin, Phys. Rev. **D84**, 084010 (2011). arXiv:1101.0931
18. G. Amelino-Camelia, L. Freidel, J. Kowalski-Glikman, L. Smolin, Gen. Relativ. Gravit. **43**, 2547 (2011). arXiv:1106.0313
19. S. Majid, H. Ruegg, Phys. Lett. **B334**, 348 (1994)
20. S. Majid, arXiv:hep-th/0604130

21. D. Oriti, arXiv:0912.2441
22. G. Amelino-Camelia, Nature **408**, 661 (2000). arXiv:gr-qc/0012049
23. M. Henneaux, C. Teitelboim, *Quantization of Gauge Systems* (Princeton University Press, 1992)

Chapter 14
A Chicken-and-Egg Problem: Which Came First, the Quantum State or Spacetime?

Torsten Asselmeyer-Maluga

Abstract In this essay I will discuss the question: Is spacetime quantized, as in quantum geometry, or is it possible to derive the quantization procedure from the structure of spacetime? All proposals of quantum gravity try to quantize spacetime or derive it as an emergent phenomenon. In this essay, all major approaches are analyzed to find an alternative to a discrete structure on spacetime or to the emergence of spacetime. Here I will present the idea that spacetime defines the quantum state by using new developments in the differential topology of 3- and 4-manifolds. In particular the plethora of exotic smoothness structures in dimension 4 could be the corner stone of quantum gravity.

Basic Assumptions in Quantum Gravity

General relativity (GR) has changed our understanding of spacetime. In parallel, the appearance of quantum field theory (QFT) has modified our view of particles, fields and the measurement process. The usual approach for the unification of QFT and GR, to a quantum gravity, starts with a proposal to quantize GR and its underlying structure, spacetime. There is an unique opinion in the community about the relation between geometry and quantum theory: The geometry as used in GR is classical and should emerge from a quantum gravity in the limit (Planck's constant tends to zero). Most theories went a step further and try to get a spacetime from quantum theory. But what happens if this prerequisite is wrong? Is it possible to derive the quantization procedure from the structure of space and time? My own research program [1–13] seems to imply a simple answer: Yes. But in this essay I will discuss it from a more general perspective and analyze the basic assumptions of quantum gravity[1] first:

---

[1] There are many books about quantum gravity, for instance [14], and the original papers which I omit to cite.

T. Asselmeyer-Maluga (✉)
German Aerospace Center, Berlin, Germany
e-mail: torsten.asselmeyer-maluga@dlr.de

© Springer International Publishing Switzerland 2015
A. Aguirre et al. (eds.), *Questioning the Foundations of Physics*,
The Frontiers Collection, DOI 10.1007/978-3-319-13045-3_14

1. Spacetime has dimension 4 (also true in superstring theory after compactification).
2. Classical spacetime, i.e. the spacetime of GR or the spacetime as a limit of quantum gravity, is a smooth, non-compact 4-manifold with Lorentz structure and (trivial) codimension-1 foliation of the form $\Sigma \times \mathbb{R}$ (Σ a smooth 3-manifold, the space).
3. In the process of quantization, spacetime admits a discrete structure and one obtains the continuum only in the limit of large scales.
4. The quantum state in quantum gravity is realized as element (or vector) in some abstract state space (the space of all connections, the space of all spin networks etc.) with no direct reference to the classical spacetime.
5. Quantum gravity (as containing GR in some limit) must be background-independent, i.e. it does not explicitly depend on a concrete shape of the spacetime (diffeomorphism invariance).
6. Classical spacetime is induced (as emergent structure) by the quantum state in quantum gravity in the semi-classical limit.

All other explicit or implicit assumptions in concrete quantum gravity theories (like action principles or the dynamics) are very specific for the corresponding approach to quantum gravity. Let us analyze its relevance. The first assumption[2] is merely known from our experience but we must realize that four-dimensional spaces have an exceptional property among other spaces which is important for physics, exotic smoothness. The second assumption is motivated by GR and the research on the implementation of causality in GR. It is the starting point of Loop quantum gravity. The form of causality which was used to establish $\Sigma \times \mathbb{R}$ (see [15, 16]) is certainly too strong. There is a unique path into the past but not an unique path in the future, otherwise one needs the net of many-worlds. So, the second assumptions can be weakened to consider every kind of codimension-1 foliation on a smooth, non-compact 4-manifold. It is known that a Lorentz structure will be a result of this foliation. The third assumption is the starting point of some approaches to quantum gravity (Causal nets, Regge calculus etc.) but all approaches are looking for discrete structures (spin networks and area quantization of Loop quantum gravity, finite string length in String theory) or to explicitly construct them (Dynamical triangulation etc.). Interestingly, all these approaches have problems in reaching the continuum limit. But is it really necessary to reduce the continuous spacetime to a discrete space? From GR we know that (by using diffeomorphism invariance) a point has no meaning in GR (see the next section for the whole discussion). But one can use relative techniques to relate two subspace of the spacetime to each other. Then only statements like "the two submanifolds intersect transversally" can be decided without introducing a concrete coordinate system. But statements of this kind are part of a mathematical theory called (differential) topology with discrete structures used to classify topological spaces. I will discuss these in more detail in Sect. "Differential Topology Unveils the Quantum Nature of Spacetime". The fourth assumption is the most problematic one. Each approach to quantum gravity can be

---

[2] Current experiments at the LHC do not give any sign for extra dimensions (see the Particle Data Group).

distinguished by its state space. So, the question must be: what is the "natural" state space of quantum gravity? Geometrodynamics started with the original superspace, i.e. the space of 3-metrics, with the Wheeler-deWitt equation as dynamics. This program was superseded by the introduction of Ashtekar variables (densitized coframe and $SU(2)$ connection) leading to Loop quantum gravity. The state space is the space of spin networks but the solution of the Hamilton constraint (as analog to the Wheeler-deWitt equation) is an unsolved problem. All state spaces are rather artificial and the relation to the geometry of the spacetime is poor. Part of the problem is the answer to the question: Is the quantum state real? If the answer is yes one has to consider quantum states in spacetimes. In particular one has to interpret the superposition of states. I will discuss this problem also in Sect. "Differential Topology Unveils the Quantum Nature of Spacetime". The fifth assumption is induced directly from GR. It is one of the crucial points in GR: the theory has to be formulated without the reference to coordinates or the choice of a coordinate system does not influence the result. The implementation of this assumption is usually done by connecting it with the sixth assumption: there is no spacetime at the quantum level, the classical spacetime emerges from the deeper quantum level. Theories with a fuzzy spacetime including also noncommutative geometry are promising candidates for a direct implementation of quantum properties into spacetime.

In summary, especially the last four assumptions are questioned in this essay. In the following I will argue that the spacetime has the right properties for a spacetime-picture of QFT. Quantum gravity should be also part of this picture.

The First Sign: Spacetime Is More Than Classical

When Einstein developed GR, his opinion about the importance of general covariance changed over the years. In 1914, he wrote a joint paper with Grossmann. There, he rejected general covariance by the now famous hole argument. But after a painful year, he again considered general covariance now with the insight that there is no meaning in referring to "the spacetime point A" or "the event A", without further specifications. Therefore the measurement of a point without a detailed specification of the whole measurement process is meaningless in GR. The reason is simply the diffeomorphism-invariance of GR which has tremendous consequences. Physical observables have to be diffeomorphism-invariant expressions. In most cases, this demand is impossible. The momentum of a moving particle is not diffeomorphism-invariant but it is physically meaningful. Therefore this momentum must be seen in the context of a measurement device which includes a concrete coordinate system. But generally, GR is background-independent of any coordinate system and, as usually thought, from the topology of the spacetime. So, if one fixes the topology then GR depends only on the (diffeomorphism-class) metric. But in Sect. "The Secret Revolution: Our Understanding of 3- and 4-Manifolds", I will discuss another possibility. Let M be a smooth 4-manifold, say M is the topologically $\mathbb{R}^4$ but with Lorentz structure, i.e. the Minkowski space. M is flat and the GR vacuum equations are trivially fulfilled. The (smooth) atlas of M is called the smoothness structure unique up to diffeomorphisms. One would expect that there is only one smooth atlas for M, all other possibilities can be transformed into each other by a diffeomorphism. But this is not true, see Sect. "The Secret Revolution: Our Understanding of 3- and 4-Manifolds ". In fact, there are infinitely many non-equivalent smoothness structures on M with no heuristic to distinguished one above the others as physically relevant.

The Concept of Spacetime in the Main Approaches to Quantum Gravity

Quantum gravity as the unification of GR and QFT is currently an open problem but with a long list of possible candidates. Among them are Superstring theory and Loop quantum gravity. For each of these there are two classes: background-dependent and background-independent, respectively. But all proposals of a quantum gravity theory have one assumption in common: the spacetime has a foam-like structure or is discrete from the beginning. In string theory, the string has the extension of one Planck length and all structures below this length are unimportant. In particular the whole spacetime is seen as an emergent phenomenon which must be derived from the full theory. The background-dependence is a problem but it will be resolved in the conjectured M-theory. The relation to GR is interesting. At the non-quantized level, the world surface of the string is embedded in some background. If this embedding is fixed then the curvature of the world surface is determined by the curvature of the embedding space and vice versa. On the quantum level, one usually argues with the help of the β function, encoding the dependence of a coupling parameter in the renormalization group. For a conformal invariant theory (conformal invariance of the world surface), this function has to vanish. In the case of string theory, one obtains the vanishing of the scalar curvature of the embedding space or the Einstein equation for the vacuum. In Loop quantum gravity, the central theme is background-independence installed by the vanishing of three constraints. One of the main results of the theory is the quantization of the area and the volume. It seems that spacetime must have a discrete structure. But the result was obtained by using an eigenvalue equation. As in the case of a harmonic oscillator, the underlying space is continuous but the spectrum of the area or volume operator is discrete. In Sect. "Differential Topology Unveils the Quantum Nature of Spacetime ", I will give an example of a class of spaces (hyperbolic 3-manifolds) with this property. Many other proposals start with a discrete structure from the scratch: Causal Sets, Quantum Causal Histories, Dynamical Triangulations etc. Therefore in all current quantum gravity proposals, the spacetime has a discrete structure or emerges from a discrete structure. But is the spacetime model of a smooth manifold dead? Mathematicians have found many interesting properties of 3- and 4-manifolds.

The Secret Revolution: Our Understanding
of 3- and 4-Manifolds

According to GR, spacetime is a smooth 4-manifold carrying a Lorentz-structure. The existence of a Lorentz-structure is closely related to the existence of a codimension-1 foliation. Therefore one also has to consider the spatial component, a smooth 3-manifold, as leaf of the foliation. From the mathematical point of view, one has to look into the theory of 3- and 4-manifolds (see [3] for more details and the references).

The central concept is the smooth manifold as a generalization of a surface. Riemann was the first one to do extensive work generalizing the idea of a surface to higher dimensions. The name manifold comes from Riemann's original German term, Mannigfaltigkeit. In his Göttingen inaugural lecture ("Ueber die Hypothesen, welche der Geometrie zu Grunde liegen"), Riemann described the set of all possible values of a variable with certain constraints as a Mannigfaltigkeit, because the variable can have many values. By induction, Riemann was able to define an $n-$times extended manifold. Then Poincare gave a definition of a (differentiable) manifold (variété) as subset of some Euclidean space which served as a precursor to the modern concept of a manifold. During the 1930s Hassler Whitney and others clarified the foundational aspects of the subject. In particular, a manifold was defined intrinsically with charts and transition functions and the equivalence of this definition with Poincare's Euclidean subsets was shown (Whitney's embedding theorem). The development of algebraic topology gave the right frame to investigate the manifolds (classification of surfaces by homology, the fundamental group etc.). The first surprise came in 1957. John Milnor constructed the first seven exotic 7-spheres (7-dimensional spheres which are homeomorphic but not diffeomorphic to each other). Therefore there must be a difference between a smooth manifold (all charts as maps to a Euclidean space and transition functions are smooth maps) and a topological manifold (the maps are now only continuous). The new topic "differential topology" was born. The (smooth) atlas (all smooth charts and transitions function to cover the manifold) was called *smoothness structure*. But how many in-equivalent smoothness structures on a $n-$manifold exists? Using powerful methods (like the h-cobordism theorem), Kervaire and Milnor were able to determine it for the spheres. Number of exotic $n-$spheres up to dimension 13: (Table 14.1).

In particular, it was shown that the number of in-equivalent smoothness structure on a $n-$manifold is only finite for $n > 4$. The uniqueness in low dimensions $n < 4$ is a classical result. Therefore only the 4-dimensional case was open. All methods developed for 2- and 3-manifolds or for higher-dimensional ($n > 4$) manifolds were useless in this dimension. In 1973, Andrew Casson gave lectures about a new construction in dimension four, which he called flexible handles (now known

Table 14.1 Number of exotic $n-$spheres up to dimension 13

Dimension n	1	2	3	4	5	6	7	8	9	10	11	12	13
$\#Diff(S^n)$	1	1	1	?	1	1	28	2	8	6	992	1	3

as Casson handles). The mimeographed notes of these three lecture came to Micheal Freedman and occupied him over the next 7 years. Eventually, he successfully classified simply-connected, compact 4-manifolds. Meanwhile, Simon Donaldson (a student of M. Atiyah) started the investigation of anti-self-dual connections of a $SU(2)$ gauge theory, in physics known as instantons. As a surprising result, he proved that not all topological, compact 4-manifolds are smoothable. With this result in mind, Kirby, Freedman and Gompf constructed the first examples of an exotic $\mathbb{R}^4$, i.e. a space homeomorphic to $\mathbb{R}^4$ but not diffeomorphic. The second surprise came in the form of the number of inequivalent smoothness structures: there are countably infinite many for most compact 4-manifolds and uncountably infinite many for most non-compact 4-manifolds including $\mathbb{R}^4$. The development of this topic is not complete. In particular for the simplest compact 4-manifold, the 4-sphere, we do not know the number of in-equivalent smoothness structures. Most mathematicians conjecture that there are countably infinite many structures on the 4-sphere. So, *dimension 4 is exceptional for smoothness*!

One point is also important for quantum gravity. With Milnor's discovery of exotic 7-spheres, one started to look for the existence of other structures on manifolds. In particular the existence question for a triangulation (or piecewise-linear structure, PL-structure) was very interesting. The surprising result of Cerf for manifolds of dimension smaller than seven was simple: PL-structure (or triangulations) and smoothness structure are the same. This implies that every PL-structure can be smoothed to a smoothness structure and vice verse. But also for manifolds of dimension higher than three, there is a big difference between the topological and PL/smooth structure, i.e. not every topological manifold is smoothable/triangulable or uniquely smoothable/triangulable. Therefore *the discrete approach (via triangulations) and the smooth manifold are the same*!

The theory of 3-manifolds also changed its view in the 80s. The theory of 3-manifolds was strongly influenced by the Poincaré conjecture but the progress was slow before 1957. In 1957, Papakyriakopoulos proved three technical theorems (the sphere and loop theorem as well Dehn's lemma) with tremendous impact on the whole 3-manifold theory. Next, Milnor proved that every compact 3-manifold can be split into an unique decomposition of prime manifolds (3-manifolds which are not splittable by using the connected sum). In 1979, Jaco, Shalen and Johannson found a finer decomposition of some prime manifolds by cutting them into pieces along embedded tori (JSJ decomposition). But the real breakthrough came from Thurston around 1980. His work was inspired by the construction of geometric structures (metric of constant curvature simulated by a homogeneous space) for the complement of knots in the 3-sphere. Based on this examples, he conjectured that any compact 3-manifold can be cut into pieces so that every piece admits a geometric structure (Geometrization conjecture). If this conjecture is true then Poincare's conjecture is also settled. So, the topological structure of 3-manifolds has a lot to do with its geometry! In the 80s, Hamilton developed the Ricci flow technique to prove this conjecture with very interesting results. But in 2002 and 2003, Perelman submitted three papers to arxiv.org in which he proposed a proof of the Geometrization conjecture. The arguments were checked by many mathematicians but no error was found. According to Hamilton and Perelman, the a 3-manifold can be determined topologically by considering the long-time behavior of the Ricci flow or *a 3-manifold is determined by a flow in the space of metrics.*

Differential Topology Unveils the Quantum Nature of Spacetime

As mentioned above, differential topology is the mathematical theory of smooth manifolds including the (smooth) relations between submanifolds. In the first section, the basic assumptions of quantum gravity were discussed. The usage of a smooth 4-manifold as spacetime was not questioned. However the choice of the smoothness structure is not unique and I will discuss it now. As starting point consider a toy model to visualize the changes of the smoothness structure. Start with a torus (or doughnut) $T^2 = S^1 \times S^1$. Now one cuts the torus along one circle to obtain a cylinder, twist one end of the cylinder by 2π and glue the two ends together. This process is called a Dehn twist and one obtains a torus again but now with a twist. But by the classification of closed surfaces, the usual and the twisted torus are diffeomorphic to each other. This fact is amazing but both tori are located in different components of the diffeomorphism group, i.e. in two different isotopy classes. In particular there is no coordinate transformation (i.e. a diffeomorphism connected to the identity) which transforms the twisted torus to the usual torus. This toy model shows the difference between coordinate transformations (diffeomorphisms connected to the identity) and global diffeomorphisms. A similar effect is the change of the smoothness structure (to a non-equivalent one) but it cannot be visualized in this easy way. As an example of this change I will consider a compact 4-manifold M (topologically complicated enough, i.e. a K3 surface or more advanced) containing a special torus T_c^2 (so called c-embedded torus). Now cut out a neighborhood $D^2 \times T_c^2$ of this torus (with boundary a 3-torus T^3) and glue in $(S^3 \setminus (D^2 \times K)) \times S^1$ (having also the boundary T^3) where $S^3 \setminus (D^2 \times K)$ denotes the complement of a knot K in the 3-sphere S^3. Then one obtains

$$M_K = \left(M \setminus \left(D^2 \times T_c^2\right)\right) \cup_{T^3} \left((S^3 \setminus (D^2 \times K)) \times S^1\right) \qquad (14.1)$$

a new 4-manifold M_K which is homeomorphic to M (Fintushel-Stern knot surgery [17]). If the knot is non-trivial then M_K is not diffeomorphic to M. One calls M_K an exotic 4-manifold, a misleading term. Nothing is really exotic here. Nearly all smoothness structures on a 4-manifold are exotic.

Now consider the physically significant non-compact examples of exotic 4-manifolds like $\mathbb{R}^4$ and $S^3 \times \mathbb{R}$. Start with $S^3 \times \mathbb{R}$. This non-compact 4-manifold has the usual form used in GR. There is a global foliation along $\mathbb{R}$, i.e. $S^3 \times \{t\}$ with $t \in \mathbb{R}$ are the (spatial) leafs. $S^3 \times \mathbb{R}$ with this foliation is called the "standard $S^3 \times \mathbb{R}$". I will denote an exotic version by $S^3 \times_\theta \mathbb{R}$. The construction of $S^3 \times_\theta \mathbb{R}$ is

rather complicated (see [18]). As a main ingredient one needs a homology 3-sphere
Σ (i.e. a compact, closed 3-manifold with the homology groups of the 3-sphere)
which does not bound a contractable 4-manifold (i.e. a 4-manifold which can be
contracted to a point by a smooth homotopy). Interestingly, this homology 3-sphere
Σ is smoothly embedded in $S^3 \times_\theta \mathbb{R}$ (as cross section, i.e. $\Sigma \times \{0\} \subset S^3 \times_\theta \mathbb{R}$).
From the geometrical point of view, this 3-manifold is also very interesting. One can
choose Σ so that it admits a hyperbolic structure, i.e. a homogeneous metric of constant negative curvature. Hyperbolic 3-manifolds have a special property: Mostow
rigidity [19]. Every diffeomorphism (especially every conformal transformation) of
a hyperbolic 3−manifold is induced by an isometry. Therefore the volume and the
curvature are topological invariants for hyperbolic 3-manifolds. In particular there
are surfaces in hyperbolic 3-manifolds (incompressible surfaces) having a special
(not divisible) volume. Then one obtains also a kind of quantized areas by purely
topological methods.

What about the foliation of $S^3 \times_\theta \mathbb{R}$? There is no foliation along $\mathbb{R}$ but there is
a codimension-one foliation of the 3-sphere S^3 (see [20] for the construction). So,
$S^3 \times_\theta \mathbb{R}$ is foliated along S^3 and the leafs are $S_i \times \mathbb{R}$ with the surfaces $\{S_i\}_{i \in I} \subset S^3$.
But what happens with the 3-spheres in $S^3 \times_\theta \mathbb{R}$? There is no smoothly embedded
S^3 in $S^3 \times_\theta \mathbb{R}$ (otherwise it would have the standard smoothness structure). But
there is a wildly embedded S^3! Let $i : K \rightarrow M$ be an embedding of K (with
dim $K <$ dim M). One calls the embedding i *wild* if $i(K)$ is not a finite polyhedron
(or $i(K)$ is not triangulated by a finite number of simplices). See the example of wildly
embedded circles in the 3-space in Fig. 14.1, the famous Fox-Artin wild knot. In [7],
we considered wildly embedded submanifolds as models of quantum D-branes. The
prominent example of a wildly embedded submanifold is Alexanders horned sphere.
Wild embedded submanifolds are fractals in a generalized manner. Now I will argue
in the following that this wild embedding is a geometric model for a quantum state.

As discussed in the first section, all approaches of quantum gravity have problems
with the construction of the state space. If I assume that the spacetime has the right
properties for a spacetime picture of quantum gravity then the quantum state must be
part of the spacetime or must be geometrically realized in the spacetime. Consider
(as in geometrodynamics) a 3-sphere S^3 with metric g. This metric (as state of GR)
is modeled on S^3 at every 3-dimensional subspace. If g is a metric of a homogeneous
space then one can choose a small coordinate patch. But if g is inhomogeneous then
one can use a diffeomorphism to "concentrate" the inhomogeneity at a chart. Now

Fig. 14.1 Examples of wild
knots

one combines these infinite charts (I consider only metrics up to diffeomorphisms) into a 3-sphere but without destroying the infinite charts by a diffeomorphism. Wild embeddings are the right structure for this idea. A wild embedding cannot be undone by a diffeomorphism of the embedding space. Secondly, this wild embedding of a 3-sphere into $S^3 \times_\theta \mathbb{R}$ is determined by its complement. So, if one understands the complement of the wild embedding then one understands the wild embedding itself. Interestingly, one can construct an operator algebra and a Hilbert space from a wild embedding (see Appendix A, below). This operator algebra can be also obtained from the foliation (by using the noncommutative geometry approach), see [10, 21]. It is the hyperfinite factor III_1 von Neumann algebra having the structure of the local algebras in a relativistic QFT with one vacuum vector. Then one obtains a background-independent approach but how does the classical spacetime appears? A wild embedding is defined by the infinite polyhedron. So, if I reduces the wild embedding to a finite polyhedron by contracting the smaller parts to zero then one obtains a finite polyhedron. But by definition, a finite polyhedron is not a wild embedding (or comes from a wild embedding). In [7], the process was discussed for quantum D-branes. In particular, the classical action of a D-brane was obtained. But if we are able to reduce the wild embedding (quantum state) to a tame embedding (classical state) then we have to show that a wild embedding is the quantization of a tame embedding. In [22]we showed even this fact:

The wild embedding can be obtained by a deformation quantization (Drinfeld-Turaev quantization) from a tame embedding. Furthermore, we constructed one of the main ingredients for a quantum field theory (C*—algebra and the observablen algebra of von Neumann type) for Alexanders horned sphere and identifying it with the knwon results (so-called factor III_1).

Decoherence and Inflation

If our approach to identify the wild embedding with a quantum state will be successful then we have to present an interpretation of decoherence or the measurement process in quantum mechanics. As an example [23] we will consider the quantum state at the beginning of the universe, i.e. a wildly embedded 3-sphere. In our model of the exotic $S^3 \times_\theta \mathbb{R}$ we made a rescaling so that the 3-sphere at $t = -\infty$ is the initial state of the cosmos (at the big bang), i.e. we assume that the cosmos starts as a small 3-sphere (of Planck radius). But in an exotic $S^3 \times_\theta \mathbb{R}$, every 3-sphere (at every time) is a wildly embedded 3-sphere S^3_θ. In the model of $S^3 \times_\theta \mathbb{R}$ we have a topologically complicated 3-manifold Σ (a homology 3-sphere) at a later time step (say at $t = 0$). As Freedman [18] this 3-manifold Σ is smoothly embedded, i.e. it represents a classical state. Therefore we obtain the transition

$$\text{quantum state} \quad S^3_\theta \xrightarrow{\text{decoherence}} \text{classical state} \quad \Sigma \qquad (14.2)$$

and we studied this process in [24] more carefully. This process has an exponential rate of expansion, a process usually called inflation today. The reason of this inflation is the transition from the quantum state (wildly embedded 3-sphere) to the classical state (complicated, smoothly embedded 3-manifold = tame embedding). At this stage, our model is very explicit: an infinite polyhedron (wildly embedded manifold) is reduced to a finite polyhedron (tame embedding) which can be part of the infinite polyhedron. Mathematically we obtain a projection or state reduction or the collapse of the wave function.

This particular example showed the main features of the decoherence process. Our model is general enough to explain also the decoherence process for wildly embedded subsystems. But I will point to one interesting result: the smoothness structure determines the classical state in our model above. If one generalize this result then (differential) topology has to be included in the discussion of the measurement process.

In summary, we obtained a state space as operator algebra of the wild embedding induced by exotic smoothness structures. The state space is

1. background-independent (diffeomorphism invariant)
2. with countable infinite basis (discrete structure)
3. and contains the classical spacetime as limit.

Of course, the whole approach is very theoretical up to now. For instance I do not start with a concrete action or list of fields. But sometimes, things went better than expected. In [12], we considered the Fintushel-Stern knot surgery above to obtain (14.1), the exotic M_K. If one started with the Einstein-Hilbert action on M then we obtained the combined Einstein-Hilbert-Dirac-Yang-Mills system. The knot complement is directly related to the fermions whereas the bosons appear as torus bundles (the pieces between the knot complements). In an extension of this work [25], the Higgs mechanism was also included. A lot of work has to be done but it is a beginning.

Conclusion

I have presented a certain number of ideas and results:

1. There is a freedom in the definition of the spacetime coming from the choice of the smoothness structure.
2. There are an infinity of exotic smoothness structures to choose from. For example the foliation of an exotic spacetime like $S^3 \times_\theta \mathbb{R}$ can be very complicated.
3. For the usual foliation $S^3 \times \{t\}$ with $t \in \mathbb{R}$ of $S^3 \times_\theta \mathbb{R}$ the 3-sphere must be a wildly embedded submanifold (represented by an infinite polyhedron).
4. A quantum state can be defined on the spacetime as wild embedding.
5. A glimpse of an action to obtain a full QFT and quantum gravity is also obtained.

Before concluding, I must add that the views expressed are only partly original. I have partially drawn from the ideas of Carl H. Brans, Jerzy Król and Helge Rosé.

Appendix A: C^*—algebras Associated to Wild Embeddings

Let $I : K^n \to \mathbb{R}^{n+k}$ be a wild embedding of codimension k with $k = 0, 1, 2$. In the following we assume that the complement $\mathbb{R}^{n+k} \setminus I(K^n)$ is non-trivial, i.e. $\pi_1(\mathbb{R}^{n+k} \setminus I(K^n)) = \pi \neq 1$. Now one defines the C^*—algebra $C^*(\mathcal{G}, \pi)$ associated to the complement $\mathcal{G} = \mathbb{R}^{n+k} \setminus I(K^n)$ with group $\pi = \pi_1(\mathcal{G})$. If π is non-trivial then this group is not finitely generated. The construction of wild embeddings is given by an infinite construction[3] (see Antoine's necklace or Alexanders horned sphere). From an abstract point of view, we have a decomposition of $\mathcal{G}$ by an infinite union

$$\mathcal{G} = \bigcup_{i=0}^{\infty} C_i$$

of "level sets" C_i. Then every element $\gamma \in \pi$ lies (up to homotopy) in a finite union of levels.

The basic elements of the C^*—algebra $C^*(\mathcal{G}, \pi)$ are smooth half-densities with compact supports on $\mathcal{G}$, $f \in C_c^\infty(\mathcal{G}, \Omega^{1/2})$, where $\Omega_\gamma^{1/2}$ for $\gamma \in \pi$ is the one-dimensional complex vector space of maps from the exterior power $\Lambda^2 L$, of the union of levels L representing γ, to $\mathbb{C}$ such that

$$\rho(\lambda \nu) = |\lambda|^{1/2} \rho(\nu) \qquad \forall \nu \in \Lambda^2 L, \lambda \in \mathbb{R} .$$

For $f, g \in C_c^\infty(\mathcal{G}, \Omega^{1/2})$, the convolution product $f * g$ is given by the equality

$$(f * g)(\gamma) = \int_{\gamma_1 \circ \gamma_2 = \gamma} f(\gamma_1) g(\gamma_2)$$

with the group operation $\gamma_1 \circ \gamma_2$ in π. Then we define via $f^*(\gamma) = \overline{f(\gamma^{-1})}$ a $*$operation making $C_c^\infty(\mathcal{G}, \Omega^{1/2})$ into a $*$algebra. Each level set C_i consists of simple pieces (for instance tubes in case of the Alexanders horned sphere) denoted by T. For these pieces, one has a natural representation of $C_c^\infty(\mathcal{G}, \Omega^{1/2})$ on the L^2 space over T. Then one defines the representation

$$(\pi_x(f)\xi)(\gamma) = \int_{\gamma_1 \circ \gamma_2 = \gamma} f(\gamma_1) \xi(\gamma_2) \qquad \forall \xi \in L^2(T), \forall x \in \gamma.$$

---

[3] This infinite construction is necessary to obtain an infinite polyhedron, the defining property of a wild embedding.

The completion of $C_c^\infty(\mathcal{G}, \Omega^{1/2})$ with respect to the norm

$$||f|| = \sup_{x \in \mathcal{G}} ||\pi_x(f)||$$

makes it into a C^*algebra $C_c^\infty(\mathcal{G}, \pi)$. The C^*-algebra $C_c^\infty(K, I)$ associated to the wild embedding I is defined to be $C_c^\infty(K, j) = C_c^\infty(\mathcal{G}, \pi)$. The GNS representation of this algebra is called the state space.

References

1. T. Asselmeyer-Maluga, Exotic smoothness and quantum gravity. Class. Quant. Gravity **27**:165002 (2010). arXiv:1003.5506v1 [gr-qc]
2. T. Asselmeyer-Maluga, C.H. Brans, Cosmological anomalies and exotic smoothness structures. Gen. Relativ. Gravit. **34**, 597–607 (2002)
3. T. Asselmeyer-Maluga, C.H. Brans, *Exotic Smoothness and Physics* (World Scientific Publication, Singapore, 2007)
4. T. Asselmeyer-Maluga, J. Król, Small exotic smooth R^4 and string theory, in *International Congress of Mathematicians ICM 2010 Short Communications Abstracts Book*, ed. R. Bathia (Hindustan Book Agency, 2010), p. 400
5. T. Asselmeyer-Maluga, J. Król, Constructing a quantum field theory from spacetime (2011). arXiv:1107.3458
6. T. Asselmeyer-Maluga, J. Król, Exotic smooth R^4 and certain configurations of NS and D branes in string theory. Int. J. Mod. Phys. A **26**, 1375–1388 (2011). arXiv:1101.3169
7. T. Asselmeyer-Maluga, J. Król, Topological quantum d-branes and wild embeddings from exotic smooth R^4. Int. J. Mod. Phys. A **26**, 3421–3437 (2011). arXiv:1105.1557
8. T. Asselmeyer-Maluga, J. Król, On topological restrictions of the spacetime in cosmology. Mod. Phys. Lett. A **27**, 1250135 (2012). arXiv:1206.4796
9. T. Asselmeyer-Maluga, J. Król, Quantum D-branes and exotic smooth $\mathbb{R}^4$. Int. J. Geom. Methods Mod. Phys. **9**, 1250022 (2012). arXiv:1102.3274
10. T. Asselmeyer-Maluga, R. Mader, *Exotic R^4 and quantum field theory, in 7th International Conference on Quantum Theory and Symmetries (QTS7)*, ed. by C. Burdik et al. (IOP Publishing, Bristol, 2012), p. 012011. doi:10.1088/1742-6596/343/1/012011, arXiv:1112.4885
11. T. Asselmeyer-Maluga, H. Rosé, Dark energy and 3-manifold topology. Acta Phys. Pol. **38**, 3633–3639 (2007)
12. T. Asselmeyer-Maluga, H. Rosé, On the geometrization of matter by exotic smoothness. Gen. Relativ. Gravit. **44**, 2825–2856 (2012). doi:10.1007/s10714-012-1419-3, arXiv:1006.2230
13. T. Asselmeyer, Generation of source terms in general relativity by differential structures. Class. Quant. Gravity **14**, 749–758 (1996)
14. C. Rovelli, *Quantum Gravity*. Cambridge Monographs on Mathematical Physics (Cambridge University Press, Cambridge, 2004). www.cpt.univ-mrs.fr/~rovelli/book.pdf
15. A.N. Bernal, M. Saánchez, Smoothness of time functions and the metric splitting of globally hyperbolic space times. Commun. Math. Phys. **257**, 43–50 (2005). arXiv:gr-qc/0401112
16. A.N. Bernal, M. Saánchez, Globally hyperbolic spacetimes can be defined as "causal" instead of "strongly causal". Class. Quant. Gravity **24**, 745–750 (2007). arXiv:gr-qc/0611138
17. R. Fintushel, R. Stern, Knots, links, and 4-manifolds. Inven. Math. **134**, 363–400 (1998) (dg-ga/9612014)
18. M.H. Freedman, A fake $S^3 \times R$. Ann. Math. **110**, 177–201 (1979)
19. G.D. Mostow, Quasi-conformal mappings in n-space and the rigidity of hyperbolic space forms. Publ. Math. IHS **34**, 53–104 (1968)

20. T. Asselmeyer-Maluga, J. Król, Abelian Gerbes, generalized geometries and exotic R^4. J. Math. Phys. (2009). arXiv:0904.1276

21. T. Asselmeyer-Maluga, J. Król, Exotic smooth $\mathbb{R}^4$, noncommutative algebras and quantization (2010). arXiv:1001.0882

22. T. Asselmeyer-Maluga, J. Król, Quantum geometry and wild embeddings as quantum states. Int. J. Geom. Methods Mod. Phys. **10**(10) (2013). arXiv:1211.3012

23. T. Asselmeyer-Maluga, J. Król, Decoherence in quantum cosmology and the cosmological constant. Mod. Phys. Lett. A **28**, 350158 (2013). doi:10.1142/S0217732313501587, arXiv:1309.7206

24. T. Asselmeyer-Maluga, J. Król, Inflation and topological phase transition driven by exotic smoothness. Adv. HEP **2014**, 867460 (2014). http://dx.doi.org/10.1155/2014/867460, arXiv:1401.4815

25. T. Asselmeyer-Maluga, J. Król, Higgs potential and confinement in Yang-Mills theory on exotic $\mathbb{R}^4$ (2013). arXiv:1303.1632

Chapter 15
Gravity Can Be Neither Classical Nor Quantized

Sabine Hossenfelder

Abstract I argue that it is possible for a theory to be neither quantized nor classical. We should therefore give up the assumption that the fundamental theory which describes gravity at shortest distances must either be quantized, or quantization must emerge from a fundamentally classical theory. To illustrate my point I will discuss an example for a theory that is neither classical nor quantized, and argue that it has the potential to resolve the tensions between the quantum field theories of the standard model and general relativity.

To Quantize or not to Quantize Gravity

Gravity stands apart from the other three interactions of the standard model by its refusal to be quantized. To be more precise, quantizing gravity is not the actual problem; gravity can be perturbatively quantized. The problem is that the so quantized theory cannot be used at energies close by and above the Planck energy, and thus cannot be considered a fundamental theory; it is said to be 'non-renormalizable', meaning it has no predictive power in the extremely high energy regime.

This mismatch between the quantum field theories of the standard model and classical general relativity is more than an aesthetic problem: It signifies a severe shortcoming of our understanding of nature. This shortcoming has drawn a lot of attention because its resolution it is an opportunity to completely overhaul our understanding of space, time and matter. The search for a consistent theory of quantum gravity that could be applied also at Planckian energies, or strong curvature respectively, has thus lead to many proposals. But progress has been slow and in the absence of experimental evidence, our reasons for the necessity of quantizing gravity are theoretical:

S. Hossenfelder (✉)
Nordita, Roslagstullsbacken 23, 106 91 Stockholm, Sweden
e-mail: hossi@nordita.org

© Springer International Publishing Switzerland 2015
A. Aguirre et al. (eds.), *Questioning the Foundations of Physics*,
The Frontiers Collection, DOI 10.1007/978-3-319-13045-3_15

1. Classical general relativity predicts the formation of singularities under quite general circumstances. Such singularities are unphysical and should not occur in a fundamentally meaningful theory. It is expected that quantum gravity is necessary to prevent the formation of singularities.
2. Applying quantum field theory in a curved background at small curvature leads to the evaporation of black holes, as first shown by Hawking [1]. This black hole evaporation however seems to violate unitary which is incompatible with quantum mechanics. It is widely believed that quantum gravitational effects restore unitarity and information is conserved.
3. All quantum fields carry energy so they all need to couple to the gravitational field, but we do not know a consistent way to couple a quantum field to a classical field. As Hannah and Eppley have argued [2], the attempt to do such a coupling leads either to a violation of the uncertainty principle (and thus would necessitate a change of the quantum theory) or to the possibility of superluminal signaling, which brings more problems than it solves. While Mattingly has argued [3] that Hannah and Eppley's thought experiment can not be carried out in our universe, that does not address the problem of consistency.

These issues have all been extensively studied and discussed in the literature and are familiar ground. The most obvious way to address them seems to be a non-perturbation theory in one or other form, and several attempts to construct one are under way. I will use the opportunity of the essay contest to stray from the well-trodden ground and argue that we should instead reinvestigate the apparent tension between the quantized matter and non-quantized gravity. It is worthwhile for the following to recall the problems with coupling a classical to a quantum field.

The first problem, as illuminated by Hannah and Eppley is that the classical and the quantum fields would have different uncertainty relations, and their coupling would require a modification of the quantum theory. Just coupling them as they are leads to an inconsistent theory. The beauty of Hannah and Eppley's thought argument is its generality, but that is also its shortcoming, because it does not tell us how a suitable modification of quantum theory could allow such a coupling to be consistent.

The second problem is that it is unclear how mathematically the coupling should be realized, as the quantum field is operator-valued and the classical field is a function on space-time. One possible answer to this is that any function can be identified with an operator on the Hilbert space by multiplying it with the identity. However, the associated operators would always be commuting, so they are of limited use to construct a geometrical quantity that can be set equal to the operator of the stress-energy-tensor (SET) of the quantum fields.

Another way to realize the coupling is to construct classical field from the operator of the SET by taking the expectation value. The problem with this approach is that the expectation value may differ before and after measurement, which then conflicts with the local conservation laws of general relativity. Coupling the classical field to the SET's expectation value is thus usually considered valid only in approximation when superpositions carry negligible amounts of energy.

Because of these difficulties to make sense of the theory, leaving gravity classical while the other interactions are quantized is not a very promising option. However, this theoretical assessment should be supported by experimental test; recent proposals for this have been put forward in [4, 5].

How to Be Neither Classical Nor Quantized

Let us carefully retrace the logic of the arguments in the previous section.

We have experimental evidence that matter is quantized in the energy regimes that we have tested. We cannot leave gravity unquantized if it couples to quantized matter. Thus gravity has to be quantized in the energy regimes we have tested. We can quantize gravity perturbatively. This theory does make sense in the energy regimes that we have tested, but does not make sense in the strong curvature regime. We have no experimental evidence for the existence and properties of singularities or black hole evaporation, or the behavior of matter in the strong curvature regime.

To conclude from the previous paragraph that we need a non-perturbative completion of quantum gravity necessitates a further assumption, that is that the quantization procedure itself is independent of the energy range at which we apply the theory. It is this assumption that I argue should be given up.

We normally think of a theory as either being quantized or classical, but let us entertain the possibility that quantization is energy-dependent. Concretely, consider that Planck's constant $\hbar$ is a field whose value at high energies goes to zero. In four space-time dimensions, Newton's constant is $G = \hbar c / m_{\text{Pl}}^2$, so if we keep mass units fix, G will go to zero together with $\hbar$, thereby decoupling gravity. If gravity decouples, there's no reason for singularities to form. If gravity becomes classical, there's no problem with the perturbative expansion. So this possibility seems intriguing, if somewhat vague. I will now make this idea more concrete and then explain how it addresses the previously listed problems with quantizing gravity.

The starting point is that Planck's constant is a massless scalar field over space time $\hbar(x, t)$, and the equal time commutation relations for all fields, including, Planck's constant itself, are proportional then to $\hbar(x, t)$. Since we have no experimental evidence for the variation of Planck's constant, the most conservative assumption is that the $\hbar$-field is presently in its ground state, and difficult to excite with energies that we have access to. This suggests that we think about quantization as the consequence of a spontaneous symmetry breaking, and we have to add a suitable potential for $\hbar$ to the Lagrangian to achieve this. We are presently experiencing $\hbar(x, t)$ as having a non-zero vacuum expectation value that we will denote with $\hbar_0$. This is the measured value of Planck's constant. But at high temperature, presumably close by the Planck energy, the symmetry can be restored, resulting in a classical theory.

Gravity and matter then have a quantized phase and an unquantized phase, and are fundamentally neither quantized nor classical in the same sense that water is fundamentally neither liquid nor solid. Quantization, in this case, is also not emergent from a classical theory because the condition for second quantization does always contain the $\hbar(x, t)$.

A New Look at Old Problems

Let us now come back to the three problems mentioned in the first section that a theory for quantum gravity should address.

First, there is the formation of singularities. We know of two types of singularities that we should worry about, the Big Bang singularity and the singularities inside black holes.

If we move backwards in time towards the early universe, the temperature of matter increases and will eventually exceed the Planck energy. This is the standard scenario in which symmetry restoration takes place [6], so the expectation value of $\hbar$ goes to zero, gravity becomes classical, and matter decouples. If matter decouples, it cannot collapse to a singularity.

Collapse to a black hole is somewhat more complicated because it's not a priori clear that the temperature of the collapsing matter necessarily increases, but it plausibly does so for the following reason.[1] If matter collapses to a black hole, it does so rapidly and after horizon formation lightcones topple inward, so no heat exchange with the environment can take place and the process is adiabatic. The entropy of the degenerate Fermi gas is proportional to $T n^{-2/3}$, where T is the temperature and n is the number density. This means that if the number density rises and entropy remains constant, the temperature has to rise [7]. So again, matter decouples and there is nothing left to drive the formation of singularities.

Note that the $\hbar$-field makes a contribution to the source term, necessary for energy conservation.

Second, there is the black hole information loss. It was argued in [8] that the problem is caused by the singularity, not the black hole horizon, and that removing the singularity can resolve the information loss problem. This necessitates the weak interpretation of the Bekenstein-Hawking entropy so that a stable or quasi-stable Planck scale remnant, or a baby-universe, can store a large amount of information. There are some objections to the existence of such remnants, but they rely on the use of effective field theory in strong curvature regimes, the validity of which is questionable [9]. Thus, unitarity in black hole evaporation can be addressed by the first point, avoiding the formation of singularities.

Third, the difficulty of coupling a quantum field to a classical field and the non-renormalizability of perturbatively quantized gravity. In the here proposed scenario, there is never a classical field coupled to a quantum field. Instead, gravity and matter are of the same type and together either in a quantum phase or a classical phase. In the quantum phase, gravity is quantized perturbatively. It then needs to be shown that the perturbation series cleanly converges for high energy scattering because $\hbar$ is no longer a constant. This is a subtle point and I can here only give a rough argument.

To see how this would work, first note that we can rewrite the equal time commutation relation into a commutation relation for annihilation and creation operators of the fields. The commutator between annihilation and creation operators is then

---

[1] I acknowledge helpful conversation with Cole Miller on this issue.

proportional to the Fourier-transform of $\hbar(x, t)$, which I will denote $\tilde{\hbar}$. The same is true for the annihilation and creation operators of $\hbar(x, t)$ (though the prefactors differ for dimensional reasons).

Now consider an arbitrary S-matrix transition amplitude with some interaction vertices. We evaluate it by using the commutation relations repeatedly until annihilation operators are shifted to the very right side, acting on the vacuum, which leaves c-numbers, or the Feynman rules respectively. If Planck's constant is a field, then every time we use the commutation relation, we get a power of the $\hbar$-field, and the S-matrix expansion is a series in expectation value of powers of $\tilde{\hbar}$ times the other factors of the transition amplitudes. Then, we use the commutation relations on $\hbar$, or its annihilation and creation operators respectively. Now note that exchanging two of these will only give back one $\tilde{\hbar}$. Thus, we can get rid of the expectation value of powers, so that in the end we will have a series in powers of vacuum expectation values of $\tilde{\hbar}$ (as opposed to a series of expectation values of powers, note the difference).

If we consider the symmetry breaking potential to be induced by quantum corrections at low order, the transition to full symmetry restoration may be at a finite value of energy. In this case then, the quantum corrections which would normally diverge would cleanly go to zero, removing this last problem with the perturbative quantization of gravity.

Summary

I have argued that the fundamental theory can be neither classical nor quantized, but that quantization may be a phase that results from spontaneous symmetry breaking. Needless to say, this proposal is presently very speculative and immature. Some more details can be found in [10], but open questions remain. However, I hope to have convinced the reader that giving up the assumption that a theory is either classical or quantized can be fruitful and offers a new possibility to address the problems with quantum gravity.

References

1. S.W. Hawking, Particle creation by black holes. Commun. Math. Phys. **43**, 199–220 (1975)
2. K. Eppley, E. Hannah, The necessity of quantizing the gravitational field. Found. Phys. **7**, 5165 (1977)
3. J. Mattingly, Why Eppley and Hannah's thought experiment fails. Phys. Rev. D **73**, 064025 (2006) [gr-qc/0601127]
4. D. Giulini, A. Grossardt, Gravitationally induced inhibitions of dispersion according to the Schrödinger-Newton equation. Class. Quant. Gravity **28**, 195026 (2011). arXiv:1105.1921 [gr-qc]
5. J.R. van Meter, Schrodinger-Newton 'collapse' of the wave function, Class. Quant. Gravity. **28**, 215013 (2011). arXiv:1105.1579 [quant-ph]
6. J.I. Kapusta, *Finite-temperature Field Theory* (Cambridge University Press, Cambridge, 1993)

7. D.S. Kothari, Joule-Thomson effect and adiabatic change in degenerate gas. Proc. Natl. Inst. Sci. India **4**, 69 (1938)
8. S. Hossenfelder, L. Smolin, Conservative solutions to the black hole information problem, Phys. Rev. D **81**, 064009 (2010). arXiv:0901.3156 [gr-qc]
9. J.F. Donoghue, When effective field theories fail. PoS EFT **09**, 001 (2009). arXiv:0909.0021 [hep-ph]
10. S. Hossenfelder, A possibility to solve the problems with quantizing gravity. arXiv:1208.5874 [gr-qc]

Chapter 16
Weaving Commutators: Beyond Fock Space

Michele Arzano

Abstract The symmetrization postulate and the associated Bose/Fermi (anti)-commutators for field mode operators are among the pillars on which local quantum field theory lays its foundations. They ultimately determine the structure of Fock space and are closely connected with the local properties of the fields and with the action of symmetry generators on observables and states. We here show that the quantum field theory describing relativistic particles coupled to three dimensional Einstein gravity as topological defects *must* be constructed using a deformed algebra of creation and annihilation operators. This reflects a non-trivial group manifold structure of the classical momentum space and a modification of the Leibniz rule for the action of symmetry generators governed by Newton's constant. We outline various arguments suggesting that, at least at the qualitative level, these three-dimensional results could also apply to real four-dimensional world thus forcing us to re-think the ordinary multiparticle structure of quantum field theory and many of the fundamental aspects connected to it.

Introduction

Quantum field theory (QFT), the theoretical framework at the basis of our understanding of particle physics, lays its foundations on a set of fundamental assumptions whose "raison d'etre" is intimately related with the existence of a fixed and highly symmetric background space-time. When gravity enters the quantum stage one is faced with a series of conceptual tensions which are the basis of the formidable challenge that the formulation of a quantum theory of geometry and matter has posed to theoretical physicists in the past eighty years [1]. The extent of this tension is dramatically evident already in the most celebrated effect in semiclassical gravity: black hole quantum radiance. In this context a free quantum field living on a black hole background produces a steady thermal emission of quanta from the horizon.

M. Arzano (✉)
Dipartimento di Fisica, Sapienza University of Rome, P.le A. Moro 2,
00185 Roma, Italy
e-mail: michele.arzano@roma1.infn.it

© Springer International Publishing Switzerland 2015
A. Aguirre et al. (eds.), *Questioning the Foundations of Physics*,
The Frontiers Collection, DOI 10.1007/978-3-319-13045-3_16

Assuming an initial pure quantum state for the system, after a crude implementation of back-reaction, such *evaporation* would end with a final mixed state thus violating the basic postulate of unitarity of quantum evolution [2].

This blatant paradox, a quantum phenomenon which predicts a violation of one of the principles of quantum theory itself, forces us to pass under scrutiny all the tacit assumptions that enter the derivation of the effect. Factorization of the Hilbert space of the quantum field states described by a Fock space is essential in the characterization of field modes inside and outside the horizon (which in turn is closely related to locality and microcausality) and is at the basis of the assumption that the use of low-energy effective field theory is reliable in the derivation of such effect [3]. In this essay we will argue, *without making any assumptions about the nature of a yet-to-be-formulated theory of quantum gravity*, that three-dimensional semiclassical gravity in the presence of the simplest form of "topological" back-reaction leads to the demise of the usual formulation of Fock space. In particular, multiparticle states will no longer be constructed from (anti)-symmetrized tensor product of one-particle states but by a "coloured" generalization of them reflecting a deformed algebra of creation and annihilation operators. Newton's constant (Planck's mass in three dimensions) enters as a "deformation parameter" which governs the non-Leibniz action of translation generators on the quantum multiparticle states. Such deformation is a consequences of the non-trivial group manifold structure of the momentum space of the corresponding classical particle which is coupled to gravity as a topological defect. Such unconventional quantization of the field modes signals a departure from several assumptions at the basis of ordinary quantum field theory from additivity of quantum charges associated to space-time symmetries to departures from locality. The fact that ordinary QFT is smoothly recovered once the Newton's constant/deformation parameter is set to zero suggests that these models can be regarded as a natural extension of the conventional field theoretic paradigm which might open new avenues in attacking the quantum gravity problem "from below".

Curved Momentum Space in "Flatland"

As it is very well known Einstein gravity in three space-time dimensions does not admit local degrees of freedom [4]. Point particles are coupled to the theory as topological defects [5]. The space-time describing a single particle will be flat everywhere except at the location of the particle where one has a conical singularity. Indeed the length of a circular path centered at the location of the particle divided by its radius will be less than 2π. The deficit angle is $\alpha = 8\pi Gm$, proportional to the mass of the particle m and Newton's constant G. For the description of the phase space of the particle we need a characterization of its three-positions and three-momenta. This can be achieved mapping the conical space-time into three-dimensional Minkowski space with a cylindrical boundary and a wedge "cut-off" representing the deficit angle of the cone [6].

In ordinary (non-gravitational) relativistic mechanics in three-dimensions, Minkowski space is isomorphic as a vector space to the Lorentz algebra and the *extended* phase space is a *vector space* given by the direct product of two copies of $\mathfrak{sl}(2, \mathbb{R})$ i.e. $\Upsilon \equiv \mathfrak{sl}(2, \mathbb{R}) \times \mathfrak{sl}(2, \mathbb{R}) \simeq \mathbb{R}^{2,1} \times \mathbb{R}^{2,1}$. Positions and momenta are parametrized by coordinates in such space i.e. $\mathbf{x} = \vec{x} \cdot \vec{\gamma}$ and $\mathbf{p} = \vec{p} \cdot \vec{\gamma}$ where γ_a are 2×2 traceless matrices.

When the particle is coupled to gravity we can map the simply connected part of the conical space-time to Minkowski space and thus coordinates will be given by a map $\mathbf{q}$ from the "cone" to $\mathfrak{sl}(2, \mathbb{R})$. To complete the embedding we need to map the local frames at each point of the cone into a fiducial inertial frame on the Minkowski side. Thus to each point, besides $\mathbf{q}$, we also need to associate an element $\mathbf{U} \in SL(2, \mathbb{R})$ which provides the information regarding the type of Lorentz rotation needed to match the local frame to the background Minkowski frame. The pair $(\mathbf{q}, \mathbf{U})$ provides an isometric embedding of the bundle of local frames on the simply connected part of the manifold into Minkowski space. The freedom of choosing the background inertial frame is reflected in the freedom of transforming the functions $\mathbf{q}$ and $\mathbf{U}$ via a global Poincaré transformation $(\mathbf{n}, \mathbf{U})$, respectively a translation and a Lorentz rotation. Likewise the values of the embedding functions on the two faces of the wedge $\mathbf{q}_{\pm}$ and $\mathbf{U}_{\pm}$ should be identified via a generic Poincaré (gauge) transformation which we denote by $(\mathbf{v}, \mathbf{P})$.

The three-position of the particle i.e. the location of its worldline in the auxiliary Minkowski space is given by the values of the function $\mathbf{q}$ on the cylindrical boundary which regularizes the singularity at the "tip" of the cone. Such function, which we denote by $\bar{\mathbf{q}}$, in principle depends on time and on an angular variable however, if we want the cylindrical boundary to look like a worldline, we must impose the additional condition that $\bar{\mathbf{q}}$ depend only on time [6] i.e. $\bar{\mathbf{q}} \equiv \mathbf{x}(t)$. Thus the three-positions of the particle will be still given by a vector, namely the "coordinates" of an element of $\mathfrak{sl}(2, \mathbb{R})$. The values of the "worldline" function $\bar{\mathbf{q}}$ at the left and right boundary of the wedge, $\bar{\mathbf{q}}_{\pm}$, are subject to the "matching condition" $\bar{\mathbf{q}}_{+} \to \mathbf{P}^{-1}(\bar{\mathbf{q}}_{-} - \mathbf{v})\mathbf{P}$.

Since $\bar{\mathbf{q}}_{+} = \bar{\mathbf{q}}_{-} = \mathbf{x}(t)$ is the location of the particle, taking the derivative with respect to time of the equation above we obtain that *the velocity of the particle must commute with the group element* $\mathbf{P}$ (remember that $\mathbf{v}$ is constant). This implies that three-momentum vectors have to be proportional to the *projection* of the group element $\mathbf{P} \in SL(2, \mathbb{R})$ on its Lie algebra $\mathfrak{sl}(2, \mathbb{R})$, i.e. if we write $\mathbf{P}$ in its matrix expansion $\mathbf{P} = u\mathbb{1} + 4\pi G\vec{p} \cdot \vec{\gamma}$, we discard the part of $\mathbf{P}$ proportional to the identity and take $\mathbf{p} = \vec{p} \cdot \vec{\gamma}$. Notice that now the components of the momentum vector are *coordinates on a group manifold*, indeed the condition $\det \mathbf{P} = 1$ implies that $u^2 - 16\pi^2 G^2 \vec{p}^{\,2} = 1$, the equation of a hyperboloid embedded in $\mathbb{R}^4$. The phase space in the presence of topological "gravitational backreaction" is thus $\Upsilon_G = \mathfrak{sl}(2, \mathbb{R}) \times SL(2, \mathbb{R}) \simeq \mathbb{R}^{2,1} \times SL(2, \mathbb{R})$.

In the following sections we will discuss the dramatic consequences that this "structural" modification of the phase space of a relativistic particle has for the corresponding (quantum) field theory.

From "Conical" Particles to "Braided" Commutators

One Particle

We turn now to the description of the field theory describing the quantization of the relativistic particle coupled to three-dimensional gravity discussed in the Introduction. Before we do that it will be useful to make a short digression on the definition of the physical phase space and the associated mass-shell relation. In order to determine the mass shell we need to find a characterization of the mass of the particle and relate it to the notion of generalized momentum. As we saw above the mass of the particle is proportional to the deficit angle of the conical space. A way to measure the deficit angle is to transport a vector along a closed path around the boundary, as a result this will be rotated by the angle $\alpha = 8\pi Gm$. Physical momenta will be thus characterized by "holonomies" $\bar{\mathbf{P}}$ which represent a rotation by $\alpha = 8\pi Gm$. Such requirement imposes the restriction

$$\frac{1}{2}\mathrm{Tr}(\bar{\mathbf{P}}^2) = \cos(4\pi Gm) \quad \longrightarrow \quad \vec{p}^{\,2} = -\frac{\sin^2(4\pi Gm)}{16\pi^2 G^2}\,, \tag{16.1}$$

on the "physical" holonomies giving us a "deformed" mass-shell condition. From a mathematical point of view such on-shell condition is equivalent to imposing that physical holonomies/momenta lie in a given *conjugacy class* of the Lorentz group (see [7] for a pedagogical discussion). Roughly speaking if in ordinary Minkowski space the mass-shell hyperboloid describing the physical momenta of a massive particle is given by an orbit of the Lorentz group in flat momentum space, physical momenta belonging to a given conjugacy class can be seen as "exponentiated" orbits of the group on its manifold.

In analogy with ordinary Minkowski space we can consider complex functions on the momentum group manifold described above. It turns out that when restricted to momenta belonging to a given conjugacy class, such space of functions carries a unitary irreducible representation of the semi-direct product of the momentum group manifold and the Lorentz group [8] and thus is analogous (modulo a choice of polarization [9]) to the ordinary one-particle Hilbert space for a quantum field.

Without loss of generality and to keep our considerations at the simplest level we now switch to Euclidean signature allowing the "phase space" of the particle to be $\Upsilon_G = \mathfrak{su}(2) \times SU(2)$. In analogy with ordinary field theory we consider plane waves labelled by group elements belonging to a given conjugacy class as representatives of a one-particle "wave function". Momenta are now coordinates on $SU(2)$, in particular we work with "cartesian" coordinates[1]

$$\mathbf{P}(\vec{p}) = p_0\,\mathbb{1} + i\,\frac{\vec{p}}{\kappa}\cdot\vec{\sigma}, \tag{16.2}$$

---

[1] For simplicity we restrict to functions on $SO(3) \simeq SU(2)/\mathbb{Z}_2$.

where $\kappa = (4\pi G)^{-1}$, $p_0 = \sqrt{1 - \frac{\vec{p}^2}{\kappa^2}}$ and $\vec{\sigma}$ are Pauli matrices. Plane waves can be written in terms of a Lie algebra element $\mathbf{x} = x^i \sigma_i \in \mathfrak{su}(2)$ as

$$e_{\mathbf{P}}(x) = e^{\frac{i}{2\kappa}\mathrm{Tr}(\mathbf{xP})} = e^{i\vec{p}\cdot\vec{x}}. \tag{16.3}$$

with $\vec{p} = \frac{\kappa}{2i}\mathrm{Tr}(\mathbf{P}\vec{\sigma})$. The main effect of the group structure of momentum space is that the composition of plane waves is *non-abelian* indeed we can define a $\star$-product for plane waves

$$e_{\mathbf{P_1}}(x) \star e_{\mathbf{P_2}}(x) = e^{\frac{i}{2\kappa}\mathrm{Tr}(\mathbf{xP_1})} \star e^{\frac{i}{2\kappa}\mathrm{Tr}(\mathbf{xP_2})} = e^{\frac{i}{2\kappa}\mathrm{Tr}(\mathbf{xP_1P_2})}, \tag{16.4}$$

differentiating both sides of this relation one can easily obtain a non-trivial commutator for the x's

$$[x_l, x_m] = i\kappa\epsilon_{lmn}x_n, \tag{16.5}$$

i.e. the "coordinates" x_i can be seen as equipped with a non-commutative algebra structure. Functions of these coordinates will inherit a non-abelian algebra structure and the corresponding field theory will be a *non-commutative field theory*. Most importantly momenta will obey a non-abelian composition rule due to the non-trivial group structure

$$\vec{p}_1 \oplus \vec{p}_2 = p_0(\vec{p}_2)\vec{p}_1 + p_0(\vec{p}_2)\vec{p}_2 + \frac{1}{\kappa}\vec{p}_1 \wedge \vec{p}_2 = \vec{p}_1 + \vec{p}_2 + \frac{1}{\kappa}\vec{p}_1 \wedge \vec{p}_2 + \mathcal{O}(1/\kappa^2). \tag{16.6}$$

Since plane waves are eigenfunctions of translation generators the non-abelian composition of momenta will correspond to a non trivial action of translation generators on multiparticle states, in particular one can easily derive the following generalization of the Leibniz rule on the tensor product of two one-particle states

$$\Delta P_a = P_a \otimes \mathbf{1} + \mathbf{1} \otimes P_a + \frac{1}{\kappa}\epsilon_{abc}P_b \otimes P_c + \mathcal{O}(1/\kappa^2). \tag{16.7}$$

Notice that $\frac{1}{\kappa} = 4\pi G$ can be seen as a *deformation parameter* and in the limit $\kappa \to \infty$ one recovers the usual action of translations as an abelian Lie algebra. As discussed in [10] such behaviour signals a departure from one of the basic postulates of quantum field theory namely the additivity of the charges associated with space-time symmetry generators which in turn is deeply connected with the locality properties of the field operators [11].

More Particles

So far we have seen that at the one particle level the mass-shell condition which in the ordinary case is given by orbits of the Lorentz group on the vector space $\mathbb{R}^{3,1}$ is replaced now by "orbits" of the Lorentz group on a *non-linear* momentum space or, more properly, *conjugacy classes*. In analogy with ordinary field theory let us label one particle states by elements of these conjugacy subspaces of $SU(2)$ and denote them by $|\mathbf{P}\rangle$. As for any quantum system the space of states of a composite object is built from the tensor product of its constituents Hilbert spaces. Since at the quantum level we are dealing with indistinguishable particles one *postulates* [12] that n-particle states are constructed from (anti)-symmetrized n-fold tensor products of one-particle Hilbert spaces for particles with (half)-integer spin. Focusing on the simplest case of a two-particle state one notices that naively adopting a standard symmetrization prescription (we assume for definiteness that we are dealing with a spinless particle) the candidate state $1/\sqrt{2}\,(|\mathbf{P}_1\rangle \otimes |\mathbf{P}_2\rangle + |\mathbf{P}_2\rangle \otimes |\mathbf{P}_1\rangle)$ *is not* an eigenstate of the translation operator due to the non-abelian composition of momenta (reflecting the non-trivial Leibniz action of translation generators discussed above). This problem can be bypassed if one resorts to the following "momentum dependent" symmetrization [13, 14]

$$|\mathbf{P}_1\mathbf{P}_2\rangle_L \equiv 1/\sqrt{2}\left(|\mathbf{P}_1\rangle \otimes |\mathbf{P}_2\rangle + |\mathbf{P}_1\mathbf{P}_2\mathbf{P}_1^{-1}\rangle \otimes |\mathbf{P}_1\rangle\right). \qquad (16.8)$$

We used the subscript L above because one could also choose

$$|\mathbf{P}_1\mathbf{P}_2\rangle_R \equiv 1/\sqrt{2}\left(|\mathbf{P}_1\rangle \otimes |\mathbf{P}_2\rangle + |\mathbf{P}_2\rangle \otimes |\mathbf{P}_2^{-1}\mathbf{P}_1\mathbf{P}_2\rangle\right). \qquad (16.9)$$

Notice how now both two-particle states are eigenstates of the generators P_a and have total on-shell momentum $\mathbf{P}_1\mathbf{P}_2 \equiv \vec{p}_1 \oplus \vec{p}_2$. In analogy with the standard case we can introduce creation and annihilation operators so that

$$a_{L,R}^\dagger(\mathbf{P}_1)a_{L,R}^\dagger(\mathbf{P}_2)|0\rangle \equiv |\mathbf{P}_1\mathbf{P}_2\rangle_{L,R}. \qquad (16.10)$$

The action of the Lorentz group on the kets will be given by conjugation [15] i.e.

$$\Lambda_{\mathbf{H}} \triangleright |\mathbf{P}_i\rangle \equiv |\mathbf{H}^{-1}\mathbf{P}_i\mathbf{H}\rangle. \qquad (16.11)$$

It is straightforward to check that both "left" and "right" symmetrizations above are *covariant* under such transformations [14]. Moreover "L" and "R"-symmetrized states are connected by a Lorentz transformation

$$(\Lambda_{\mathbf{P}_1} \otimes \Lambda_{\mathbf{P}_2}) \triangleright |\mathbf{P}_1\mathbf{P}_2\rangle_L = |\mathbf{P}_1\mathbf{P}_2\rangle_R. \qquad (16.12)$$

In order to determine the algebra satisfied by such operators we start by noticing the useful relation

$$(\Lambda_{\mathbf{P}_2^{-1}} \circ \Lambda_{\mathbf{P}_1}) \otimes \mathbf{1} \rhd |\mathbf{P}_1\mathbf{P}_2\rangle_L = |\mathbf{P}_2\mathbf{P}_1\rangle_L. \tag{16.13}$$

Defining $\mathcal{R}_L^{-1}(\mathbf{P}_1, \mathbf{P}_2) \equiv (\Lambda_{\mathbf{P}_2^{-1}} \circ \Lambda_{\mathbf{P}_1}) \otimes \mathbf{1}$ we can then write the following *braided* commutators

$$a_L^\dagger(\mathbf{P}_1)a_L^\dagger(\mathbf{P}_2) - \mathcal{R}_L^{-1}(\mathbf{P}_1, \mathbf{P}_2)a_L^\dagger(\mathbf{P}_2)a_L^\dagger(\mathbf{P}_1) = 0 \tag{16.14}$$

$$a_L(\mathbf{P}_1)a_L(\mathbf{P}_2) - \mathcal{R}_L(\mathbf{P}_1, \mathbf{P}_2)a_L(\mathbf{P}_2)a_L(\mathbf{P}_1) = 0. \tag{16.15}$$

One can proceed in an analogous way for the *right* operators to find similar commutation relations with $\mathcal{R}_L^{-1}(\mathbf{P}_1, \mathbf{P}_2)$ replaced by $\mathcal{R}_R^{-1}(\mathbf{P}_1, \mathbf{P}_2) \equiv \mathbf{1} \otimes (\Lambda_{\mathbf{P}_1} \circ \Lambda_{\mathbf{P}_2^{-1}})$. The cross-commutators between $a(\mathbf{P})$ and $a^\dagger(\mathbf{P})$ will be similarly [14] given by

$$a_L(\mathbf{P}_1)a_L^\dagger(\mathbf{P}_2) - \hat{\mathcal{R}}_L(\mathbf{P}_2)a_L^\dagger(\mathbf{P}_2)a_L(\mathbf{P}_1) = \delta(\mathbf{P}_1^{-1}\mathbf{P}_2) \tag{16.16}$$

where $\hat{\mathcal{R}}_L(\mathbf{P}_2) = \mathbf{1} \otimes \Lambda_{\mathbf{P}_2}$ and $\delta(\mathbf{P}_1^{-1}\mathbf{P}_2)$ is the Dirac delta function on the group [16]. One can proceed in an analogous way for the "R" operators and obtain

$$a_R(\mathbf{P}_1)a_R^\dagger(\mathbf{P}_2) - \hat{\mathcal{R}}_R(\mathbf{P}_2)a_R^\dagger(\mathbf{P}_2)a_R(\mathbf{P}_1) = \delta(\mathbf{P}_1^{-1}\mathbf{P}_2) \tag{16.17}$$

where now $\hat{\mathcal{R}}_R(\mathbf{P}_1) = \Lambda_{\mathbf{P}_1} \otimes \mathbf{1}$.

We arrived to a modification of the usual algebra of creation and annihilation operators which is quite suggestive. It is reminiscent of the algebra of q-deformed oscillators or "quons" [17] but with "colour"-dependent q-factors given by $\mathcal{R}_{L,R}(\mathbf{P}_1, \mathbf{P}_2)$ and $\hat{\mathcal{R}}_{L,R}(\mathbf{P})$. We leave it open to speculation whether such deformed commutators can be interpreted as the quantum counterpart of the braiding of the worldlines of classical point-defects.

Discussion

The familiar form of the algebra of creation and annihilation operators that we are accustomed to from quantum field theory textbooks is intimately related to the quantization condition one imposes on fields and their conjugate momenta. The latter is *assumed* on the basis of the analogy with ordinary quantum mechanical commutators between position and momenta of a non-relativistic particle. The results we presented show that Einsten gravity in three space-time dimensions *clearly* indicate a possible relaxation of such assumption and a departure from the basic structures underlying our familiar formulation of local quantum field theory. The most immediate consequence of the deformed algebra of oscillators, as we showed above, is that the Fock space of the theory loses its simple structure in terms of (anti)-symmetrized tensor products of given one-particle states. It has been suggested [18] that these types of departures from ordinary Fock space might reflect a new kind of uncertainty on the measurement of momenta of multiparticle states namely that measuring the total momentum of a system precludes complete knowledge of the total momenta of its components and vice-versa. Besides this observation what is evident now is that due to the "braided" nature of the multiparticle states the question of decoupling of the low energy degrees of freedom form the high energy ones must be handled with care. This could suggest a weak link in the assumptions at the basis of the derivation leading to the information paradox, namely the use of low energy effective field theory in the presence of backreaction.

Another key aspect that is put at stake in these models is locality. In the discussion above we briefly touched upon the fact that the Leibniz action of symmetry generators on quantum states is deeply connected with the local properties of the fields. It turns out that allowing a non-trivial geometry for the momentum space of a classical particle has been subject to recent investigations in the context of the "relative locality" paradigm [19]. The phase space of a particle coupled to three dimensional gravity can indeed be seen as an example of a relative locality theory [20]. The conceptual breakthrough of such models lies in the observer-dependent notion of crossing of particle worldlines. The far reaching implications of this new feature have been widely discussed in the literature [21, 22]. In the perspective of our discussion it will be useful to investigate the behaviour of field operators constructed via the deformed operators above in order to check whether "classical" relative locality translates at the quantum level into departures from the ordinary local field paradigm.

Of course all the discussion so far is very specific to three dimensional gravity and its topological nature. What about the more realistic four-dimensional world? Obviously in four space-time dimensions Einstein's gravity *is not* a topological theory and thus in general similar arguments would not hold. Surprisingly though there exist suggestive results on Planckian scattering in quantum gravity that appear to hint in the right direction. Early work by Hooft [23] and by Verlinde and Verlinde in the early 90s [24] showed that forward scattering at Planckian center of mass energies in 3+1 quantum gravity can be treated semiclassically and gravity splits in a weakly coupled sector and a strongly coupled sector whose quantum dynamics can be described by a topological field theory. Could we be dealing with a similar state of affairs also in this four dimensional regime? As of today the question remains open.

The recent framework of *piecewise flat* gravity in $3 + 1$ dimensions [25] proposed as a model for gravity which displays only a finite number of degrees of freedom per compact regions of space-time could also provide a bridge to the real four dimensional world. Indeed this model is based on a straightforward extension of the picture of a system of particles described as defects which is found in three dimensional gravity. To our knowledge nobody has attempted a study of the phase space of these particles/strings in the same spirit of [6]. It would be not surprising if one would end up finding non-trivial structures analogous to the ones we discussed in this essay.

Finally, following the "relative locality" framework mentioned above one could argue that a curved momentum space is just a feature of a regime of four dimensional *quantum gravity* in which the Planck length is negligible while the Planck mass remains finite [22]. This formally means that both quantum and local gravitational effects become negligible, while their ratio remains finite and governs the non-trivial geometry of momentum space. If this assumptions are correct then our arguments would qualitatively hold true in four dimensions and they would indicate that "first order" quantum gravity corrections to local QFT would be exactly the kind described above.

In our opinion and in the light of the observations above, large part of the conceptual apparatus of local QFT is ripe for re-thinking and the three dimensional world is there to point us the way to go beyond the various assumptions that lie their roots in the very structure of Minkowski space. What we find remarkable is that the simple combination of *ordinary* classical gravity and quantum theory (via a topological coupling), without any reference to a specific "quantum gravity" model, suggests that departures from local QFT become quite natural when gravity enters the game. This suggests that the "humble" framework of semiclassical gravity has still a lot to teach us on various puzzling aspects of the marriage between gravity and the quantum world.

Acknowledgments I would like to thank J. Kowalski-Glikman and V. De Carolis for discussions. This work is supported by a Marie Curie Career Integration Grant within the 7th European Community Framework Programme and in part by a grant from the John Templeton Foundation.

References

1. S. Carlip, Rep. Prog. Phys. **64**, 885 (2001). [gr-qc/0108040]
2. S.W. Hawking, Phys. Rev. D **14**, 2460 (1976)
3. S.D. Mathur, Lect. Notes Phys. **769**, 3 (2009). arXiv:0803.2030 [hep-th]
4. S. Carlip, (Cambridge University Press, Cambridge, 1998), p. 276
5. S. Deser, R. Jackiw, G. 't Hooft, Ann. Phys. **152**, 220 (1984)
6. H.J. Matschull, M. Welling, Class. Quant. Gravity **15**, 2981–3030 (1998). [gr-qc/9708054]
7. B.J. Schroers, PoS QG -PH, 035 (2007). arXiv:0710.5844 [gr-qc]
8. T.H. Koornwinder, F.A. Bais, N.M. Muller, Commun. Math. Phys. **198**, 157 (1998). [q-alg/9712042]
9. M. Arzano, Phys. Rev. D **83**, 025025 (2011). arXiv:1009.1097 [hep-th]
10. M. Arzano, Phys. Rev. D **77**, 025013 (2008). arXiv:0710.1083 [hep-th]
11. R. Haag, J.T. Lopuszanski, M. Sohnius, Nucl. Phys. B **88**, 257 (1975)
12. A.M.L. Messiah, O.W. Greenberg, Phys. Rev. **136**, B248 (1964)
13. M. Arzano, D. Benedetti, Int. J. Mod. Phys. A **24**, 4623 (2009). arXiv:0809.0889 [hep-th]
14. M. Arzano, J. Kowalski-Glikman, to appear
15. Y. Sasai, N. Sasakura, Prog. Theor. Phys. **118**, 785 (2007). arXiv:0704.0822 [hep-th]
16. N.J. Vilenkin, A.U. Klimyk, *Representation of Lie Groups and Special Functions—1991*, vols. 3 (Kluwer, 1993)
17. O.W. Greenberg, Phys. Rev. D **43**, 4111–4120 (1991)
18. G. Amelino-Camelia, A. Marciano, M. Arzano, in *Handbook of Neutral Kaon Interferometry at a Phi-factory*, vol. 22, ed. by A. Di Domenico (2007), pp. 155–186
19. G. Amelino-Camelia, L. Freidel, J. Kowalski-Glikman, L. Smolin, Phys. Rev. D **84**, 084010 (2011). arXiv:1101.0931 [hep-th]

20. G. Amelino-Camelia, M. Arzano, S. Bianco, R.J. Buonocore, Class. Quant. Gravity **30**, 065012 (2013) arXiv:1210.7834 [hep-th]
21. G. Amelino-Camelia, M. Arzano, J. Kowalski-Glikman, G. Rosati, G. Trevisan, Class. Quant. Gravity **29**, 075007 (2012). arXiv:1107.1724 [hep-th]
22. G. Amelino-Camelia, L. Freidel, J. Kowalski-Glikman, L. Smolin, Gen. Relativ. Gravity **43**, 2547 (2011) [Int. J. Mod. Phys. D **20**, 2867 (2011)]. arXiv:1106.0313 [hep-th]
23. G. Hooft, Phys. Lett. B **198**, 61 (1987)
24. H.L. Verlinde, E.P. Verlinde, Nucl. Phys. B **371**, 246 (1992). [hep-th/9110017]
25. G. Hooft, Found. Phys. **38**, 733 (2008). arXiv:0804.0328 [gr-qc]

Chapter 17
Reductionist Doubts

Julian Barbour

Abstract According to reductionism, every complex phenomenon can and should be explained in terms of the simplest possible entities and mechanisms. The parts determine the whole. This approach has been an outstanding success in science, but this essay will point out ways in which it could nevertheless be giving us wrong ideas and holding back progress. For example, it may be impossible to understand key features of the universe such as its pervasive arrow of time and remarkably high degree of isotropy and homogeneity unless we study it holistically—as a true whole. A satisfactory interpretation of quantum mechanics is also likely to be profoundly holistic, involving the entire universe. The phenomenon of entanglement already hints at such a possibility (Somewhat more technical material that appeared as endnotes in my essay entry now appear as footnotes. I have also added some further footnotes, identified by 2014 at their beginning, to bring my original essay up to date, and an addendum at the end. In a few cases, I have made very minor changes to the original text for the sake of clarity or to correct a type. These are not indicated.).

Reductionism's Strengths and Weaknesses

Nature does not begin with elements as we are obliged to begin with them. It is certainly fortunate that we can, from time to time, turn aside our eyes from the overpowering unity of the All and allow them to rest on individual details. But we should not omit ultimately to complete and correct our views by a thorough consideration of the things which for the time being we left out of account" (Ernst Mach, 1883)

To get an idea where reductionism's strong and weak points lie, let's go to its source in Newton's worldview: utterly simple laws that govern the motions of bodies in space and time.

The law of inertia is the most basic: a force-free body moves uniformly in a straight line forever. Things are almost as simple if bodies interact. When far apart, they move with near perfect inertial motions, but when closer they can, through forces

Julian Barbour (✉)
College Farm, The Town, South Newington, Banbury, OX15 4JG, Oxon, UK
e-mail: barbourj@physics.ox.ac.uk

© Springer International Publishing Switzerland 2015
A. Aguirre et al. (eds.), *Questioning the Foundations of Physics,*
The Frontiers Collection, DOI 10.1007/978-3-319-13045-3_17

like gravitation, begin to change each other's motions. If body A has a greater mass than body B, then A affects B more than B does A. With allowance for the masses, action and reaction are equal.

The behaviour of a large system of many bodies, in principle the whole universe, can, according to Newton, be entirely explained by the inherent tendency of each body to move inertially modified merely by the forces exerted by all the other bodies in the universe. Just as reductionism proposes, the whole truly is the sum of its parts. Or is it?

The weak spot in Newton's scheme is the very thing that makes it reductionist. The position and motion of individual bodies such as atoms, the simple entities, are defined relative to invisible space and time, the framework in which simple laws can be formulated. Mach [1] and others argued that in reality the position of any one object is defined relative to every other object in the universe. That is obviously a far more complicated approach and it is clearly holistic. Mach argued that nevertheless it could still reproduce all of Newton's successes because force-free bodies are observed to move rectilinearly relative to the stars, which suggests that, in their totality, they exert a powerful causal effect on individual objects. Newton could have mistaken this empirical fact as evidence for what he called absolute space.

It is easy to see how this could have led to seriously wrong ideas about the universe. Barely over 100 years ago, most scientists thought we lived in an 'island universe' of a few million stars (our Galaxy) with nothing but space outside it. According to Newton's laws, such an island universe could exist and have angular momentum, L, about an axis, making it oblate like the rapidly rotating Jupiter. This was to be expected.

However, a simple implementation of Mach's ideas [2] rules it out. A Machian island universe must have angular momentum *exactly equal to zero*: $L = 0$, even though subsystems of bodies in small regions of the Galaxy can behave as Newton's laws predict and have $L \neq 0$. It is merely necessary for the values of L for all the subsystems to add up to zero. Seeing Newton's laws confirmed in the solar system, which does have $L \neq 0$, astronomers had no reason to question any of the predictions of Newtonian theory, which in no way is capable of forcing an island universe to have $L = 0$.

But suppose astronomers had, on Machian grounds, been convinced around 1900 that the universe must have $L = 0$ and had found this was not so for the Galaxy. Far from giving up the Machian theory, they would have *predicted* that our Galaxy cannot be the entire universe. They would have predicted a universe with more matter whose angular momentum exactly balances the Galaxy's.

How does the universe we can now observe meet this expectation? Extraordinarily well. It is full of galaxies and clusters of galaxies. Each and every one manifestly possesses some angular momentum, but there is no indication that the individual angular momenta add up to give a significant non-zero angular momentum of the universe as a whole.

In fact, my example is too simplistic, being based on Newtonian theory and not its superior successor in the form of Einstein's general relativity. But I think it makes my point. A reductionist standpoint may be very misleading.

Einstein, Mach, and General Relativity

Mach's idea that Newton's invisible absolute space could be replaced by an effect of the entire universe had a deep influence on Einstein, who called the idea *Mach's principle* [3]. It was the single greatest stimulus to the creation of his wonderful theory of gravity, general relativity (GR).

If GR implements Mach's principle, reductionism will be challenged. Simple parts and their interactions will not determine the whole; the whole will determine the way the parts behave. In fact, we shall see that the whole to a considerable degree determines what we perceive as parts.

So does GR implement Mach's principle? This has been a matter of controversy ever since Einstein created GR, mainly because he set about implementing Mach's principle in an indirect manner that left its status in GR obscure [4]. In fact, his definition of Mach's principle [3] was inconsistent [4], p. 93. I have given what I believe is the correct definition of Mach's principle [5] and argued that if the universe is closed up in three dimensions like the earth's surface in two then GR does implement Mach's principle [5]. If the universe is spatially infinite, the answer is equivocal. It is Machian however far you can imagine, but infinity is unreachable, and one can never establish a complete sense in which the whole determines the part.

In the previous section I showed how reductionism can mislead if we conceive the universe as a mere sum of its parts, but it did not suggest any need to change the conception of the parts, the individual stars, but only the way they interact. The Machian interpretation of general relativity is much more sophisticated and changes radically the way we conceive the parts and not just the way they interact. In this section I show how the standard representation of GR presents the universe as a sum of parts. In the next section, I give the alternative Machian interpretation, in which GR appears much more holistic. In the final section, I discuss possible implications of a holistic quantum view of the universe.

By parts in GR I mean little (infinitesimal) regions of spacetime. Imagine a two-dimensional surface that, at each point, is smooth and curved in accordance with a definite local law: measurements of the curvature in the immediate vicinity of every point on the surface would confirm the law holds at that point. Any surface for which this is true may exist. This is a reductionist situation: the parts (infinitesimal elements of surface) and the local law they satisfy determine what wholes (extended two-dimensional surfaces) can exist.

In general relativity, in the simplest situation in which no matter is present, the infinitesimal elements have not two but four dimensions: one of time and three of space. Otherwise the situation is very similar to what I described. Infinitesimal regions of spacetime satisfy Einstein's famous field equation $R_{\mu\nu} = 0$, and his theory permits any spacetime in which this local law is satisfied everywhere. Presented in these terms, it is a great triumph of reductionism; the predictions of GR have been very well confirmed for subsystems of the universe (the solar system and binary pulsars) and are not in conflict with cosmological observations. But in fact these present numerous puzzles whose solution may well call for a truly holistic approach that restricts the solutions GR allows and puts them in a different perspective. I now turn to that.

Gravity, Angles, and Distances

In the previous section, I identified the 'parts' in GR as infinitesimal pieces of space-time. Their structure is determined by the fundamental quantity in GR: the metric tensor $g_{\mu\nu}$. Being symmetric ($g_{\mu\nu} = g_{\nu\mu}$) it has ten independent components, corresponding to the four values the indices μ and ν can each take: 0 (for the time direction) and 1, 2, 3 for the three spatial directions. Of the ten components, four merely reflect how the coordinate system has been chosen; only six count. One of them determines the four-dimensional volume, or *scale*, of the piece of spacetime, the others the *angles* between curves that meet in it. These are angles between directions in space and also between the time direction and a spatial direction.

Now Mach's attitude to physics was not only holistic but also that all theoretical concepts should stay as close as possible to directly observable quantities. Using grand philosophical terms, the gap between epistemology—what can be observed—and ontology—what is assumed to exist—should be as small as possible. Ideally, there should be no gap at all. From this perspective, Einstein's ontology as expressed through the metric tensor $g_{\mu\nu}$ can be questioned on two grounds.

First, time is taken to be just as real as space. But Mach argued that time cannot be a primary notion; it must be derived from motion. In the first of these fqxi essay competitions, I showed how Newton's notion of an independent absolute time can be derived from motion. In Einstein's theory, Newton's global absolute and nondynamical time is replaced by local proper time, which interacts with space. This is a major change, but time, like space, is still taken to be a primary ontological concept, not something to be derived from more basic concepts.

Second, there is a big difference between, on one hand, quantities with scale, like lengths and volumes, and, on the other, angles: the former cannot be specified without the human choice of some unit—say metres or yards. That is arbitrary. In contrast, an angle can be specified in dimensionless terms as a fraction of a radian, which itself is a dimensionless quantity.

Let me spell this out to underline the Machian aspect. Suppose yourself at the centre of a circle of radius r looking at two points on the circumference separated by r. The angle that you see between them is by definition one radian. Key here is that the angle is the same whatever the radius r. This brings out the difference between angles and lengths.

Now imagine yourself on a clear night at high altitude in a desert. What do you see? Thousands of stars studding the black sky, all with definite angles between them.

It's beautiful. The scientific point is that you directly see the angles between the stars. But you cannot tell their distances.[1]

There are two possible reactions to the difference between dimensionful and inherently dimensionless quantities like angles. The standard one in science is to say that it is not a big deal; one must simply express everything in terms of ratios. Astronomers, for example, express interplanetary distances as ratios, taking the semi-major axis of the Earth's orbit as unit. Moreover, these distances have *dynamical* effect: Newton's gravitational force is inversely proportional to the square of the distance.

The alternative attitude to dimensionful quantities is to deny them any fundamental role. Is this feasible? I think so. Note first that, despite being defined through ratios of distances, angles can be truly said to belong to a point. The radius of the circle taken to define a radian can always be taken as small as one likes. In contrast, a distance is necessarily associated with two points, the ends of the considered interval. In the standard formulation of GR, angles and distances both have ontological—dynamical—status. Distances do because the curvature of spacetime plays a central role in GR, and curvature is defined by comparisons over distances even though they can be taken arbitrarily small.[2]

Curvature, and with it distance, is so deeply embedded in the conceptual and technical foundations of GR it would seem only a mad man would try to deny it a deep dynamical significance. General relativity without curvature seems like Hamlet without the Prince of Denmark. Remarkably, a new formulation of GR, called *shape dynamics* [6–8], suggests that, in a well defined sense, it is only angles that count. Because it shows how reductionism could be misleading us, I will sketch how the significance of angles emerged.

Dirac [9] and Arnowitt, Deser, and Misner (ADM) [10] took the first step over 50 years ago by replacing Einstein's vision of spacetime as a four-dimensional 'block' universe by a picture in which a three-dimensional entity evolves, building up space-time in the process. This is known as the ADM formalism. The hope, still held by many, is that the Dirac–ADM approach will one day lead, as its creators intended, to a quantum theory of gravity.

I need to say what the Dirac–ADM three-dimensional 'entity' is. It is a Riemannian three-geometry (named after the great mathematician Riemann, who introduced the concept). A two-geometry is easy to visualize: it is like the two-dimensional curved surface of the Earth, which closes up on itself. A closed geometry is needed to model a universe as a whole. A closed three-geometry is much harder to imagine, but is mathematically possible. In the ADM formalism, angles and infinitesimal distances are, dynamically, on an equal footing.

However, a hint that angles might have priority emerged from subsequent work related to the technically important ADM initial-value problem. There is no need

---

[1] Even when distances are determined by parallax, it is changes of observed angles that determine the distances, which are themselves ratios of the trigonometric base length.

[2] Technically, curvature involves second derivatives of the metric tensor, whereas angles are defined by the undifferentiated metric tensor.

for me to give details except to say that without its solution it is simply impossible to make any practical use of GR to establish, for example, what happens when two black holes coalesce. Among other things, this work, by James York [11] with input from his student Niall Ó Murchadha [12], established the amount of information that is needed to determine a physically sensible solution of GR.

The required information can be expressed in terms of a 3-geometry and the rate at which it is changing. As with spacetime, at each point of a 3-geometry there is encoded angle and scale information, 2 and 1 numbers respectively (compared with $5 + 1$ for four-dimensional spacetime). The rates at which these are changing are also characterized by $2 + 1$ numbers. York showed that the purely angle part of this information, i.e., $2 + 2 = 4$ numbers at each point, is sufficient to solve the initial value problem and hence determine a complete spacetime that satisfies Einstein's equations.

This suggested that GR is a theory of the way angle data change and that distance data play no dynamical role. However, this possibility was never taken too seriously because York's technique seemed to violate a foundational principle of GR—that there is no distinguished definition of simultaneity in the universe. For this reason, York's work, which does single out a notion of simultaneity, has usually been regarded as a merely technical device, albeit very valuable, for solving problems but not settling issues of principle.

Shape Dynamics

To what extent does shape dynamics (SD), which I mentioned earlier and for which the above has been a preparation, change this?

Imagine a 'toy' model universe of just three particles in two dimensions. Picture them as dots on an infinite sheet of grid paper. Then two coordinates determine the position of each. The six coordinates together define a *Newtonian configuration*, the sheet playing the role of absolute space. But one can ignore the positions of the dots on the sheet and regard as real only the three distances between each particle pair. That defines a Machian *relative configuration*. Defined in its own intrinsic terms, it does not exist anywhere in space. The final step is to say only the shape of the triangle, defined by two of its angles, is real.

The progressive elimination of non-Machian data takes us from Newtonian configurations through relative configurations to *shapes*. Now comes a key fact. Mathematically it is vastly easier to describe change in terms of coordinates, at the level of Newtonian configurations. That is why Newton introduced absolute space—and with it reductionism. Because shapes are irreducibly holistic, it is much harder to work with them and achieve the gold standard of dynamical theory: the ability to determine the future from an initial state of the system. In Newton's dynamics, a configuration and its rate of change define an initial state. In shape dynamics, we

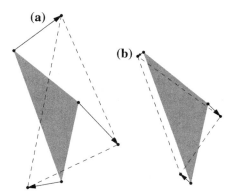

Fig. 17.1 a An arbitrary relative positioning of two triangles determines an initial Newtonian configuration (vertices of the *grey triangle*) and its rate of change (length of the *arrows* to the vertices of the *dashed triangle*). In the best-matched positioning **b**, shifting, rotating and scaling has brought the *dashed triangle* as close as possible to perfect overlap (congruence) with the *grey one*. An appropriate mathematical expression [5] measures the residual incongruence, the determination of which depends only on the shapes, not on their sizes or positions in space. The procedure is holistic, since the shapes alone determine the outcome, which is moreover expressed tractably through the coordinates of the vertices of the *dashed triangle*

want a shape and its rate of change to determine an initial state.[3] That's where the problem, the passage from reductionism to holism, lies. It is solved by *best matching*, which is explained in Fig. 17.1.

Once the shape-dynamic problem of defining initial states has been solved, it is easy to determine how shapes change according to a law that in no way relies on absolute duration, on position in an frame like grid paper, or on an external scale that measures size. The idea can be applied very generally, in particular to Riemannian geometries from which all information except that relating to angles between curves has been abstracted. Then we are dealing with *shapes of closed geometries*, just as York did, but now with some differences.

First, SD has led to an improved understanding of York's technique. This is a technical matter that I will relegate to a footnote[4] for experts.

---

[3] I need to say something about what rate of change means in shape dynamics. It is not with respect to any external time but relates to the expansion of the universe, which cosmology indicates is a reality. In the context of shape dynamics, this introduces a single overall scale ratio: one can say the universe is twice as large now as it was in an earlier epoch. This scale ratio, which is why shape dynamics is based on volume-preserving conformal transformations, has nothing to do with angles but provides a single global parameter that defines the rate of change of the angles. Thus, the angles do not depend on time but on the scale ratio. I do find the need for a scale ratio mysterious. Perhaps it has to be there to provide a substitute for time. 2014: I discuss this further in the Addendum.

[4] York based his technique, for a spatially closed universe, on full conformal transformations (which change local sizes but leave the angle information unchanged), whereas SD is based on volume-preserving conformal transformations. This is a tiny restriction, but it explains a bizarre feature of York's method that seemed completely *ad hoc*. I am referring to the scaling law that York adopted

Second, and relevant for us, York did not solve the initial-value problem following any clearly formulated first principles but by exploring different mathematical techniques until he found one that worked, supremely well in fact. In contrast, SD derives not only Einstein's equations but also York's method for solving the initial-value problem on the basis of clearly formulated Machian first principles, of which there is a 'holy trinity': duration, position, and size must all be relative. The first two of these were foundational principles of Einstein's original derivation,[5] but the third was not. In the shape-dynamic reformulation of GR, Einstein's relativity of simultaneity is traded for relativity of size.

The generalization of best matching to 'shapes of geometries' is direct but so far reaching it is worth outlining. One starts from $2+2$ given pieces of angle information at each point and considers how one could add to them in all possible ways $1+1$ pieces of 'trial' information that specify distances and their rates of change. The extra $1+1$ pieces of information are then varied in all possible ways until best matching is achieved. The mathematics is sophisticated but conceptually fully analogous to adjusting the dashed triangle to the best-matched positioning in Fig. 17.1. As in that case, there is always just one unique way, for given $2+2$ angle information, in which the best-matching condition can be satisfied. One then has an initial state that determines the future uniquely.[6]

Prior to the best matching, there is *nothing* in the specification of the initial $2+2$ pieces of angle information at each point of space that determines the distance between nearby points, the rate at which time flows at each space point, or the path of an inertially moving body. What I find really remarkable is that best matching applied to the distance-free $2+2$ pure angle information leads to complete determination of distances, rates of time flow, and inertial motion everywhere.[7] It is a triumph of Mach's intuition, going far beyond his original tentative suggestions.

Best matching is profoundly holistic. The procedure based on it that determines how the extra $1+1$ pieces of distance information are added at each point has to take into account the angle information *everywhere*—at all points. What is fixed *here* is determined by everything that is *there*.[8] There is a remarkable delicate interdependence of everything.

(Footnote 4 continued)

for the trace part of the momentum in the Lichnerowicz–York equation, which seemed incompatible with the law adopted for the trace-free part. SD provides a simple explanation for the law [6].

[5] 2014: I see that what I wrote originally was a bit confusing. Relativity of simultaneity, rather than of duration, was a foundational principle for Einstein. However, the latter is a consequence of the former in conjunction with Einstein's other assumptions. The key difference in SD is then, as stated, that relativity of simultaneity is traded for relativity of size as a foundational principle.

[6] My characterization of shape dynamics emphasizes its conceptual aspects. My collaborators Henrique Gomes, Sean Gryb, and Tim Koslowski also use the name shape dynamics for a specific gauge-theoretical implementation of the underlying ideas in a form that is likely to be best suited for technical developments.

[7] Soon after its completion, York's work suggested to Isenberg and Wheeler [13] a formulation of Mach's principle in somewhat similar terms but without a clear shape-dynamic underpinning.

[8] Best matching in the case of 'shapes of geometries' involves the solution of elliptic partial differential equations. As York and Ó Murchadha showed, these have very good existence and uniqueness

You may well ask: has anything been achieved, or is SD merely a different way of interpreting GR? The real test will be whether SD helps in the creation of a quantum theory of gravity. I shall consider that in the next section. It is however noteworthy that GR allows a great number of solutions that seem physically most implausible; this has long been a concern. The ADM formalism, if regarded as the true form of the theory, already considerably restricted the allowed solutions. Shape dynamics goes significantly further. It requires the universe to be compatible with generation from York initial data and to be spatially closed. This is what the truly holistic approach requires. In this respect (in postulating a spatial closure), it runs counter to much current thinking in cosmology. It is always good for a theory to make bold predictions and 'live dangerously'.

The main aim of this section has been to show that a reductionist approach can give a seriously misleading intuition about 'what the world is made of' and how it 'works'. In certain key respects, the standard representation of GR remains close to Newton's world view. Time and space are fused together, but locally their ontological nature and inertial motion in them are barely changed. Distance in space and duration in time are both real and foundational. Shape dynamics suggests that only angles are real.[9]

Quantum Implications

Consider the archetypal quantum two-slit experiment in which, in Dirac's famous words, "each photon interferes with itself". This implies a particle that somehow splits in two and manages to 'go through' both slits at once. This obviously stretches the imagination.

Could a reductionist mindset be misleading us in trying to interpret quantum experiments in terms of particles moving in space and time? It's a very Newtonian picture. We see the Moon but never a photon. All we see is equipment and an event that we attribute to a photon. We see a photon-generating laser, a filter to ensure that

---

properties. In the case of spatially closed universes, there are no boundary conditions to be specified; everything is determined intrinsically and holistically.

[9] I should add two caveats here. First, the discussion so far has ignored matter. That can be treated by York's method and hence added to shape dynamics without difficulty. Indeed it adds conceptual clarity to the picture. This is because matter fields define curves in space. The angles between any two curves are real (ontological), but distances along the curves are not; they are gauge. Second, all relativists would agree that a spacetime is determined by the specification of $2 + 2$ pieces of information at each point in a three-geometry. However, a majority would probably deny that these pieces of information must be exclusively and exhaustively angle data; they could be a mixture of angle and distance data. If relativity of simultaneity is taken to be sacrosanct, that is undeniable but it leaves one with a very indeterminate situation. What, if any, mixture of angle and distance data is correct? Three arguments speak for pure angle data: angles are conceptually more fundamental, the choice is clean (no indeterminate mixture), no other general and robust method apart from York's pure-angle method has been found to solve the initial-value problem.

only one photon at a time reaches the slits, and then the screen on which spots appear one after another.

The observed events are certainly discrete, but are we right to try to interpret them in terms of point-like particles? We do so because the experiment results in those spots on the screen or some other localized responses in apparatus. Of course, Bohr above all emphasized the complementary nature of quantum phenomena: particle-like or wavelike features are manifested depending on the macroscopic arrangement of the experiment. Change the arrangement and complementary aspects of reality are manifested. One can measure either momentum or position of a particle but never both at once.

This led Bohr to insist on the impossibility of formulating any meaningful statements about quantum phenomena without the framework provided by the classical, non-quantum world of macroscopic instruments. In this view quantum theory is not about 'what is' but about 'what we can say'.

But perhaps quantum theory *is* about 'what is' and the problem is that reductionism has led us to the wrong picture of 'what is'. Instead of thinking of particles in space and time, we should perhaps be thinking in terms of complete shapes of the universe. Then a photon would not be a particle that moves from one place to another but a law telling us how one shape of the universe becomes another. After all, that is what happens in the two-slit experiment. A spot appears on a screen that was not there before. In the first place, this tells us the shape of the universe has changed. Of course, it is only one change among an immense multitude, but it is a part of the totality of change. We may be doing violence to the universe—Mach's overpowering unity of the All—by supposing individual events are brought about by individual things, by parts. Hard as it may be, I think we need to conceptualize in terms of wholes.

The Machian interpretation of GR gives us hints of how this could change things. It suggests that what we observe locally has a part that is 'really here'—in the first place angle information—and a framework and basic laws that 'seem to be here'— local distances, times and inertial motion—but are in reality the effect of the complete universe.

In many ways, Bohr had a Machian mindset: the interpretation of quantum events depends crucially on the relationships that define the experimental layout. But where does that end and what determines it? Quantum entanglement measurements are now performed between the Canary Islands and the west coast of Africa and would be impossible without clocks that measure photon travel times. But if the Machian interpretation of GR is correct, the very distances over which travel times can be measured and even the times themselves are in a real sense created by the universe.

Thus, I think Bohr was on the right track but that one needs to go much further along it. Where will it take us? I don't know but will hazard a conjecture or two.

First, the quantum mechanics of the universe, like the Machian interpretation of GR, is holistic: it is about shapes. However, in contrast to shape dynamics, in which one shape follows another, in the quantum universe *all* shapes are present with different probabilities. But then whence comes our tremendously strong impression that time passes and we have a unique history?

In *The End of Time* [14], I attempted to answer this question with the notion of a *time capsule*: a very special configuration of the universe whose structure suggests that it is the outcome of a process that has taken place in time in accordance with definite laws. In [14] I still took distance to be real. The ideas that later led to shape dynamics were only briefly mentioned (in Box 3). I now distinguish carefully between relative configurations, in which both distances and angles are fundamental, and shapes defined by angles alone.

In [14] I conjectured a 'timeless quantum competition' between all possible configurations of the universe in which time capsules get the greatest quantum probabilities.[10] We experience them as instants of time bearing every appearance of the transition from a definite past to an indefinite future. I justified this conjecture by the pronounced asymmetry of what I called *Platonia*, the collection of all possible relative configurations of the universe. But with distance removed from the foundations, we are left with shapes. Platonia is *shape space*.

In this view, the quantum universe is *doubly holistic*: the possible shapes of the universe are wholes, each a unity in itself, but there is also shape space, the higher whole formed by the totality of possible wholes. Shape space is the ultimate quantum arena.[11] That is my second conjecture.

Seen from a 'god's eye' perspective, the striking asymmetry of Platonia that I noted in [14] becomes even more pronounced and suggestive: in any space of shapes there is always one unique shape, appropriately called Alpha, that is *more uniform* than any other possible shape.[12]

It interesting to note in this connection that the theory of cosmological inflation has some significant achievements to its credit but nevertheless must assume, as a

---

[10] 2014: As I will explain in the Addendum below, I still believe that the concept of time capsules is correct, but am now inclined to believe that one can speak meaningfully of evolution of the wave function of the universe.

[11] In the ADM approach to quantum gravity, the wave function of the universe for vacuum gravity is defined on superspace, the space of Riemannian three-geometries. The problem with this arena is the supreme difficulty, unresolved despite over 50 years of attempts, of implementing the ADM Hamiltonian constraint. I suspect that the root of the problem is the indeterminate mixture of angle and distance information encoded in superspace, which, as noted in the main text, is unavoidable as long as relativity of simultaneity (refoliation invariance) is held to be sacrosanct. In shape dynamics, the wave function of the universe is defined on conformal superspace. This brings undoubted conceptual clarity, as York already pointed out, but severe problems still remain. The ADM Hamiltonian may be conceptually hybrid but it is at least local; the shape-dynamic Hamiltonian is non-local.

[12] 2014: During work on the paper [15], my collaborators Tim Koslowski and Flavio Mercati pointed out to me that this statement is not true for shape space, on which a very natural measure of uniformity (and, simultaneously, complexity—see footnote 13) puts a plateau-like structure within which there are narrow infinitely deep potential wells. A uniquely distinguished 'peak' Alpha is not present. Fortunately, my slip in the essay does not affect my argument. Indeed, in [15] we present encouraging support for it. In an addendum at the end, I comment on the main results of [15], which follow already from the plateau-like structure.

hitherto unexplained initial condition, that the presently observed universe emerged from a state that was highly isotropic and homogeneous: in a word, very uniform. This suggests to me, as my third conjecture, that we should not be looking to explain the isotropy and homogeneity by a special initial condition but through the dominant status of Alpha in the universe's shape space. This is reflected in the fact that every shape has a degree of complexity that ranges from the most uniform possible to shapes of ever increasing complexity.[13] One never reaches Omega.

If there is no Omega in shape space, there must still be an end to this essay. Let it be my final conjecture, which sharpens the one in [14]. Our experiential life is dominated by time and a great asymmetry: the transition from past to future, from birth to death. Science has so far failed to explain the asymmetry and has been forced to attribute it to a special—*highly uniform*—initial state of the universe. I suspect we need look no further than the structure of shape space. Its highly uniform Alpha,[14] which holistic shape dynamics tells us must be deeply significant, seems to match perfectly what we take to be the uniform initial state of the universe.

Mathematics, the tool of theoretical physics, can only describe becoming through differences of structures that simply are—they exist in a timeless Platonic realm. Becoming is the ever mysterious moving reflection of being. We cannot explain the indubitable asymmetry of becoming that we experience unless being is asymmetric. It is.

Addendum (2014)

Since this essay was completed in 2012, my collaborators and I have had several insights that, I believe, significantly strengthen the plausibility of the conjectures made in my essay. These are set out in detail in [15], which builds on [17, 18]. The first key insight, achieved in [17, 18], relates to the nature of time and its relation to scale. It was in part anticipated in the submitted essay in footnote 3 and can be explained in simple Newtonian terms.

Consider N point particles with masses m_i and suppose they represent an 'island universe'. If we treat it in Machian shape-dynamic terms as suggested in the text, we

---

[13] Complexity is a surprisingly difficult notion to capture in a definition. However, in the case of the Newtonian N-body problem, there is a uniquely appropriate measure of the complexity of an N-body system that I call the scale-invariant gravitational potential V_{si}. It is the product of the ordinary Newtonian potential and the square root of the centre-of-mass moment of inertia $I_{cm} := (1/M) \sum_{i<j} m_i m_j r_{ij}^2$, where r_{ij} is the distance between particles i and j and M is the total mass. It is a function on shape space and takes its maximum on the most uniform shape the system can have. For large N, the corresponding distributions of the particles are remarkably uniform [16]. As the distributions become more clustered, V_{si} takes ever larger negative values. Given the close analogy between the Newtonian N-body problem when recast in Machian terms and GR, I anticipate the existence of a function on conformal superspace (the shape space for GR) that is analogous to V_{si}.

[14] 2014: In line with the correction noted in footnote 12, this should be replaced by 'Its striking plateau-like structure'.

would like to insist that only the shapes formed by the set of points have dynamical significance. Within Newtonian theory, this is *almost* the case if, as is reasonable, we insist that the universe as a whole has vanishing total energy and angular momentum. Then position and orientation in space of the universe as a whole have no meaning. However, if one sticks with the Newtonian gravitational potential energy, it is not possible to eliminate completely the role of scale. At a given instant, it has no physical significance—it would be changed by the human convention of measuring in inches rather than centimetres. However, a *ratio* of sizes at different instants, which is a dimensionless number and independent of the choice of units, does have meaning.

Now recall what I said about looking at the stars: you can directly see angles between stars but not their individual distances. This is closely related to my conviction, implicit in the distinction that I made between angles and distance, that shape is something far more fundamental than size. One sees this already in the case of a triangle, for which its shape is intuitively different from its size and entirely independent of any units of measurement. What is particularly striking is that one needs $3N - 7$ numbers to fix the shape of N points embedded in three-dimensional space but only one to fix its size, which, moreover, depends on the unit of measurement. One can have arbitrarily many shape degrees of freedom, but only one size degree of freedom. In the N-body problem, the Newtonian theory of N point particles interacting gravitationally, there is a (mass-weighted) measure of size that is uniquely distinguished by the dynamics. It is the (centre-of-mass) moment of inertia I.

In [15], my collaborators and I draw attention to a fact discovered well over a century ago by Lagrange, which is this. If the total energy E of the system is non-negative, $E \geq 0$, then (except in very special cases that need not concern us here) the curve of I (as a function of the Newtonian time) is U-shaped upward. It passes through a minimum and tends to infinity in both time directions. Now suppose you start 'at infinity' on one side of the U. It has a slope, we call it D in [15], which is negative and infinitely large in magnitude—the curve points vertically downward. As you move further down the U, the slope (with allowance for its sign) steadily increases from $-\infty$ until it reaches the bottom of the U, where it passes through zero, $D = 0$, after which it goes on increasing all the way to $+\infty$ 'at the top' of the other side of the U. Thus, the slope of D is always increasing. Moreover, if you track the behaviour of D going in the opposite direction, you find exactly the same thing: with allowance for its sign, the slope of D always increases. To find such behaviour of a degree of freedom in any dynamical system is remarkable. It is given a special name and called a Lyapunov variable.

In [18], we draw attention to an almost uncanny similarity between the time t in Newtonian theory and this behaviour of D. Both are invisible, and their role in dynamics can only be inferred from the behaviour of things that we can see. Moreover, both t and D are *monotonic*: they either always increase or decrease.[15] In fact, in

---

[15] Of course, one normally thinks that t always increases, but Newton's equations take the same form if one reverses the direction of time, which corresponds to reversing all the velocities in a given solution. The form of the law does not imprint a direction of time on the solution. Newton's laws, like Einstein's, are time-reversal symmetric.

[18] we argue that, if one is considering a complete Newtonian universe, a variable related to D should be regarded as 'time' and not t. It is particularly interesting that in general relativity there is no analogue of Newtonian time but there is a close analogue of D called the York time. It emerges naturally from the method, discussed in Sect. "Gravity, Angles, and Distances", by which York solved the initial-value problem in GR.

The recognition of the special status of D and its strong similarity to Newton's time might have great significance for the creation of quantum gravity, which has long been plagued by the so-called *problem of time*. This arises because quantum mechanics seems to need a well-defined time variable with properties like Newton's t. But in general relativity as created by Einstein with relativity of simultaneity taken as a fundamental postulate it is not possible to identify such a time. The case is altered if one is prepared, as we do in shape dynamics, to trade relativity of simultaneity for relativity of size. In the classical theory, this does not lead to any conflict with known observations, but it could have far-reaching implications in the quantum theory. In particular, if the York time is identified as the 'time' that quantum gravity seems to need, one would no longer have a frozen wave function of the universe and face the acute problem of how to recover our vivid experience of the passage of time.[16]

For me, one of the most interesting things about the U-shaped curve of the moment of inertia is the light it may cast on the origin of the arrow of time. As noted in footnote 15, the known laws of nature do not distinguish a direction of time. However, the great bulk of the actual processes that we observe taking place in the universe exhibit a marked directionality [19]. It is particularly striking that qualitatively different kinds of processes all have 'arrows' that point in the same direction. Moreover, all observations suggest that the direction is the same everywhere in the universe and at all times in its history. It has been a mystery for well over a century how this behaviour can arise from time-symmetric laws.

The most common suggestion is known as the past hypothesis, which is that at the big bang the universe was created in a special state of extremely low entropy. If this was so, then there would be no conflict with the second law of thermodynamics, according to which entropy always increases. It started very low and has been increasing ever since. Such a scenario would also be compatible with the growth of structure that we see everywhere on the earth and in the universe: there is no conflict with the second law provided the decrease of entropy associated with growth of structure in one region is offset by growth of entropy elsewhere.

The past hypothesis relies on a fundamental fact about physical laws: they all allow a great many different solutions. In a laboratory, the solution one actually encounters is determined by some initial (and boundary) conditions. There is, thus,

---

[16] Readers familiar with my earlier work, especially *The End of Time* [14], will see that I have moved from advocacy of the position that there is nothing in the 'external universe' corresponding to change and the passage of time to serious consideration of the possibility that difference of scale could play a role somewhat like, but certainly not identical to, Newton's time. While it is a bit embarrassing to make such a major change, I am happy that it came about through systematic study of the role of scale in gravitational dynamics very much in line with relational principles of the kind advocated by Leibniz and Mach.

a dual explanation for any given phenomenon: the underlying law and an initial condition. The past hypothesis implies that one of the most all-pervasive phenomena we observe both near and far in the universe—the arrow of time—is not a consequence of the law that governs the universe but of a very special initial condition. Why such an initial condition should have been realized is a complete mystery. This is hardly a satisfactory state of affairs, and in [15] we question it.

Our key observation relates to the fact that effectively *all* solutions of the N-body problem with zero energy and angular momentum divide into the two halves either side of the bottom of the U-shaped curve described above. Reflecting the time-reversal symmetry of Newton's law, there is *qualitative* symmetry of each solution about the central point. However, when one examines the halves in detail, they are quite different. Moreover, in all the typical solutions one finds that near the bottom of the U, where $D = 0$, the motion is basically chaotic but that with increasing distance from $D = 0$ structures, especially orbiting pairs of particles, tend to form. This is clearly reflected in the behaviour of the complexity/uniformity function that is defined in footnote 13 as a measure of how structured the system of particles is. Its value fluctuates but grows steadily between bounds. If, as is very natural, one defines a direction of time to match the growth of structure, then 'time' increases in both directions away from the point $D = 0$. One can say that each solution consists of two 'histories' that have a single common past, at $D = 0$, but separate futures, at the two infinite 'tops' of the U. Moreover, as we argue in [15], any observer living in such a solution must be in one of the halves and will take that to be effectively a single history which emerges from a relatively uniform but rather chaotic initial state.

It is far too soon to claim that our observation will help to crack the great problem of the origin of the arrow of time. However, to us at least, it does suggest that the law which governs the universe might well, by virtue of its form and without any special initial conditions, create manifest arrows of time for observers within the universe.

References

1. E. Mach. Die Mechanik in ihrer Entwicklung Historisch-Kritisch Dargestellt (1883)
2. J. Barbour, B. Bertotti, Mach's principle and the structure of dynamical theories (downloadable from platonia.com). Proc. R. Soc. Lond. A **382**, 295–306 (1982)
3. A. Einstein, Prinzipielles zur allgemeinen Relativitätstheorie. Annalen der Physik **55**, 241–244 (1918)
4. J. Barbour, H. Pfister (eds.), *Mach's Principle: From Newton's Bucket to Quantum Gravity, Einstein Studies*, vol. 6 (Birkhäuser, Boston, 1995)
5. J. Barbour, Mach's principle: a response to Mashhoon and Wesson's paper Annalen der Physik (2011). arXiv:1106.6036, arXiv:1108.3057
6. E. Anderson, J. Barbour, B.Z. Foster, B. Kelleher, N.Ó. Murchadha, The physical gravitational degrees of freedom. Class. Quant. Gravity, vol. 22:1795–1802, (2005), gr-qc/0407104
7. J. Barbour, Shape dynamics: an introduction. in *Quantum Field Theory and Gravity. Proceedings Conference at Regensburg 2010*, ed. by F. Finster et al. (Birkhäuser, 2012)
8. H. Gomes, S. Gryb, T. Koslowski, Einstein gravity as a 3D conformally invariant theory. Class. Quant. Gravity **28**, 045005 (2011) (arXiv:1010.2481)

9. P.A.M. Dirac, Fixation of coordinates in the Hamiltonian theory of gravitation. Phys. Rev. **114**, 924–930 (1959)

10. R. Arnowitt, S. Deser, C.W. Misner, The dynamics of general relativity, in *Gravitation: An Introduction to Current Research*, ed. by L. Witten (Wiley, New York, 1962), pp. 227–265

11. J.W. York, The role of conformal 3-geometry in the dynamics of gravitation. Phys. Rev. Lett. **28**, 1082–1085 (1972)

12. N.Ó. Murchadha, J. York, Existence and uniqueness of solutions of the Hamiltonian constraint of general relativity on compact manifolds. J. Math. Phys. **4**, 1551–1557 (1973)

13. J. Isenberg, J.A. Wheeler, Inertia here is fixed by mass-energy there in every W model universe, in *Relativity, Quanta, and Cosmology in the Development of the Scientific Thought of Albert Einstein*, ed. by M. Pantaleo, F. deFinis (Johnson Reprint Corporation, New York, 1979), pp. 267–293

14. J. Barbour, *The End of Time* (Oxford University Press, Oxford, 1999)

15. J. Barbour, T. Koslowski, F. Mercati, A gravitational origin of the arrows of time. arXiv:1310.5167 [gr-qc]

16. R. Battye, G. Gibbons, P. Sutcliffe, Central configurations in three dimensions. Proc. Roy. Soc. Lond. **A459**, 911–943 (2003)

17. J. Barbour N.Ó. Murchadha, Conformal superspace: the configuration space of general relativity. arXiv:1009.3559

18. J. Barbour, T. Koslowski, F. Mercati, The solution to the problem of time in shape dynamics (Accepted for publication in Class. Quant. Gravity 2014). (2013) arXiv:1302.6264 [gr-qc]

19. H.D. Zeh, *The Physical Basis of the Direction of Time* (Springer, New York, 2007)

Chapter 18
Rethinking the Scientific Enterprise: In Defense of Reductionism

Ian T. Durham

Abstract In this essay, I argue that modern science is *not* the dichotomous pairing of theory and experiment that it is typically presented as, and I offer an alternative paradigm defined by its functions as a human endeavor. I also demonstrate how certain scientific debates, such as the debate over the nature of the quantum state, can be partially resolved by this new paradigm.

The Scientific Enterprise

> I have begun to enter into companionship with some few men who bend their minds to the more solid studies, rather than to others, and are disgusted with Scholastic Theology and Nominalist Philosophy. They are followers of nature itself, and of truth, and moreover they judge that the world has not grown so old, nor our age so feeble, that nothing memorable can again be brought forth. –Henry Oldenberg, as quoted in [4].

Science is a living, breathing—and very human—enterprise. As such, it has always been a malleable process. Indeed, that is one of its enduring traits: not only does science prescribe a system by which its predictions may be refined by additional knowledge, but its *very nature* changes as our understanding of the world and ourselves broadens. Nevertheless, there is an over-arching paradigm to modern science whose origins are rooted in the works of Alhazen[1] who flourished during the Islamic Golden Age, circa 1000 CE. In its simplest form, this paradigm consists of the posing of questions and the subsequent testing of those questions [12]. This process is, of course, cyclic, as the testing of the original questions very often leads to new ones. But the asking of a question is really at the root of all scientific endeavors and stems from humanity's innate curiosity about itself and the world around us. In a sense, we all remain children, continually asking 'why?' In more modern scientific terms, the act of questioning forms the basis of a scientific theory that is "a well-substantiated

---

[1] Abū 'Alī al-Ḥasan ibn al-Ḥasan ibn al-Haytham (965 CE—c. 1040 CE), also known as Ibn al-Haytham and sometimes al-Basri.

I.T. Durham (✉)
Department of Physics, Saint Anselm College, Manchester, NH 03102, USA
e-mail: idurham@anselm.edu

© Springer International Publishing Switzerland 2015
A. Aguirre et al. (eds.), *Questioning the Foundations of Physics*,
The Frontiers Collection, DOI 10.1007/978-3-319-13045-3_18

explanation of some aspect of the natural world, based on a body of facts that have been repeatedly confirmed through observation and experiment" [13]. In other words, Alhazen's paradigm breaks science into two equal parts: theory and experiment.

While there have been modern refinements to Alhazen's basic framework, notably the adoption of the hypothetico-deductive[2] model, the basic division into theory and experiment remains. Victoria Stodden has recently proposed that computational science be recognized as a third division and, indeed, this is an attractive suggestion [14]. But it would fail to address certain persistent problems with both theory and experiment that raise deeper questions about the overall methodology of science. Clues to a solution to these problems can be found in the origins of that methodology.

While a precise formulation of the history of modern scientific methodology is not only lengthy but somewhat subjective, it is generally agreed that the revolution it sparked began in 17th century Europe and many of its principles were codified in the early documents and practices of the Royal Society of London, arguably the world's oldest scientific organization[3] [4]. As Thomas Sprat wrote, the Royal Society's purpose was "not the Artifice of Words, but a bare knowledge of things" expressed through "Mathematical plainness" [4]. This early scientific community developed a highly mechanistic approach to science that, while applied with equal vigor to anything tangible (and thus encompassing the modern fields of astronomy, chemistry, biology, physiology, et. al.), was decidedly grounded in the *physical*. The modern field that we recognize as physics has been called "the most fundamental and all-inclusive of the sciences" [5]. Arguably a portion of that inclusivity stems from the fact that all the other sciences are constrained by physical laws. This is one way in which scientific reductionism can be interpreted—a 'reduction' of the other sciences to physics. But physics is also inclusive by dint of its methods. Physics, throughout its history, has hewn most closely to the mechanistic approach developed in the 17th century and, indeed, this is the other way in which scientific reductionism is traditionally interpreted—a 'reduction' of a *system* to its constituent parts in an effort to better comprehend the whole.

This interpretation of reductionism is closely related to the notion of causality and, as a view of science, has been challenged in recent years as a result of work on emergence and complex systems [1, 9, 10, 16]. As Jonah Lehrer[4] wrote in a recent article

> [t]his assumption—that understanding a system's constituent parts means we also understand the causes within the system—is not limited to the pharmaceutical industry or even to biology. It defines modern science. In general, we believe that the so-called problem of causation can be cured by more information, by our ceaseless accumulation of facts. Scientists refer to

---

[2] The term 'hypothetico-deductive' has been attributed to William Whewell, though evidence for this is lacking as the term does not appear in any of his works on the inductive sciences.

[3] The history of the Royal Society is tightly linked with a number of organizations that arose in the mid–17th century including Académie Monmor, the Académie des sciences, and Gresham College [4].

[4] The ideas for the present essay were in large part developed as a rejoinder to Lehrer *prior* to his resignation from the *New Yorker* after admitting to fabricating quotes. That incident should have no bearing on what is written and discussed here.

this process as reductionism. By breaking down a process, we can see how everything fits together; the complex mystery is distilled into a list of ingredients [10].

Lehrer's article, however, focused almost exclusively on a single aspect of scientific methodology that is not necessarily mechanistic and that is misunderstood, even by scientists themselves: statistics and mathematical modeling. If reductionism is indeed what Lehrer claims it is, then statistical methods and mathematical modeling are most definitely *not* reductionist since they only seek to find mathematical structures that explicitly match existing data. This point is perhaps the most misunderstood in all of science. As an example, consider first the relationship between statistics and probability.

Mathematical Models

Statistics often accompanies probability (at least in textbook titles and encyclopedia entries). But this belies a subtle but important difference between the two. Both are indeed disciplines in their own right that fall under the larger umbrella of mathematics and logic. But only statistics is an actual *tool* of science. Probability is a logico-mathematical description of random processes. Statistics, on the other hand, is a methodology by which aggregate or 'bulk' information may be analyzed and understood. It loses its meaning and power when applied to small sample sizes. And there's the rub. If reductionism is the act of breaking down a process in order to understand its constituent parts, as Lehrer claims, statistics is the *antithesis* of reductionism because it makes no such effort.

Why then do we stubbornly persist in thinking that statistical methods in science can masquerade as some kind of stand-in for reductionism? Why do we expect more from statistics than we have a right to? Statistics is a very—*very*—important tool in science, but it is often misapplied and its results are often misinterpreted. Few understood this better than E.T. Jaynes. Jaynes spent the better part of his career attempting to correct one of the more egregious misconceptions, one that is intimately related to the difference between probability and statistics.

Roughly speaking, statistics generally describe information we *already know* or data we've *already collected*, whereas probability is generally used to *predict what might happen in the future*. As Jaynes astutely noted, if we imagine data sampling as an exchangeable sequence of trials,

> the probability of an event at one trial is not the same as its frequency in many trials; but it is numerically equal to the *expectation* of that frequency; and this connection holds whatever correlation may exist between different trials ...The probability is therefore the "best" estimate of the frequency, in the sense that it minimizes the expected square of the error [7].

In other words, probabilities can only be *accurately* formulated from statistical data if that data arose from a *perfectly repeatable* series of experiments or observations. This is the genesis of the interpretational debate over the meaning of the word 'probability,' with the frequentists on one side claiming a probability assignment is really nothing more than an assignment of the frequency of occurrence of a given outcome of a trial, and the Bayesians on the other side claiming that a probability assignment is a state of knowledge. As Jaynes clearly notes, the frequency interpretation is only valid under strictly enforceable conditions whereas the Bayesian view is more general.

What does the Bayesian interpretation of probability tell us about reductionism? The key to the Bayesian interpretation is the notion that, if probabilities represent our states of knowledge, measurements *update* these states of knowledge. Thus knowledge is gained in an incremental manner[5] which is the essence of reductionism. Thus probabilities, in a Bayesian context, are absolutely reductionist. As Jaynes points out, it *is* possible to give probabilities a frequentist interpretation, in which case they connect to the more aggregate descriptions provided by statistics, but only under certain strict conditions.

All of this does not necessarily obviate the need for the broader generalizations provided by statistics. In fact, as the foundational basis for thermodynamics, statistics as understood in the sense of distributions of measured quantities, has been very successful in explaining large-scale phenomena in terms of the bulk behavior of microscopic processes. Similar arguments can be made in terms of fluid dynamics, atmospheric physics, and like fields. As Jaynes pointed out,

> [i]n physics, we learn quickly that the world is too complicated for us to analyze it all at once. We can make progress only if we dissect it into little pieces and study them separately. *Sometimes, we can invent a mathematical model which reproduces several features of one of these pieces, and whenever this happens we feel that progress has been made* [8], [emphasis added].

Thus statistics is one of the primary methods by which larger-scale patterns are discovered. These patterns as such *emerge* in aggregate behavior from the underlying pieces. However, it is wrong to assume that such patterns can emerge *completely independently* of the underlying processes. This is tantamount to assuming that macroscopic objects can exist independently of their underlying microscopic structure. The melting of an ice cube clearly refutes this notion.

Of course, very few true anti-reductionists would argue this fairly extreme view. Instead they argue an intermediate position such as that proposed by P.W. Andersen [1]. Andersen fully accepts reductionism, but argues that new principles appear at each level of complexity that are not merely an extension of the principles at the next lower level of complexity. In another words, Andersen is suggesting that were we to be endowed with a sufficiently powerful computer *and* were we to have a full and complete understanding of, say, particle physics, we *still* would not be able to 'derive' a human being, for example, or, at the very least, the basic biological laws governing human beings. Biology and chemistry, to Andersen, are more than just applied or extended physics. This is precisely the point Lehrer is trying to make. But there is at least one fundamental problem with this argument: it assumes that no amount of additional knowledge can bridge the gap between levels of complexity,

---

[5] This is not necessarily the same thing as *sequential*, as is clearly demonstrated by certain quantum states.

i.e. it takes as *a priori* that reductionism (or 'constructionism,' as Andersen calls it) is either wrong or incomplete. But this is *logically unprovable*. As Carl Sagan wrote, "[y]our inability to invalidate my hypothesis is not at all the same thing as proving it true" [11]. In fact, this is precisely the same argument that proponents of creationism and intelligent design employ in claiming the universe (and, in particular, the biological life therein) is too complex to arise from simpler, less complex rules [2].

This does *not* mean that there aren't fundamental limits to our ability to 'know' the universe. For example, Gödel's incompleteness theorems place an inescapable limit on our ability to mathematically describe any physical system representable in an axiomatic manner [6]. Consider two physical systems, X and Y, each independently described by the same set of mathematical structures, M, that we take to be the minimum set that fully describes each system. Now suppose that completely combining these physical systems gives rise to a *third* physical system, Z, that is described by a set of mathematical structures, N, where $M \neq N$. We assume that N is taken to be the minimum set of structures that fully describes Z. In this scenario, X and Y are more 'fundamental' than Z and thus M must necessarily be a more restrictive set of structures than N. If M and N are formally distinct then Gödel's theorems tell us that there will exist truths in M, for instance, that cannot be proven from within M itself. Likewise for N. Thus it might be that N cannot be derived from M alone. In fact, it implies that there are structural elements of N that cannot be derived from *any* more primitive set of structures. Is this evidence for the anti-constructionist view? Not necessarily. While this is a valid limit to our knowledge, it only applies to any complete axiomatic description of the systems M and N. The universe as a whole may not be fully axiomatic. In fact, in the history of science, axiomatization was largely the realm of deductivism. But science is equal parts deductive and *inductive* and Gödel's theorems say nothing about inductive methods. In other words, the limits on our knowledge apply to certain methods. As yet, there does not appear to be a known limit to *all* methods. Thus it may be more instructive to think about science in terms of *methodologies*.

A New Scientific Paradigm

Recall that Alhazen's paradigm breaks science into two equal parts: theory and experiment. In this paradigm, experiments 'describe' the universe and theories 'explain' it. In this light, consider the development of Newtonian gravity in the 17th century. We can assign Galileo the role of experimenter/observer for his work with falling bodies, bodies on an inclined plane, and his observations of the moons of Jupiter, the latter of which importantly showed that celestial objects could orbit other celestial objects aside from the earth. This final point emphasizes the fact that a full theory of gravity had to take into account the movement of *celestial* bodies as well as terrestrial ones. Where, then, in this historical context, can we place Kepler? The data used by Kepler in the derivation of his three laws of planetary motion was largely taken by Tycho Brahe. They were not *explained* until nearly six decades after Kepler's death (in 1630) when Newton published his *Philosophiæ Naturalis Principia Mathematica* in 1687.[6] Thus, Kepler was neither the one who performed the original observations nor was he the one who discovered the explanation for the patterns exhibited by the observational data. He was, in fact, performing precisely the same general function as statisticians, climate scientists, and anyone performing clinical drug trials: he was fitting the data to a mathematical structure; *he was modeling*. This is *neither* theory *nor* experiment. It is a *methodology* for ascertaining how the universe works. In a sense, it is a *function*.

To some extent we have, as scientists, successfully ignored this problem for four centuries largely because it didn't seem to matter. After all, the dichotomy of theory and experiment was only a rough guide anyway and didn't have much of an impact (if any) on the science itself. But now, in certain areas of science and particularly in physics, this dichotomy does not appear to be working as it should. The most obvious example of this may be quantum mechanics where we have more than a century's worth of reliable experimental data, a well-established mathematical structure fit to that data, but no universally agreed upon *interpretation* of this data and its mathematical structure. Conversely, with string theory we have a well-established mathematical structure and a generally agreed-upon theory, but *no data*. In climate science, on the other hand, we have a consensus theory concerning climate change and we have a vast amount of experimental data, but we have no universally agreed upon mathematical model taking all of this data into account (i.e. we haven't reduced climate change to a self-contained set of equations yet). These examples appear to suggest that Stodden is on the right track in suggesting that there is a third division to science.

But how would adding a third division of science to the usual two solve the problems raised by Lehrer, Andersen, and others? To answer this question, let us first re-examine the purpose of each division's methods. What is it that experimentalists are *really* doing? Are they actually describing the universe or is their aim something else? I would argue that the aim of experimental science is, in fact, *not* to merely describe the universe. Even Aristotle described the universe. What Aristotle *didn't* do was describe it in a precise and consistent manner. His interpretation of what he saw had to fit pre-conceived philosophical notions. The revolution that marked the advent of modern *experimental* science aimed at measuring quantities free from pre-conceived notions of what those quantities *should* be. In other words, experimental science does not describe things, it *measures* things. Inherent in this aim is *precision* since measurement without precision is meaningless. Achieving a measure of precision itself requires repeatability—experimental results must be repeatable and independently verifiable. In fact, this latter point is so crucial that it is often *more* important for experimentalists to describe their procedures as opposed to their data. The data will often speak for itself but the procedure must be comprehensible if it is to be repeated and verified.

---

[6] Robert Hooke famously claimed priority in the formulation of the inverse square law, but, as Alexis Clairaut wrote in 1759 concerning this dispute, there is a difference "between a truth that is glimpsed and a truth that is demonstrated" (quoted and translated in [3]).

The aim of theory, on the other hand, has always been to *explain* the world around us and not merely to describe it. What sets *modern* theoretical science apart from Aristotelianism and other historical approaches is that it aims for logical self-consistency with the crucial additional assumption that science, as a whole, is ultimately universal. This last point implies that all of science is intimately *connected*. Thus we fully expect that biological systems, for example, will still obey physical and chemical laws. Crucially, modern theoretical science also aims to *predict* the future behavior of systems. Thus a 'good' scientific theory is both explanatory as well as predictive.

Description, then, is the realm of mathematics. Mathematics is ultimately how we describe what we 'see' in the experimental data. However, since mathematics is *such* an integral part of science, neither theorists nor experimentalists can carry out their work entirely free of it. It is this all-pervasive nature of mathematics that then leads to confusions and mis-attributions of the kind argued by Lehrer as well as interpretational problems vis-à-vis probability theory and its relation to statistics. As we noted earlier, roughly speaking, statistics generally is applied to prior knowledge (collected data) whereas probability theory is predictive in nature. As such, statistics is generally descriptive whereas probability theory is predictively explanatory. Thus I would argue that some of these issues could be cleared up if, rather than thinking of science in the way Alhazen did, perhaps with the added 'third division' suggested by Stodden, we instead should think of science as being divided into three functions: **measurement, description**, and **predictive explanation**. These functions, of course, are the *essence* of reductionism.

Now consider the rather sticky example of quantum mechanics which appears to be lacking a single, unifying 'interpretation' (i.e. 'theory' in the sense we have discussed above). In our parlance, it would seem that there are multiple predictive explanations that exist for quantum mechanics. But, in fact, most of the differences in the various interpretations of quantum mechanics differ in their interpretation of the quantum state. Thus consider a generic quantum state,

$$|\Psi\rangle = c_1|\psi_1\rangle + c_2|\psi_2\rangle.$$

If we interpret this statistically, then the values c_1 and c_2 are arrived at only by making repeated measurements. Instead, we can interpret this as a state of knowledge about the system that can be updated with a subsequent measurement. In other words, it can be interpreted as being predictive, at least in a probabilistic sense. On the other hand, if we take the state to be ontological, then it actually exists in the form given by $|\Psi\rangle$ and thus the state is merely descriptive. Thus these three interpretations of the quantum state correspond exactly to the three 'functions' of science and, when viewed in that light, do not necessarily contradict one another. Perhaps, instead of requiring *no* interpretation, as Brukner has suggested [15], quantum mechanics actually requires *multiple* interpretations.

What does this suggested shift in the description of science imply for complexity and emergence? If science is to be considered universal, connective, and self-consistent, perhaps the problem is not that reductionism is a broken paradigm, but rather that we are mis-ascribing some of our activities to the wrong scientific function, e.g. perhaps some of our so-called theories are actually more descriptive than predictively explanatory. Or perhaps they're built on the wrong description. Either way, without a formal proof that reductionism is incapable of certain descriptions of nature, it would seem a bit premature to declare one of the most successful methods of human inquiry dead. In fact it may simply be that, since the time of Alhazen, we have simply been missing a key ingredient. In order to maintain science as a productive, respected, and vital discipline we must ensure that it remains true to its foundational functions while always allowing room for introspection. Otherwise, science risks being ignored and too much is at stake for us to let that happen.

Acknowledgments The ideas discussed in this essay were tested on a few unsuspecting audiences over the course of a little more than a month. Thus, for helping me feel my way through these ideas, I would like to thank the following for giving me a pulpit from which to preach: the Clemson University Symposium for Introduction to Research in Physics and Astronomy (SIRPA); the Kennebunk Free Library's 'Astronomy Nights,' co-hosted by the Astronomical Society of Northern New England; and the Saint Anselm College Philosophy Club.

References

1. P.W. Andersen, More is different: broken symmetry and the nature of the hierarchical structure of science. Science **177**(4047), 393–396 (1972)
2. P. Atkins, Atheism and science, in *The Oxford Handbook of Religion and Science*, ed. by P. Clayton, Z. Simpson (Oxford University Press, Oxford, 2006)
3. W.W. Rouse Ball, *An Essay on Newton's Principia* (Macmillan and Company, London, 1893)
4. D.J. Boorstin, *The Discoverers: A History of Man's Search to Know his World and Himself* (Vintage Books, New York, 1983)
5. R.P. Feynman, R.B. Leighton, M. Sands, *The Feynman Lectures on Physics*, vol. I (Addison-Wesley, Reading, 1963)
6. Kurt Gödel, Über formal unentscheidbare sätze der principia mathematica und verwandter systeme, i. Monatshefte für Mathematik und Physik **38**, 173–198 (1931)
7. E.T. Jaynes, Where do we stand on maximum entropy? in *Papers on Probability, Statistics and Statistical Physics*, ed. by R.D. Rosencrantz, E.T. Jaynes (Kluwer Academic Publishers, Dordrecht, 1989), pp. 211–314
8. E.T. Jaynes, *Probability Theory: The Logic of Science* (Cambridge University Press, Cambridge, 1998)
9. S.A. Kauffman, Beyond reductionism: reinventing the sacred. Edge.org (2006)
10. J. Lehrer, Trials and errors: why science is failing us. Wired, December 2011
11. C. Sagan, The dragon in my garage, in *The Demon-Haunted World: Science as a Candle in the Dark* (Ballantine, New York, 1996)
12. S. Sambursky, *Physical Thought from the Presocratics to the Quantum Physicists: An Anthology* (Pica Press, New York, 1974)
13. Steering Committee on Science and Creationism, National Academy of Sciences. *Science and Creationism: A View from the National Academy of Sciences* (The National Academies Press, Washington, 1999)
14. V. Stodden, The Scientific Method in Practice: Reproducibility in the Computational Sciences (MIT Sloan Research Papers, 2010), Working paper no. 4773–10
15. B. Swarup, The end of the quantum road? Interview with Caslav Brukner (2009). http://fqxi.org/community/articles/display/114
16. Robert E. Ulanowicz, *Ecology: The Ascendent Perspective* (Columbia University Press, New York, 1997)

Chapter 19
Is Life Fundamental?

Sara Imari Walker

> *One can best feel in dealing with living things how primitive physics still is.*
>
> Albert Einstein

Although there has been remarkable progress in understanding some pieces of the puzzle, the emergence of life is still a mystery, presenting what is arguably one of the greatest unsolved questions in science. For the physicist or biologist, this may seem a problem for chemistry, and that the difficulty is simply that we don't have the know-how to engineer chemical networks quiet as complex as life, at least not yet. However, current challenges and limitations in chemical synthesis and the design of complex chemical networks may be only part of the story. The central challenge is that we don't know whether life is "just" very complex chemistry,[1] or if there is something fundamentally distinct about living matter. Until this issue is resolved, real progress in understanding how life emerges is likely to be limited.

What's at stake here is not merely an issue of how chemical systems complexify; the question of whether life is fully reducible to just the rules chemistry and physics (albeit in a very complicated manner) or is perhaps something very different forces us to assess precisely what it is that we mean by the very nature of the question of the emergence of life [1]. Stated most acutely, if a fully reductionist account is sufficient,

---

[1] This is not to imply that life is any less remarkable if a full account of biological organization turns out to indeed reduce to nothing more than the underlying rules of chemistry and physics subject to the appropriate boundary conditions and no additional principles are needed.

S.I. Walker (✉)
School of Earth and Space Exploration and Beyond Center for Fundamental
Concepts in Science, Arizona State University,Tempe, AZ, USA
e-mail: sara.i.walker@asu.edu

S.I. Walker
Blue Marble Space Institute of Science, Seattle, WA, USA

© Springer International Publishing Switzerland 2015
A. Aguirre et al. (eds.), *Questioning the Foundations of Physics*,
The Frontiers Collection, DOI 10.1007/978-3-319-13045-3_19

and life is completely describable as the nothing other than very complicated sets of chemical reactions, what then can we say originated? Taken to the extreme, the "all life is just chemistry" viewpoint advocates in a very real sense that life does not exist and as such that there is no transition to be defined. While this may very well be the case, when cast in these terms, even the avid reductionist might be unwilling, or at least hesitant, to accept such an extreme viewpoint. At the very least, although it is an open question whether this viewpoint is fundamentally correct, it is counterproductive to think in such terms—without a well-defined distinction between the two, there is no constructive mode of inquiry into understanding the transition from nonliving to living matter. As much as (or perhaps more than) any other area of science, the study of the emergence of life forces us to challenge our basic physical assumptions that a fully reductionist account is adequate to explain the nature of reality.

An illustrative example may be in order. It is widely appreciated that the known laws of physics and chemistry do not necessitate that life should exist. Nor do they appear to explain it [2]. Therefore in lieu of being able to start from scratch, and reconstruct 'life' from the rules of the underlying physics and chemistry, most are happy to avert the issue nearly entirely. We do so by applying the Darwinian criterion and assuming that if we can build a simple chemical system capable of Darwinian evolution the rest will follow suit and the question of the origin of life will be solved [3]. Accordingly, the problem of the origin of life has effectively been reduced to solving the conceptually simpler problem of identifying the origin of Darwinian evolution. Although this methodology has been successful in addressing specific aspects of the puzzle, it is unsatisfactory in resolving the central issue at hand by stealthily avoiding addressing when and how the *physical transition* from nonlife to life occurs. Therefore, although few (barring the exception of our avid reductionist) are likely to be willing to accept a simple molecular self-replicator as living, the assumption goes that Darwinian evolution will invariably lead to something anyone would agree is "alive". The problem is that the Darwinian criteria is simply too general, applying to *any* system (alive or not) capable of replication, selection, and heritage (e.g. memes, software programs, multicellular life, non-enzymatic template replicators, etc.). It therefore provides no means for distinguishing complex from simple, let alone life from non-life. In the example above, the Darwinian paradigm applies to both the precursor of life (i.e. a molecular self-replicator) and the living system it is assumed to evolve into, yet most might be hesitant to identify the former as living. It is easy to see why Darwin himself was trepidatious in applying his theory to explain the emergence of life.[2] If we are satisfied to stick with our current picture decreeing that "all life is chemistry" with the caveat "subject to Darwinian evolution", we must be prepared to accept that we may never have a satisfactory answer to the question of the origin of life and in fact that the question itself may not be well-posed.

---

[2] Darwin is famously quoted as stating, "It is mere rubbish thinking, at present, of the origin of life; one might as well think of the origin of matter" [4].

The central argument of this is essay is that we should not be satisfied with this fully reductionist picture. If we are going to treat the origin of life as a solvable scientific inquiry (which we certainly can and should), we must assume, at least on phenomenological grounds, that life is nontrivially different from nonlife. The challenge at hand, and I believe this is a challenge for the physicist, is therefore to determine what—*if anything*—is truly distinctive about living matter. This is a tall order. As Anderson put it in his essay *More is Different*, "The ability to reduce everything to simple fundamental laws does not imply the ability to start from those laws and reconstruct the universe" [5]. From this perspective, although an explanation of the physics and chemistry underlying the components of living systems is fully reducible to known physics, for all practical purposes we just can't work in the other direction and expect to really nail the issue down. If we can't work from the bottom-up, then we must work from the top-down by identifying the most distinctive features of the organizational and logical architecture of known living systems, which set them apart from their nonliving counterparts. We must therefore assume, right at the outset, that the "all life is chemistry" picture is inadequate to explain the phenomenon of life. We must ask, if life is not just complex chemistry, then *what is life*?

Despite the notorious difficulty in identifying precisely what it is that makes life seem so unique and remarkable, there is a growing consensus that its informational aspect is one key property, and perhaps the key property. If life is more than just complex chemistry, its unique informational aspects may therefore be the crucial indicator of this distinction. The remainder of this essay focuses on an illustrative example of how treating the unique informational narrative of living systems as more than just chemistry may open up new avenues for research in investigations of the emergence of life. I conclude with a discussion of the potential implications of such a phenomenological framework—if successful in elucidating the emergence of life as a well-defined transition—on our interpretation of life as a fundamental natural phenomenon.

"It from Bit from It"

Wheeler is quite famously quoted as suggesting that all of reality derives its existence from information, captured cleverly by his aphorism "it from bit" [6]. If Wheeler's aphorism applies anywhere in physics, it certainly applies to life, albeit in a very different context than what Wheeler had originally intended. Over the past several decades the concept of information has gained a prominent role in many areas of biology. We routinely use terminology such as "signaling", "quorum sensing" and "reading" and "writing" genetic information, while genes are described as being "transcribed", "translated", and "edited", all implying that the informational narrative is aptly applied in the biological realm. The manner in which information flows through and between cells and sub-cellular structures is quiet unlike anything else we observe in the natural world.

As we now learn it in school, the central dogma of molecular biology states that information flows from DNA → RNA → protein. In reality the situation is much more complicated than this simple picture suggests. The central dogma captures only the bit-by-bit transfer of Shannon (sequential) information. However, biology seems to employ a richer and more challenging concept of information than that tackled by Shannon, to the point that it is hotly debated what is even meant by the term "biological information". Consider as an example DNA, which acts as a digital storage repository for the cell. The human genome, for instance, contains roughly 3.2 billion base pairs, corresponding to roughly 800 MB of stored data. Compare this to rare Japanese plant *Paris Japonica*, with a genomic size of a whopping 150 billion base pairs or 37.5 GB of data—one of the largest genomes known [7]. *Paris Japonica* therefore vastly outstrips humans in terms of its genome's Shannon information content. Does this somehow imply that this slow-growing mountain flower is more complex (i.e. processes more information) than a human? Of course the answer is no. Across the tree of life, genome size does not appear to readily correlate with organismal complexity. This is because the genome is only a small part of the story: DNA is not a blueprint for an organism,[3] but instead provides a database for transcribing RNA, some (but by no means all) of which is then translated to make proteins.

The crucial point here is the action is not in the DNA, no information is actively processed in the DNA itself [8]. A genome provides a (mostly) passive access on demand database, which contributes biologically meaningful information by being read-out to produce functional (non-coding) RNAs and proteins. The biologically relevant information stored in DNA therefore has nothing to do with the chemical structure of DNA (beyond the fact that it is a digital linear polymer). The genetic material could just as easily be another variety of nucleic acid and accomplish the same task [9]. What is important is the *functionality* of the expressed RNAs and proteins. Functional information is a very strange beast, being dictated in part by the global context rather than just the local physics [10]. For example, the functionality of expressed RNA and proteins is context-dependent, and is meaningful only in the larger biochemical network of a cell, including other expressed proteins, RNAs, the spatial distribution of metabolites, etc. Sometimes very different biochemical structures (in terms of chemical composition, for example) will fill the same exact functional role—a phenomenon known as functional equivalence (familiar from cases of convergent evolution) where sets of operations perform the same functional outcome [11]. Only small subsets of all possible RNA and protein sequences are biologically functional. A priori, it is not possible to determine which will be functional in a cell based purely on local structure and sequence information alone (although some algorithms are becoming efficient at predicting structure, functionality is still determined by insertion in a cell, or inferred by comparison to known structures). Biologically functional information is therefore not an additional quality, like electric charge or spin, painted onto matter and fixed for all time. It is only definable in a relational sense, and thus must be defined only within a wider context.

---

[3] Here a blueprint is defined as providing a one-to-one correspondence between the symbolic representation and the actual object it describes.

One is left to conclude that the most important features of biological information, such as functionality, are inherently nonlocal. Biological information is clearly not solely in the DNA, or any other biochemical structure taken in isolation, and therefore must somehow be stored in the current state of the system (e.g. the level of gene expression and the instantaneous biochemical interaction network). Moreover, molecular biologists are continuing to uncover a huge variety of regulatory RNAs and proteins, which acting in concert with other cellular components, dictate the operating mode (e.g. phenotype) of a cell. Therefore, not only is the information specifying functional roles distributed, but information control is also a widely distributed and context-dependent feature of biological organization [12].

Superficially this may not seem to be anything particularly insightful or illuminating. One might argue that such distribution of information and control is an inevitable consequence of the complexity of biochemical networks. However, on closer inspection this state of affairs is really quiet remarkable for a physical system and potentially hints at something fundamentally different about how living systems process information that separates them from their nonliving counterparts. Cutting straight to the point, in biology *information* appears to have causal efficacy [11, 13]. It is the information encoded in the current state that determines the dynamics and hence the future state(s) and *vice versa* [14].

Consider a simplified example: the case of the genome and proteome systems, where the current state of the system—i.e. the relative level of gene expression—depends on the composition of the proteome, environmental factors, etc. that in turn regulate the switching on and off of individual genes. These then in turn dictate the future state of the system. An important point is that these two subsystems cannot function in isolation. Colloquially, this dynamic is often referred to as a chicken-or-egg problem, where neither the genotype nor the phenotype can exist without the other. Such a dynamic is well-known from the paradoxes of self-reference [15]; picture for example Escher's *Drawing Hands* where each of a pair of hands is drawing the other with no possibility of separating the two: it is unclear which hand is the cause and which the effect.

In biology, we cannot disentangle the genotype and phenotype because causation is distributed within the state of the system as a whole (including the relations among all of the subcomponents). Similar dynamics are at play throughout the informational hierarchies of biological organization, from the epigenome [16], to quorum sensing and inter-cellular signaling in biofilms [17], to the use of signaling and language to determine social group behavior [18]. In all of these cases where the informational narrative is utilized, we observe context (state) dependent causation, with the result that the update rules change in a manner that is both a function of the current state and the history of the organism [14]. Here casting the problem in the context of an informational narrative is crucial—the foregoing discussion may be formalized by stating that the algorithm describing the evolution of a biological system changes with the information encoded in the current state and vice versa. Contrast this with more traditional approaches to dynamics where the physical state of a system at time t_1 is mapped into the state at a later time t_2 in accordance with a *fixed* dynamical law and imposed boundary conditions. Thus, for example, Newtonian mechanics provides the algorithm that maps the state of the solar system today onto its state tomorrow by specifying a trajectory through phase space. The key distinction between this situation and that observed in biology is that information doesn't "push back" and actively influence the ensuing rules of dynamical evolution as it does in living systems. This feature of "dynamical laws changing with states" as far as we know, seems to be unique to biological organization and is a direct result of the peculiar nature of biological information (although speculative examples from cosmology have also been discussed, see e.g. [19]). It therefore serves as a contender for defining living matter.

Wheeler's dictum, as applied to the biological realm should therefore read more as *"it from bit from it"*,[4] where lower of levels of matter dictate the informational state of a system which then in turn dictates its future evolution. In this picture, *life is a dynamical phenomenon that emerges when information gains causal efficacy over the matter it is instantiated in* [20]. A situation made possible by the separation of information from its physical representation (i.e. through functional equivalence, coded intermediates, etc.). Thus, in biology the informational narrative is freed up to be almost independent of the material one and we may sensibly discuss cell-sell signaling, or sense data flowing along nerves, without specific reference to the underlying activity of electrons, protons, atoms or molecules. Of course all information requires a material substrate, but the important point here is that life cannot be understood in terms of the substrate alone. Thus it is meaningless to say that any single atom in a strand of DNA is alive. Yet, it is meaningful to state that the organism as a whole is living. "Aliveness" is an emergent global property.

Informational Efficacy and the Origin of Life

The liberation of the informational narrative from the material one potentially elicits a well-defined physical transition (even if currently not well-understood), which may be identifiable with the physical mechanism driving the emergence of life. In this picture, the origin of life effectively mediates the transition whereby *information* a "high-level" phenomenon gains causal efficacy over matter in a top-down manner[5] [20]. In physics we are used to the idea of "bottom-up" causation, where all causation stems from the most fundamental underlying layers of material reality. In contrast, top-down-causation is characterized by a higher level in an organizational hierarchy influencing a lower level by setting a context (for example, by changing some physical constraints) by which the lower level actions take place. In such cases, causation can also run downward in organizational hierarchies [21, 22]. Thus, top-down causation

---

[4] Perhaps an even better dictum might be *"it from bit from it from bit … ad infinitum"* to capture the self-referential nature of dynamical laws changing with states.

[5] In practice, 'top' and 'bottom' levels are typically not easily identified in hierarchical systems. Conceptually one may view both top-down and bottom-up causal effects as inter-level phenomenon, occurring between neighboring levels in a hierarchy, a phenomenon referred to as 'level-entanglement' by Davies (not to be confused with entanglement in quantum systems) [19].

opens up the possibility that high-level non-physical entities (i.e. information) may have causal efficacy in their own right [19, 23].

There is a vast literature suggesting top-down causation as a unifying mechanistic principle underlying emergence across the sciences, from quantum physics to computer science, to evolutionary biology, to physiology and the cognitive and social sciences (see e.g. [22]). In some areas of science, such as physiology, the existence of top-down causal effects is taken as self-evident and essential to making scientific progress. For example, it is not even a subject of debate that information control is widely distributed within living organisms (and thus that causation is also distributed). In other areas of science, such as chemistry and physics, which may be more familiar to the reader, top-down causation is not nearly as widely accepted. In particular, its role in chemistry is not well understood at all [24]. Poised at the intersection of the domains of science where top-down causation is widely accepted (biology) and where its role is not readily apparent (chemistry and physics) sits the emergence of life, suggesting that some very interesting physics may be occurring at this transition, and it may have everything to do with the appearance of genuinely new high-level causes.

Adopting this picture as constructive scientific inquiry into the emergence of life, an important question immediately presents itself: if a transition from bottom-up causation only (e.g. at the level of chemistry), to top-down (intermingled with bottom-up) causation may be identifiable with the emergence of life, what sets the origin of life apart from other areas of science where the role of top-down causation is clearly evident? As outlined by Ellis, there may in fact be several different mechanisms for top-down causation, which come into play at different hierarchical scales in nature [13]. In this regard, there may in fact be something unique to the emergence of life, which stems from the unique informational narrative of living systems as described in the previous section. Namely, biological systems (and other physical systems derivative of the biosphere such as computers and societies) seem to be unique in their implementation of top-down causation via information control [11, 13]. According to Auletta et al. who have rigorously defined this concept in the biological realm "Top-down causation by information control is the way a higher level instance exercises control of lower level causal interactions through feedback control loops, making use of functional equivalence classes of operations" [11]. The key distinction between the origin of life and other realms of science is therefore due to the onset of distributed information control, enabling context-dependent causation, where information—a high level and abstract entity, effectively becomes a cause. Cast in the language of the previous section this is just another way of stating that the origin of life might be associated with the onset of dynamical laws changing with states [20].

In contrast to other quantities attempting to capture the role of information in living systems, such as functional or semantic information, or even 'dynamical laws changing with states' (e.g. self-referential dynamics), causality is readily definable, and in principle measureable (although often difficult in practice). This is a primary reason why top-down causation is widely heralded as one of the most productive formalisms for thinking about emergence [22]. This framework therefore potentially enables a methodology for identifying a non-trivial distinction between life and nonlife, delineated by a fundamental difference in how information is processed. For the later, information is passive, whereas for the former information plays an active role and is therefore causally efficacious. The catch is that one must be willing to accept (at the very least on phenomenological grounds) the causal role of information as a defining feature in the story of life right along side the substrate narrative of the underlying chemistry. This forces new thinking in how life might have arisen on lifeless planet, by shifting emphasis to the origins of information control, rather than the onset of Darwinian evolution or the appearance of autocatalytic sets (that lack control) for example, which do not rigorously define how/when life emerges. It permits a more universal view of life, where the same underlying principles would permit understanding of living systems instantiated in different substrates (either artificial or in alternative chemistries). It may also encourage new thinking about the emergence of the apparent arrow of time in the biosphere, trending in a direction of increasing complexity with time: dynamical evolution where laws change with states is likely to not be time-reversal invariant (although this remains to be rigorously demonstrated). Once life emerges, we might therefore expect it to complexify and diversify over time, particularly as information gains causal efficacy over increasingly higher-levels of organization through major evolutionary innovations [25].

In practice, utilizing this framework as a productive paradigm for addressing the emergence of life will likely be very difficult. We currently don't have any good measures this transition. Although there is a vast literature in top-down causation, the role of a possible shift in informational efficacy (control) and thus causal structure as the key transition mediating the emergence of life has been absent in nearly all discussions of life's origins (see e.g. [20] for an exception relevant to this discussion). Part of the challenge is that we do not have the proper tools yet. Walker et al. proposed one possible measure, applying transfer entropy to study the flow of information from local to global and from global to local scales in a lattice of coupled logistic maps [25]. Nontrivial collective behavior was observed each time the dominant direction of information flow shifted from bottom-up to top-down (meant to act a toy model for the transition from independent replicators to collective reproducers characteristic of many major evolutionary transitions). However, this measure falls far short of being satisfactory. In particular, it doesn't capture true emergence where the parts do not exist without the whole (i.e. the cells in your body cannot exist outside of the multicellular aggregate that is you). Furthermore, it does not capture the causal relations among lower level entities and therefore is incapable of quantifying how the informational state of a system influences these lower level causal relations. In fact, it is not even a causal measure. In a very different context, a step in this direction may be provided by Tononi's measure of integrated information ϕ, which has been proposed as a way to quantify consciousness by measuring causal architecture based on network topology [26]. This measure effectively captures the information generated by the causal interactions of the sub-elements of a system beyond that which is generated independently by its parts. It therefore provides a measure of distributed information generated by the network as a whole due to its causal architecture. A version of the theory whereby ϕ itself is treated as a dynamical variable that has causal power in its own right might provide a way of quantifying the causal efficacy of information in the context that has been discussed here. Additional formalisms will need to also account for reliable encodings, where the same high-level phenomenon is reliably produced. In biology we have the example of the genetic code, but are far from decoding more distributed aspects of algorithmic information processing as occurs in the epigenome or the connectome.

It is an open question what will ultimately provide a useful phenomenological formalism for understanding the emergence of life. At the minimum the framework presented here provides a non-trivial distinction between life and nonlife and thus formulates the origin of life as a well-defined scientific problem, a key requirement for rigorous inquiry into life's emergence as discussed in the introduction. Life may be identified as fundamentally distinct from "just" complex chemistry due to its causal structure dictated by the causal efficacy of information. This immediately suggests several lines of inquiry into the emergence of life (which may or may not be practical at present). A top-down approach is to identify the causal architecture of known biochemical networks by applying measures (such as ϕ, or other measures of causal relationships [27]), for example by focusing on regulatory networks (information control networks). A bottom-up approach is to determine how information control emerges *ab initio* from chemical kinetics as well as how control evolves once this "information takeover" has occurred. Some of these principles will likely be testable in simple laboratory systems. A third line of inquiry could focus on the fundamental aspects of the problem, such as state-dependent dynamical laws, or the reproducibility of high-level outcomes via reliable encodings.

This is only a place to start, and it is entirely possible that additional and/or other novel physical principles will be required to pin-down what really drove the emergence of life. Whatever proper formalism emerges, we should not shy away from treating life as a distinct and novel physical phenomenon when addressing its origins. If this line of inquiry provides a productive framework for addressing the origin of life, a question, which must eventually be asked, is: *Is life fundamental?* For example, characterizing the emergence of life as a shift in causal architecture due to information gaining causal efficacy over the matter it is instantiated would mark the origin of life as a unique transition in the physical realm. Life would therefore be interpreted as logically and organizationally distinct from other kinds of dynamical systems,[6] and thus be a novel state of matter emerging at higher levels of reality. Our usual causal narrative, consisting of the bottom-up action of material entities only, would therefore be only a subset of a broader class of phenomena—including life—which admit immaterial causes in addition to material ones and which are characterized by their causal architecture. We would therefore have to consider that higher levels of reality admit the emergence of novel fundamental phenomena.

---

[6] Note this does not preclude that there may exist a gradation of states which are "almost" life with properties somewhere between completely passive and active informational dynamics, i.e. some parts might exist autonomously—an interesting question to consider in the context of astrobiology.

References

1. C.E. Cleland, C.F. Chyba, Defining life. Orig. Life Evol. Biosph. **32**, 387–393 (2002)
2. P.C.W. Davies, *The Fifth Miracle: The Search for the Origin and Meaning of Life* (Simon and Schuster, New York, 1999)
3. G. Joyce, Bit by bit: the Darwinian basis of life. PLoS Biol. **418**, 214–221 (2012)
4. C. Darwin, Letter to J.D. Hooker, in *The Correspondence of Charles Darwin 1863*, vol. 11, ed. by F. Burkhardt, S. Smith (1999), p. 278. (29 March 1863)
5. P.W. Anderson, More is different. Science **177**, 393–396 (1972)
6. J.A Wheeler, Sakharov revisited: "It from bit", ed. by M. Man'ko In: *Proceedings of the First International A.D Sakharov Memorial Conference on Physics, Moscow, USSR* (Nova Science Publishers, Commack, New York, 1991)
7. J. Pellicer, M.F. Fay, I.J. Leitch, The largest Eukaryotic genome of them all? Bot. J. Linn. Soc. **164**(1), 10 (2010)
8. D. Noble, Genes and causation. Philos. Trans. R. Soc. A **366**, 3001–3015 (2008)
9. V.B. Pinhero et al., Synthetic genetic polymers capable of heredity and evolution. Science **336**, 341–344 (2012)
10. G. Auletta, *Cognitive Biology: Dealing with Information from Bacteria to Minds* (Oxford University Press, Oxford, 2011)
11. G. Auletta, G.F.R. Ellis, L. Jaeger, Top-down causation by information control: from a philosophical problem to a scientific research programme. J. R. Soc. Interface **5**, 1159–1172 (2008)
12. U. Alon, *An Introduction to Systems Biology: Design Principles of Biological Circuits* (CRC Press Taylor & Francis, 2006)
13. G.F.R. Ellis, Top-down causation and emergence: some comments on mechanisms. J. R. Soc. Interface **2**(1), 126–140 (2012)
14. N. Goldenfeld, C. Woese, Life is physics: evolution as a collective phenomenon far from equilibrium. Annu. Rev. Condens. Matter Phys. **2**(1), 375–399 (2011)
15. D. Hofstadter, *Godel, Escher, Bach: An Eternal Golden Braid* (Basic Books Inc., New York, 1979)
16. P.C.W. Davies, The epigenome and top-down causation. J. R. Soc. Interface **2**(1), 42–48 (2012)
17. M.R. Parsek, E.P. Greenberg, Sociomicrobiology: the connections between quorum sensing and biofilms. Trends Microbiol. **13**, 27–33 (2005)
18. J.C. Flack, F. de Waal, Context modulates signal meaning in primate communication. In: Proc. Nat. Acad. of Sci. USA **104**(5) 1581–1586 (2007)
19. P.C.W. Davies, The physics of downward causation, in *The Re-emergence of Emergence*, ed. by P. Clayton, P.C.W. Davies (Oxford University Press, Oxford, 2006), pp. 35–52
20. S.I. Walker, P.C.W. Davies, The algorithmic origins of life (2012). arXiv:1207.4803
21. D.T. Campbell, *Levels of organization, downward causation, and the selection-theory approach to evolutionary epistemoloty*, ed. by G. Greenber, E. Tobach. Theories of the Evolution of Knowing. T.C. Schneirla Conference Series, (1990), pp. 1–15
22. G.F.R. Ellis, D. Noble, T. O'Connor, Top-down causation: an integrating theme within and across the sciences? J. R. Soc. Interface **2**, 1–3 (2011)
23. G.F.R. Ellis, On the nature of emergent reality, in *The Re-emergence of Emergence*, ed. by P. Clayton, P.C.W. Davies (Oxford University Press, Oxford, 2006), pp. 79–107
24. E.R. Scerri, Top-down causation regarding the chemistry-physics interface: a sceptical view. Interface Focus **2**, 20–25 (2012)
25. S.I. Walker, L. Cisneros, P.C.W. Davies, Evolutionary transitions and top-down causation, in *Proceedings of Artificial Life XIII*, pp. 283–290 (2012)
26. G. Tononi, An information integration theory of consciousness. BMC Neurosci. **5**, 42 (2004)
27. J. Pearl, *Causality* (Cambridge University Press, Cambridge, 2000)

Appendix
List of Winners

First Prize

Robert Spekkens: *The paradigm of kinematics and dynamics must yield to causal structure*[1]

Second Prizes

George Ellis: *Recognising Top-Down Causation*
Steve Weinstein: *Patterns in the Fabric of Nature*

Third Prizes

Julian Barbour: *Reductionist Doubts*
Giacomo D'Ariano: *Quantum-informational Principles for Physics*
Benjamin Dribus: *On the Foundational Assumptions of Modern Physics*
Sabine Hossenfelder: *Gravity can be neither classical nor quantized*
Ken Wharton: *The Universe is not a Computer*

Fourth Prizes

Giovanni Amelino-Camelia: *Against spacetime*
Michele Arzano: *Weaving commutators: Beyond Fock space*

---

[1] From the Foundational Questions Institute website: http://www.fqxi.org/community/essay/winners/2012.1.

© Springer International Publishing Switzerland 2015
A. Aguirre et al. (eds.), *Questioning the Foundations of Physics*,
The Frontiers Collection, DOI 10.1007/978-3-319-13045-3

Torsten Asselmeyer-Maluga: *A chicken-and-egg problem: Which came first, the quantum state or spacetime?*

Olaf Dreyer: *Not on but of.*

Ian Durham: *Rethinking the scientific enterprise: In defense of reductionism*

Sean Gryb & Flavio Mercati: *Right about time?*

Daryl Janzen: *A Critical Look at the Standard Cosmological Picture*

Israel Perez: *The Preferred System of Reference Reloaded*

Angelo Bassi, Tejinder Singh & Hendrik Ulbricht: *Is quantum linear superposition an exact principle of nature?*

Sara Walker: *Is Life Fundamental?*

Titles in this Series

Quantum Mechanics and Gravity
By Mendel Sachs

Quantum-Classical Correspondence
Dynamical Quantization and the Classical Limit
By Dr. A. O. Bolivar

Knowledge and the World: Challenges Beyond the Science Wars
Ed. by M. Carrier, J. Roggenhofer, G. Küppers and P. Blanchard

Quantum-Classical Analogies
By Daniela Dragoman and Mircea Dragoman

Life—As a Matter of Fat
The Emerging Science of Lipidomics
By Ole G. Mouritsen

Quo Vadis Quantum Mechanics?
Ed. by Avshalom C. Elitzur, Shahar Dolev and Nancy Kolenda

Information and Its Role in Nature
By Juan G. Roederer

Extreme Events in Nature and Society
Ed. by Sergio Albeverio, Volker Jentsch and Holger Kantz

The Thermodynamic Machinery of Life
By Michal Kurzynski

Weak Links
The Universal Key to the Stability of Networks and Complex Systems
By Csermely Peter

The Emerging Physics of Consciousness
Ed. by Jack A. Tuszynski

© Springer International Publishing Switzerland 2015
A. Aguirre et al. (eds.), *Questioning the Foundations of Physics*,
The Frontiers Collection, DOI 10.1007/978-3-319-13045-3

Quantum Mechanics at the Crossroads
New Perspectives from History, Philosophy and Physics
Ed. by James Evans and Alan S. Thorndike

Mind, Matter and the Implicate Order
By Paavo T.I. Pylkkanen

Particle Metaphysics
A Critical Account of Subatomic Reality
By Brigitte Falkenburg

The Physical Basis of the Direction of Time
By H. Dieter Zeh

Asymmetry: The Foundation of Information
By Scott J. Muller

Decoherence and the Quantum-To-Classical Transition
By Maximilian A. Schlosshauer

The Nonlinear Universe
Chaos, Emergence, Life
By Alwyn C. Scott

Quantum Superposition
Counterintuitive Consequences of Coherence, Entanglement, and Interference
By Mark P. Silverman

Symmetry Rules
How Science and Nature Are Founded on Symmetry
By Joseph Rosen

Mind, Matter and Quantum Mechanics
By Henry P. Stapp

Entanglement, Information, and the Interpretation of Quantum Mechanics
By Gregg Jaeger

Relativity and the Nature of Spacetime
By Vesselin Petkov

The Biological Evolution of Religious Mind and Behavior
Ed. by Eckart Voland and Wulf Schiefenhövel

Homo Novus—A Human without Illusions
Ed. by Ulrich J. Frey, Charlotte Störmer and Kai P. Willfiihr

Brain-Computer Interfaces
Revolutionizing Human-Computer Interaction
Ed. by Bernhard Graimann, Brendan Allison and Gert Pfurtscheller

Extreme States of Matter
on Earth and in the Cosmos
By Vladimir E. Fortov

Searching for Extraterrestrial Intelligence
SETI Past, Present, and Future
Ed. by H. Paul Shuch

Essential Building Blocks of Human Nature
Ed. by Ulrich J. Frey, Charlotte Störmer and Kai P. Willführ

Mindful Universe
Quantum Mechanics and the Participating Observer
By Henry P. Stapp

Principles of Evolution
From the Planck Epoch to Complex Multicellular Life
Ed. by Hildegard Meyer-Ortmanns and Stefan Thurner

The Second Law of Economics
Energy, Entropy, and the Origins of Wealth
By Reiner Köummel

States of Consciousness
Experimental Insights into Meditation, Waking, Sleep and Dreams
Ed. by Dean Cvetkovic and Irena Cosic

Elegance and Enigma
The Quantum Interviews
Ed. by Maximilian Schlosshauer

Humans on Earth
From Origins to Possible Futures
By Filipe Duarte Santos

Evolution 2.0
Implications of Darwinism in Philosophy and the Social and Natural Sciences
Ed. by Martin Brinkworth and Friedel Weinert

Probability in Physics
Ed. by Yemima Ben-Menahem and Meir Hemmo

Chips 2020
A Guide to the Future of Nanoelectronics
Ed. by Bernd Hoefflinger

From the Web to the Grid and Beyond
Computing Paradigms Driven by High-Energy Physics
Ed. by Rene Brun, Federico Carminati and Giuliana Galli Carminati

Printed in the United States
By Bookmasters